THE LAST OF THE
HUNTER GATHERERS

This 1992 oil painting by Icelandic artist Finnur Jónsson, is reproduced courtesy of the
owner, the Federation of Icelandic Vessel Owners.

Michael Wigan

THE LAST OF THE HUNTER GATHERERS

FISHERIES CRISIS AT SEA

SWAN·HILL PRESS

Copyright © 1998 Michael Wigan

First published in the UK in 1998
by Swan Hill Press, an imprint of Airlife Publishing Ltd

British Library Cataloguing-in-Publication Data
A catalogue record for this book
is available from the British Library

ISBN 1 85310 771 9

Typeset by Servis Filmsetting Ltd, Manchester, England.
Printed in England by Biddles Ltd, Guildford and King's Lynn.

Swan Hill Press
an imprint of Airlife Publishing Ltd
101 Longden Road, Shrewsbury, SY3 9EB, England

Preface

This is not a book simply describing world fisheries. Although the Bohai and Yellow Sea areas of China are amongst the most intensively fished in the world and China is an ancient state with a historic fishery, reliable data is scanty. Many other major fisheries, for one reason or the other, have not been mentioned. Some aspects of fisheries which have been described are omitted. The book picks out examples of fisheries and management philosophies which are themselves of interest or which throw interesting light on problems elsewhere. If fisheries are run for man and not for fish, man eventually eradicates himself from the equation. I have tried to show how fishing states have responded to this fact and to identify the management systems which have stood the test of time and why they have succeeded.

The degree to which those in fisheries have been willing, keen and helpful in supplying information and opinion has been a revelation. The story of fishing and fisheries is crying out to be told. There is a feeling amongst fishery scientists, if my antennae are working correctly, that they have important stories to contribute, that their field of research is as fascinating as anything in the natural world could be, and that, although people might seem to be at loggerheads, in a curious way everyone knows in their heart of hearts that they are pointing the same way. Fishermen are sometimes forgetful, and give the same impression. The co-operation I have had has been superb and I only hope my attempts to use the material does justice to the efforts of all who have been frank and generous with their expertise.

Thanks are due to the following people in particular: David Cushing, Terry Rowantree of the Scottish fisheries consultancy Enviro-Marine, Richard Miles of the publishers Fishing News Books, Serge Garcia, Brian O'Riordan, Tom Hay, Rögnvaldur Hannesson, Ragnar Arnason, Orri Vigfússon, Nick Johnstone of the International Institute for Environment and Development, Arne Kalland, John Pope, Ransom Myers, Conor Nolan of the Falkland Islands Government, Iain Sutherland, Bill Schrank, Michael Sissenwine, John Tower, fisherman and fishing writer David Thomson, Warren Wooster, Petur Bjornsson of Isberg, Ciaran Crummey of Bord Iascaigh Mara, Euan Dunn of the Royal Society for the Protection of Birds, Ken Whelan of the Burrishoole Research Station in Ireland, and Eila Grahame.

Many others have contributed and helped me to formulate a view or understand a mechanism, biological or human. Notably, these have been: Stephen Akester of fishing consultants Gowan McAllister, Hugh Allen of the Mallaig and North-West Fishermen's Association, David Armstrong, Jean Collins of the FAO, Ronald Fonteyne of the Fisheries Research Station in Ostend, the author Fred Archer, John Ashworth, R. Boddeke, Richard Brill of the University of Hawaii, John Barton of the Falkland Islands Government, the author Kevin

Crean, Sid Cook, Jim Cowie of Cowie Seafoods, William Calder, David Cummins, Tony Curran, Joe Cordero of the National Marine Fisheries Service in California, the film maker John Dollar, Peter Donald, Bob van Marlen of the Dutch Fisheries Institute, Astrid Thysegen, Tony Dickinson, Andreas Demetropoulos of the Fisheries Department in Nicosia, Yvonne L. de Reynier, Eric Edwards of the Shellfish Association of Great Britain, Bob Earll, Axel Eikimo, the fishing journalist Tom Wray, Richard Hughett, Dick Ferro, Norman Graham of the University of Humberside, Paul Goodisson, Beatrix Gorez of the Coalition of Fair Fisheries Agreements in Brussels, John Goodlad of the Shetlands Fishermen's Association, Ian Harrison formerly of the National Fishing Heritage Centre in Grimsby, Mike Heath of the Aberdeen Marine Laboratory, Ray Hilborn of the University of Washington, Dennis Haanpaa of Alaska's Department of Fish and Game, Tony Bailey of ICES, Gunnar Stefansson of the Marine Research Institute in Iceland, Jon Reynir Magnusson, Sveinn Hjörthur Hjartarson, Brynjolfur Bjarnason of Grandi, Thorsteinn Pálsson of the Icelandic Ministry of Fisheries, Stymir Gunnarson, Artúr Bogason, Reidar Thoresen, Stuart Barlow of IFOMA, Ray Gambell, Curt Johansson of Sweden's National Board of Fisheries, Thorë Jacobsen, John Koegler, Bob Lindley, Einer Lenke of the Greenland Home Rule Government, John Casey of the Marine Laboratory in Lowestoft, Joan McGinley, Philip MacMullen, fisherman Pete Mustand, fisherman Terry Norquay, Philip Atkinson of the New Zealand High Commission in London, Barry Deas of the National Fishermen's Federation Organisation, Jim Portus, Dave Pessel of the Plymouth Trawler Agents Association, Jim Clarke of Pescanova, Thorstein Pedersen of the University of Tromso in Norway, Peter Pearse, Thomas Poulsen of the Danish Institute of Fisheries Technology and Aquaculture, Mike Picken of Rubh Mor Oysters, John Pirie, Brian Riddell of the Department of Fisheries in British Columbia, Ingolf Røttingen of the Institute of Marine Research in Bergen, Robert Allan of the Scottish Fishermen's Federation, John Tumilty of the Sea Food Industry Association, Colin Smales of F. Smales and Son, J. Dale Shiveley of Texas Parks and Wildlife, John Shepherd of Southampton University, Mike Stuart of Scallop Kings, Jim Slater, Despina Symons, Roy Torres of the National Marine Fisheries Service in California, John Williams from the Ministry of Fisheries in Oslo, David Wileman of the Danish Institute of Fisheries Technology and Aquaculture, Carl Walters of the University of British Columbia, fisherman Govey Cargill, Ian Harrison, Joseph France, Richard McCormick, Howard Stock, Francisco Goulart, Andrade de Medeiros, Chesley Sanger, Quentin Bates, Phil Lockley, Chris Law, Dr Peter Henderson, and Alastair MacFarlane from the New Zealand Fishing Industry Board.

In a separate category I would like to thank Arthur Anderson who made the BBC Scotland film *Troubled Waters*, Eric Malling of Canadian CTV, Theresa King of Worldfish, Tim Oliver editor of *Fishing News* and the publishing group EMAP which also publishes *Fish Farming International* and *Fishing News International*, Frances Berrigan of Cicada Films, Tom Richardson and Barry Gibson and John Brownlee of Saltwater Sportsman, Gary Caputi editor of *Big Game Journal*, and Doug Olander editor of *Sport Fishing*.

Above all, thanks are due to Zahava Hanen of Alberta, author, founder of the David Thompson Trust, and my sponsor, without whom time to write this book would never have been made available.

Definitions of seas, oceans and species of fish I have taken unadulterated from source material. Similarly I have used measures and denominations as written of in source material, not attempting to harmonise metric and standard measurements. The metric tonne replaces the ton with the passage of time; they are in any case almost identical.

Michael Wigan

Contents

Introduction

'We come home and our very souls are purified by the sea.' (Manual Lopez Muniz)

'You're your own boss and it's up to you to do it. I love it. It's the only way to fly.' (Newfoundland cod fisherman)

'Ironically, many fishermen do not look at the competitive race for the fish in the way that economists do. Commercial fishing is an exhilarating form of high-stakes gambling and competing against other fishermen is part of the game. The free-for-all hunting lifestyle passed on from generation to generation partly explains why the argument for subsidies to 'protect the way of life' carries so much weight in the fishing industry. Both the public and the fishermen themselves see the profession as one of the last vestiges of the hunter-gatherer society.' (Laura Jones, environmental economist at the Fraser Institute; from *Fish or Cut Bait*, published by the Fraser Institute, Vancouver, 1997)

There is something about its sea fishermen that defines a nation. Those who ply the waves for a living in all weathers, for whom the horizon is a straight silvery line tipping off the curvature of the earth, whose business is beneath, in a dark, cold, pressurised habitat where man has no place, bring back with them onto land a whiff of something faraway and intangible. In our modern, laptop-dominated lives, the salt, wind and waves ring an echo. They remind us of the time when we were all hunter gatherers, hunters of finfish and gatherers of shellfish. We ventured forth to take from Nature's bounty, extractors of good things on which to subsist, predators with no conscience.

Fishermen are the last of the hunter gatherers. They have survived into a world where their practices have no equal or parallel. Dr J. Strickland many years ago at the 2nd Oceanographic Conference in Moscow remarked that our fishing methods were '10,000 years out of date'. Ingenious their contemporary refinements might be, but a harvest from the wild in which a third is jettisoned because it is the wrong species or size is surely being conducted with a cavalier disregard for the world's shrinking resources. Anyone with sense and a basic understanding of resource science can see this plainly. It is therefore extraordinary that it has taken until the last tickings away of the twentieth century for the bulk of people in the developed world to apprehend that we stand on the brink with regard to marine resources, in fact, in several instances – the anchovy fishery off Peru, the North Sea herring, Newfoundland's cod fishery – we stepped over it.

It is now becoming a truism that the extractive phase in world fisheries is long gone. The custom which perhaps started with the arrival of French and Portuguese fishing boats off Newfoundland's Grand Banks in 1504, or with the deployment of an astonishing 149 boats from eastern England in Iceland's fish-rich waters in 1528, and is continuing today, quite unremarked, with the fishing agreements between impoverished West African countries and the EU's surplus fishing fleets operating off their shores, is ending. The herring fishery, mentioned in the Domesday Book as yielding to the King of England 60,000 herrings from the little village of Dunwich, has already been forced to close in 1977. The four-century-old concept of distantwater fleets harvesting the offshore bounties of less muscular nations began to fade when nations started to declare their 200-mile limits in the 1950s. Fisheries from now on must be managed responsibly and sustainably, for despite the almost unbelievable fecundity, resilience and persistence of ocean systems, we are on the threshold of serious and sudden depletion. World catches of the fish prized by humans for eating have peaked. The take of tuna, the premium seafish, peaked in 1990, and, despite increased fishing effort, by more countries, using more sophisticated gear, it has not risen since.

Cod, another dietary staple, is in a worse condition. As several scientists from England's world leader fishery institute at Lowestoft echoed, it is remarkable, given the volume of fishing effort, that there are any cod left in the North Sea.

Several seas have already been fished to alarmingly low levels. The Food and Agriculture Organisation (FAO) of the United Nations, based in Rome, in its document 'The State of World Fisheries and Aquaculture', said that of the world's 200 major fishery resources 35 per cent show declining yields, 25 per cent are plateauing, and 40 per cent are at the development phase. The low exploitation category was empty. More recent FAO publications highlight the point that in the world's most productive oceans catches have been in long-term decline for twenty years. Many are producing similar yields of the higher-value fish as in the 1960s when far fewer boats, manipulating less deadly technology, were in action. Where once complex ecosystems existed, based on undersea productivity estimated to be equal to the whole plant production of the terrestrial surface, there is now, in some places, only seawater; few bigger creatures swim.

World fisheries have changed dramatically in the last forty years. In 1950, ninety-eight per cent of the world catch was caught in the north hemisphere; Japan alone took one eighth of it. Many of the world's most productive fishing grounds were barely exploited, the most obvious example being the fertile upwelling off Peru, in 1996 the world's biggest fishery by tonnage, but in 1950 a sardine-only resource, utilised by millions of seabirds. To the traditional seafaring nations of Europe, Norway, Iceland, Japan, Korea, Russia, Canada, America, and South Africa, have been added the apprentice fishing nations, revelling in their newly-discovered resources. These include New Zealand, Australia, the Philippines, Peru, Chile, Indonesia, Thailand, Taiwan, and Mexico. Peru in 1953 landed 113,000 tonnes of fish, compared to Japan's 4.5

million tonnes, whereas in 1994, Peru landed 8.4 million tonnes as to Japan's 8.1 million. Fish is no longer synonymous with cod, herring and mackerel. Today's largest catches are of low-grade species – anchovies, pollack and jack mackerel.

Changes in the structure of world fisheries have been huge and, in the traditional species, depletions have also been huge. It used to be estimated that an acre of sea off the English east coast yielded as much good food for human consumption as a hundred acres of Northamptonshire grassland. Yet the cod, one of the most closely-researched species, was reckoned in that same sea in 1996 to be clinging to survival at just one per cent of its estimated 'virgin', or original, biomass. One of the world's most fertile and vigorous fishes, the female of which can lay 10 million eggs, and specimens of which used to tip the scales at 200 pounds, the cod is now down to bare subsistence numbers. Its representatives are almost all small and young, and 70 per cent of cod from the North Sea are aged two or less. Yet they only mature between the ages of three and six. In the days of virgin cod stocks the sixteenth-century European seamen sailing out to the great cod banks of Canada's east coast lowered baskets over the side and raised them full of fish. The early pioneers reported cod so thick the rowing was hard. Today, between the fish lies a lot of empty seawater.

There is presently a partial moratorium on Canada's Grand Banks, a fertile wide shelf stretching south from Labrador to Newfoundland. This fishery's demise is the classic case of a contemporary fishery, under modern management, with a world famous stock, totally collapsing. Less spectacular has been the decline in the Common Fisheries Policy (CFP) area of EU member states where the North Sea, also a shallow shelf, provides another of the great fishing-grounds. Scientists' alarm bells have been ringing here for several decades. The total recorded catch fell from 7 million tonnes in 1983 to 5.5 million tonnes in 1990. The 1996 House of Lords Select Committee report on fisheries, which took evidence from all over the world and is an acknowledged defining document, expressed the overhanging crisis in fisheries generally in the strongest possible terms: 'All over the world stocks of most species of fish are being depleted by overfishing' ; 'Almost every variety of fisheries policy and management has proved inadequate to prevent overfishing' ; 'Immediate steps must be taken by all developed nations to reduce their commercial fishing effort very significantly.'

As fish stocks shrink, ways of reducing them more and more quickly develop apace. In one recent issue of the British fishing weekly *Fishing News* no fewer than four radically improved inventions for detecting and capturing fish were profiled. The race between seek-and-catch sophistications and methods of conserving non-target fish through specialised fishing gear is being won by the former. We live in the age of flow-scales, by which fish are weighed volumetrically as they come up the ramp out of the sea, fishing trips hauling home 1,000 tonnes, supertrawlers churning forward on 7,000 horsepower engines, robot jigging machines controlling tens of thousands of hooks, and ships which are harvesting and processing factories in one. 12,000 tonnes have been caught in one month in Peru's sardine fishery with over 500 tonnes being hauled in one net. On-vessel Swedish filleting machines are now capable of decapitating and filleting one fish every half second. Satellites aid fish location. The potential to empty seawaters of their denizens and put them onto dinner plates is growing quickly. Deepwater species are being jerked rapidly into the unfamiliar light from hitherto inconceivable depths. Much-used fishing-grounds, like the North Sea, are being so repeatedly trawled in the space of a year that the bottom resembles arable land after fine harrowing.

One marine species, the three-ton, vulnerable, kelp-eating Steller's sea-cow was exploited to extinction within a few years of its first discovery in 1741. The human ability to wreak this sort of damnation on saltwaters has vastly increased since. The chances, for example, of a North Sea cod reaching the end of its natural life free of interception by man is almost non-existent. It is some time since dredging destroyed most of the last natural oyster beds in Europe, but exploitation had already reached a crazy pitch back in 1843 when 70 million oysters were dredged from one bay, Cancale, in France.

There is a converse argument. The sea remains a big place, very well-known in the fish-rich shallows but less known in the depths. Even today almost no major biological oceanographic expedition goes to sea without returning with a new species. One fish of enormous size, the megamouth shark, was only discovered in 1976 when it became tangled in a cargo parachute being used as an anchor in a US Navy boat off Hawaii. The illustrious scientist Edward O. Wilson believes that other very large and unclassified sharks still swim the sea. When a Scottish fish factory worker in the north-west spotted on the processing conveyor the fifth *lamprogrammus scherbachori* to be discovered worldwide, he remarked 'It's just a fish. You quite often find strange things here.'

The list of 40,000 known species of fish is being added to. To date, no sea fishes, knowingly, have been rendered extinct, although the degradation of coral reefs, with their thousands of species, may have resulted in unlogged casualties, and, of course, it is hard to declare sea fish to be gone. Some species, like the coelocanth, supposed extinct for 70 million years and rediscovered in 1938, have re-surfaced. The deeper canyons under the big oceans previously considered devoid of fish are now turning up their own inhabitants, alien creatures of Gothic strangeness adapted to lightless, abyssal life, proof of the boundlessness of the unknown. Discoveries about the sea is the real pioneer's frontier; their findings have more relevance for humanity than anything in space.

The use of fish as food may in future be only a portion of their service. Important elements in seawater are collected and concentrated by fish – copper and zinc in oysters, and cobalt in lobsters and mussels. The role of 'industrial' fish species as the raw material for fish meal in aquaculture, nurturing bigger, more valuable fish is accelerating. Fish waste, like guts and heads, which can form a substantial part of bodyweight, can now be converted into a liquid silage feed. Fish have a recreational use which in America's sport fishery has already overtaken, in dollar values, the total commercial catch. Interest in the seas and their inhabitants has transformed those parts of the world's coastline sited to benefit from reef tourism. The chain of Sea Life Centres in Britain is flourishing and growing in number, just when zoos filled with terrestrial animals are bedevilled by lower visitor numbers and an agonising ethical and philosophical dialectic. The refinement, complexity and sheer bizarreness of some of the life forms which populate the sea has started to fascinate minds which once looked to outer space for sci-fi thrills.

In its rôle as healthy protein, fish is changing its position in world cultures. From being a cheap source of essential protein in many industrialised societies and a basic subsistence diet for coastal dwellers, fish has rocketed up the menu into top position. In Europe in the post-war period fish prices rose more slowly than those of other commodities, and for war-battered populations crying out for protein, fish was the most readily accessible resource. Today, the more affluent the society the more fish is consumed. Annual fish consumption in Canada between 1970 and 1990 rose from 4.9 kg per person to 7.2 kg. In Britain the rise was from 5.4 kg to 7.7 kg. In America it has been even faster, and in Japan, the archetypal fish-eating culture and, ironically and atypically, one in which the trend-setters are moving towards meat-eating, the average consumption is an astonishing 75 kg.

Fish, the last major harvest of the hunter gatherers, has become a luxury. Once the cheapest source of protein fish is now fashionable, health-friendly, and more suited, with its quicker digestibility, to sedentary lifestyles in the developed world. In America, fish products are the third largest import, after illegal drugs and oil. Fish imports cost America $6 billion annually. The fish trade deficit expanded from $1.8 billion to $3.5 billion in only six years, starting from 1980, spectacular growth by any standards. In Japan, where the search for the most perfect steak from a bluefin tuna is conducted daily with an almost fanatical fervour in Tokyo's frantic fish market, the freshest, fattest, and best can fetch 20,000 yen (£124) a kilo. By the time it has reached the plate the price of this segment of the ocean's pride has suffered a considerable increase.

The developed economies are busily chasing the resources of the undeveloped economies, whilst competing for each others has largely passed into history with the declaration of 200-mile limits. These Exclusive Economic Zones (EEZs) started in the 1950s and became widespread in the 1970s. Behind many of the world's international quarrels lie fishing interests. This is most obvious around Japan, the world's supremely fish-conscious nation, which has the inconvenience of neighbours' incursive EEZs.

Japan is disputing ownership of Takeshima Island with Korea and that of the Senkaku Islands with China; and the long-standing dispute with Russia about the Kurile Islands has consistently focused on fishing rights. The Falklands War between Argentina and Britain in 1982 had a substantial, though little spoken-of fisheries dimension, and since its resolution the Falkland Islands Government has enjoyed rental revenues of up to £20 million a year from squid licences. Disputes about the fishing limits between the two states are ongoing. The argument between Spain and Canada about Spanish trawlers catching fish off the edge of Canada's Grand Banks marked the first occasion when a fishing nation claimed rights over its resource even after the fish had moved, in unusual circumstances, beyond the 200-mile limit. The whole question of who owns the fish, and at what point, took on fresh energy, and it was revealing that nations with strong fishing traditions, like Britain, took Canada's side, whilst lesser fishing states in the EU adopted purely legalistic attitudes.

The threatened disappearance of the island of Kolbeinsey, a small pinnacle of rock to the north of Iceland, poses dilemmas for a fishery-dependent state facing the loss of a lump of rock which considerably enlarges, through its 200-mile orbit, the national EEZ. Never have such stony carbuncles been so precious. As the world's boundaries shrink, and the value of fish rises, fishing rights are gaining increasing significance in the lists of natural assets. For fish is the only major food produced from the wild. Fisheries have been called the most valuable renewable natural resource of all.

Renewable is the critical word. The circulation of global currents, the earth's magnetic field, and the complex oceanographic effects of melting and freezing at the ice-caps, ensure that all unenclosed oceans are regularly refreshed. Unlike on farmland, which has to be reinvigorated with field dressings, the sea, enriched at a microbial level by the sun reaching into the depths to cultivate weed-growth, restores itself. Indeed it is the presence of oceans, and water, that distinguishes this planet from all the others, which are arid.

At a localised level the actions of man have altered the constitution of sea-water. An example is the south North Sea where rivers laden with phosphate-rich sewage boosted fish populations in the short term until sewage treatments commenced in 1991; or in the eastern Mediterranean after the Suez Canal was built and the Nile's nutrients were lost, decimating Israel's sardine fishery. But on the larger scale man has had little effect on ocean currents and the rhythms and cycles of water warming and chilling. It is only on the fish that inhabit the sea, and on some coastal habitats like coral reefs, that the human influence has been drastic.

The challenge in the future is to allow the ocean's reproductive system to function to its best potential. The use of the sea and its fish resource calls for the most difficult thing in modern democracies, restraint and wise usage for the benefit of tomorrow's populations and tomorrow's voters.

Not only will the political problems have to be faced, of sharing equitably the huge fish resource, but the call on fishery scientists will be not only for guidance on catch rates, but for ingenious technology which can refine the way in which species are targeted. It is an irony that some of the most species-specific gear has arisen from sport fisheries and the knowledge gained, notably in America, has concerned bait species as well as targets like sailfish and billfish. Waste on the scale that has become customary, in order to comply with complex fishing policy regimes, will be increasingly unacceptable. It would have been a scandal long ago if publicly exposed.

Also unacceptable should be the notion of sacred cows, even if the cows are seals, or potent symbols of green activism. The green movement has rightly drawn attention to shameless and selfish exploitative fishing practices, often involving sea mammals rather than fish. Sea otters and turtles are prototype plunder victims to whose predicament everyone has been alerted. However, some species have now become inviolable totems. The enlightenment that greens urge on commercial fishermen must be pressed on themselves. The romanticisation of seals has led to an absurd impasse where the numbers of virtually all marine creatures are regulated by man's take with the exception of some of the most devastating predators of all. The green movement's uncritical espousal of marine mammals is a fine thing for nursery school but no part of a mature fisheries policy. It is doing enormous social and economic damage to the livelihoods of many small and fragile communities in the northern hemisphere, from the Canadian Inuits to the fish farmers in the Shetland Islands. Under relentless public relations pressure from self-righteous welfarists communities of traditional seal hunters in Alaska, deprived of a market for seal products, have in some cases been decimated by suicides almost on a mass scale. A responsible policy for marine exploitation can have no untouchables and no management-exempt orders.

Fisheries must not be the playthings of politicians looking for soft targets. Nature and the seasons dictate when fish are in best fettle for eating and that is when they should be landed. The calendar of the decision-making in the CFP involves fish often being caught in poor condition. This is organisationally inept, wasteful and unnecessary, and causes disillusionment within the harvesting and processing industries.

Fisheries are one of the sternest tests for international co-operation. 200-mile limits are a crude initial attempt to ring-fence the intangible; territorial demarcation by continental shelf would be more logical. Fish move in pursuit of food differently in fluctuating conditions, and what is available to be caught by one state one year may be in the neighbour's zone the next. Fish are slippery customers and their habit of ranging, most pronounced in pelagic (surface-dwelling) species, must be accommodated in sufficiently flexible and far-sighted arrangements, whatever the cost.

It also has to be recognised that international waters cannot be a free-for-all. Free for all, it is true, is free for no one. It may be that only ten per cent of the fish caught are from the 'high seas', not from the productive shelves and fishing banks, nor from the spectacularly fertile ocean upwelling areas. Nonetheless in the far out international waters are some of the keynote species like tuna and sharks. On the cusp of the third millennium the high seas are being brought into the sphere of internationally-agreed management and becoming part of fishing nations' responsibility. The formative influence was crashing stocks.

Some say the first aquaculture treatise was written by Fen Li in China in 475 AD. However, early aquaculture was of freshwater fish, cultured inland. The sorts of marine fish cultures, or mariculture, which stir most interest today are something quite different. In the case of salmon, fish farming has converted a renowned delicacy into a common food, available anywhere at a lower price than yesterday's staples such as cod. In other cases industrial production-line philosophies have been applied to species that cannot always withstand such intensification – or the ambient environment cannot. The rôle of mariculture in swabbing up coastal unemployment, in the survival of wild fish in the sea through market stabilisation, and in the sustenance of the world's human billions, has aroused many imaginations and cost bankers dear. The boom and bust cycles, both economic and biological, which have characterised modern mariculture, have hurt and broken many. Some brave entrepreneurs, living on high wires, deploying cutting-edge technology, have been involved. However, the atmosphere with environmentalists will only get more embattled while fish farming exploits the comforts of protected estuaries and accessible brackish waters. They are critical areas for many wild fish. The challenge is to farm fish in an environmentally neutral way, and in enhancing wild stocks to see at what point to hand the lead back to nature.

What has happened in Norway, the world leader in farmed Atlantic salmon, where fish farmers imperilling the wild run now argue that efforts to maintain wild stocks should be abandoned, is a frightening precedent. They say farmed and wild salmon are the same. How so? Farmed salmon migrate, and come from a very varied gene pool. Arrogant commercial interests are prepared to sweep away natural biota, from which their industry was founded initially, in the thrust for monopolistic management. It is the same as saying giant pandas are fine if some are in zoos, so why bother re-introducing them in China?

The North Sea has been called a giant natural fish farm. It is not: farm animals do not eat each other. Sea fish almost without exception live off each other, the same fish often, at different stages of growth, moving from being victim to predator of the same neighbour. Some deep-sea fish like anglers and groupers, with stomachs which massively distend, can digest fish much larger than themselves. The reason for this physical development is intriguing. Deeper water fish are few and far between, the thickest densities of plankton, the base building blocks of nourishment in the sea, being nearest the surface. Deep-dwelling fish therefore need huge gapes to ingest all the rare passing meals, regardless of size. The scope for research and discovery in this complex subject is immeasurable. I have yet to hear a fisheries biologist engaged in practical research who is not mesmerised with fascination about his subject. The unknowns are legion.

A simple example in a keynote species is the continuing mystery of what happens to salmon smolts when they leave European rivers and head into the sea. They head north, for the plentiful feeding of the North Atlantic. Their osmo-regulation, adapting from fresh to salt water, is a critical transition during which they are under stress, that we know. Once in the North Atlantic they grow fantastically fast. That is the sum of knowledge. Their diet, passage, the depth they swim, how much deeper they fall in the night, how much closer to the surface they ascend in the day, the risks they run, the mechanics of their migrating instinct, is a blank. No wonder in the books of thoughtful oceanographers, like the late Sir Alister Hardy, once Zoology Professor at Oxford University, there is an almost breathless excitement as he describes sifting through the contents of his experimental tow-nets and speculates on the intricacies of the food chain.

If the world wants to learn about the fish in the sea there are more people than greens, scientists and doom-mongers to be listened to. It is a failure of the technological age to assume that the experts can tell the whole story. The egghead bent over his logarithms will not produce workable fishing policies on his own. The fellow in the oilskins is too often forgotten or ignored. The one with roughened hands and sea-sprayed eyes telling of a different relationship with the sea and its denizens gets left out. His sort of evidence, often presented with character not algebra, in dialect not equations, may be hard to assimilate and it may be uncomfortable to stomach. But it is of a nature, and from a source, that must not be sidelined. After all, the basic data for fisheries all over the world starts with fishing catches and the performance of fishermen. The hunter on the sea like other hunters has a knowledge which has earned him a living. His hunches have had to pay. His apothegms are authentic.

Furthermore, the fishermen are being asked to carry out the policies. If they feel the regulations applied to them are illegitimate, for whatever reason, they will flout them. No number of 'eyes in the sky' will dissuade them from lawbreaking. It may be inconvenient for governments, galling for scientists to be contradicted and second-guessed by men in rough polo-necks, it may dismay the greens, but no fishery management will flourish if it is sourced only academically. For fishermen, the 'economic actors' as theorists distastefully refer to them, are a special breed. Ponder this question: what other community or industry commemorates its dead thirty years after the fateful calamity?

With the introduction of the quota system, whereby each boat has permission to catch a specific number of fish, often of a specific size, within a defined period, usually in a specific area, the whole rôle of the fisherman changed. He used to use his experience and judgement to decide when to fish and where. The modern fisherman is an agent of the catching process, hemmed round with regulations. So his public image has changed. Once known for his resourcefulness and independence of action the fisherman is more frequently characterised now as an irresponsible plunderer, eyeing up opportunities to beat the system. As Torben Vestergaard of Denmark's Aahus University put it: 'The strategies of action that used to earn them (fishermen) recognition now criminalise them.' The remark is not too strong. Many bitter fishermen feel this way, and letters columns in fishermen's magazines show this.

It has to alter. If fisheries are one of the most important renewable resources, care must be paid to get harvesters of the resource into an understanding with management and the processors. Fishermen, as all primary producers have had to do, must realise that they are working to a specific market, that the requirement to land fish in good condition is growing quickly. Until recently, scientists have regarded predictive assessments of fish stocks as a logical and achievable science. It may not be. Doubts have grown from unexplained events. Leading scientists in Europe's CFP have now conceded that stock assessments may be twenty-five per cent out. Old-timers in fishing were always wary of stock prediction, and had a more flexible way of life which allowed them to prosecute the fishery hard in times of abundance, relenting when cycles turned down. Contrary to the broad image, there were examples of fishing communities where restraint from overfishing was an agreed discipline. Informal systems of resource sharing were established. Such independent systems are difficult for governments and rely on the producer being the master of the market. Today the market rules, and in recent times the number of scientists, administrators and managers has risen as fishermen's numbers have decreased and fish stocks declined.

This book is about saltwater fisheries. The distinction between saltwater and sweetwater fisheries is a trifle partial. At the beginning of time seawater was fresh and lacked salt. It is rivers that naturally carry a salt load. Over thousands of years the miniscule proportions of salt washed from the earth's face have made seawater salty, the sea acting as the earth's salt-trap. Some freshwater fish species – for example, the powan, a type of herring, in Scotland's Loch Lomond – were originally sea-swimming fish that got trapped as ice-sheets retreated, and adapted to freshwater survival. Others, anadromous fish like salmon and the European eel, use both fresh and saltwater at different stages of their life-cycle. The legendary gamefish tarpon and red drum breed in brackish water, then turn seawards to the salt. Many commercially important species actually commence life near shore, in estuaries, saltmarshes, mangrove forests, seagrass and seaweed beds, coral reefs, bays, and wetlands.

It is in saltwater, if freshwater anglers will allow, that fish are at their optimum. They are bigger, faster, and stronger in the larger element. Any rod fisherman who has had a 4-pound bonefish streaking off the tropical flats, burning towards the deepwater zone at a seemingly impossible speed, can testify to that. It is in saltwater that fish, or sea mammals, exhibit the capacity for phenomenal growth rates. Famously, the blue whale's foetus balloons to immense proportions in a short time as the mother feeds off Antarctic krill. Less than an inch long at a few weeks after conception, they are about 23 feet long only ten months later when they are born. The growth rates of tuna are also astonishing, and tuna have metabolic rates almost equivalent to those of mammals, which is what gives them their extraordinary speed (up to 50 mph) and strength.

The superb feeding richness of the sea is based on plankton, from the Greek word meaning 'wandering'. Plankton consists of all the tiny microscopic creatures, including the eggs, larvae, and fry of many fish, and the spores of all marine algae, which float in the ocean top layer. Shallow seas have more plankton than deep seas, cold seas more than warm ones.

Diatoms are the commonest plants in the sea's phytoplankton, or the huge masses of microscopic, drifting plant life. These single-celled plants, which have neither leaves, stalks nor roots but rather a transparent shell made of silica, can divide in two in a single day, and so multiply prodigiously in a short time. As the sun mounts the sky in springtime the diatoms, which are dormant through winter, start to grow by utilising the phosphates and nitrates in seawater. Large areas of ocean may change colour, becoming green or brownish-green. Such is the efflorescence of the diatom explosion that it alters the chemistry of seawater. Diatoms bloom briefly because the warm summer sun heats only the sea's surface, failing to reach the depths from which more nutrients could come. With autumn gales, and a levelling of temperatures between shallow and deeper water, fresh nutrients are churned upwards towards the light and there provide food for the diatom's second outburst before the cooler, less illumined period of winter. Copepods feed on diatoms. Copepods are short-lived, small, and very numerous crustaceans, in turn the diet of finfish. One herring stomach has been found to contain 60,000 copepods.

In temperate waters the pelagic fish like herring and mackerel feed mostly from spring through to autumn, whilst bottom-feeders, or demersal species, like plaice, scavenge for young molluscs and crustaceans on the seabed. The feeding season is different in polar regions where there is only a single plant outburst, and also in the tropics where planktonic plants grow, under unchanging conditions, unchangingly. In warm waters dinoflagellates, tiny creatures with two whip-like swimming arms, replace diatoms, and it is these miniature life-forms which, massed together, create the luminescence on the sea reported by sailors.

In the world's upwelling areas, where deep currents meet a submerged bank, or each other, or hit a shelving shoreline, or where powerful offshore winds blow surface waters seawards producing an upthrust of nutrient-rich replacement water from below, are the world's most volumetrically spectacular fisheries, the largest being the anchovy swarm off Peru. The once-great cod fishery on the Grand Banks of Newfoundland derives from the turbulence and vertical circulation caused by the collision of the Labrador and Gulf Stream currents. The reason the remainder of the world's most productive fishing areas are on shallow grounds or banks, like Dogger Bank in the North Sea, is that light penetrates shallow water most easily, creating plankton growth which is the foundation of all animal life in the seas.

The nutritional value of plankton is most evident in its exclusive sustenance of the sea's largest dweller, larger than any known dinosaur. The diet of the blue whale, which can weigh over 150 tons, is entirely plankton, some of it invisible except through a microscope. A two-ton baby whale, after weaning, can reach maturity and a weight of 80 tons in two years. It is a fish farmer's dream conversion ratio come true! Unlike in fish farms there are no natural diseases in the sea caused by overcrowding.

What the sea cannot do, and what multiple retailing and consumer convenience require, is supply specific, regular quantities of like-sized protein on demand, regardless of season, breeding cycle, fish abundance, or ought else. Fishery policies dictated by politicians, at the behest of large multinationals attempting to universalise supply products, are doomed.

Predictably this realisation has been acted on by businessmen sooner than politicians. In the market-place, the movers and shakers are not bound hand and foot by pronouncements made last week or last year, or obliged to say things they know to be absurd. This latter predicament has been the fate of politicians trying to defend Europe's CFP who, knowing the realities, have preached fairy tales. It happened too in mid-collapse of the Newfoundland cod fishery. Retailers have a different agenda; public image is important, so is the persistence of high-profit product. Therefore it is unsurprising that Unilever, the largest international trader in fish products, moved first to preserve fish stocks, at the same time, conveniently, as purifying its public image as the seller of a declining resource. A 1996 announcement declared that the company would phase out the purchase of fish oil from unsustainable fisheries in European waters. The Unilever target was well-chosen. 'Industrial' fishing in EU waters chiefly took sandeels, a fishery prosecuted principally by Denmark, in a catch that, unusually, is unrestricted. No quotas were in place. Unilever's decision was rapidly copied by other companies whose directors took the view that to have their headquarters surrounded by environmentalists dressed as puffins – making the point that puffins, which eat sandeels, are being deprived of their staple diet – was attention they could do without. Thus, we entered the era of market distinctions about sustainable stocks.

If fisheries and the fish are to be secured, history is not auspicious. It is a little dismaying to note that the Roman writers Juvenal and Seneca reported the passion for eating fish had impoverished the Mediterranean and the Tyrrhenian Sea, although they probably meant the inshore fisheries not the great tuna migrations. On the other hand, thousands of years have passed, and the fish, though thinned down rather more, still persist.The great plunders – of whales, sea otters, commercial finfish like cod and herring – have continued almost unabated until the eleventh hour, until, often, it has simply become uneconomic to continue. In some fisheries, especially where the politics are too entangled for politicians to extricate themselves without committing electoral suicide, that will happen again. In others, particularly where a larger proportion of the population understands how fisheries tick and how they have to be sensibly exploited, where there are long fishing traditions, there will be, and already are, carefully tuned management policies. George Orwell said, 'History becomes more a race between education and catastrophe.' Crunch-time is coming.

Chapter 1

Herring, Cod, and Whales

'It is a certain maxim that all states are powerful at sea as they flourish in the fishing trade.' (William Wood, *A Survey of Trade*, London 1772)

'. . . few of the activities of men are half as wasteful as whaling has always been. Where else does man permit himself to kill so magnificent an animal for a reward so disproportionately small?' (Hawthorne Daniel and Francis Minot, *The Inexhaustible Sea*)

'And God said . . . let them have dominion over the fish of the sea.' (*The Bible*, Book of Genesis)

'The whaling industry simply represented the industrial revolution's expansion to the most remote waters of the world. This expansion occurred at a time when the acquisition of natural resources was considered to be a God-given right.' (John R. Bockstoce, *Whales, Ice and Men*)

'If every egg shed by every fish were able to develop and become a cod each year it would only take a few years to fill the seas of all the world so packed tight with cod-fish that there would be no room for any more. But this disaster will never happen to us whatever others may befall since innumerable dangers, thank heaven, surround the egg before it hatches.' (F.D. Ommanney, *North Cape*)

'The ship is rolling gently; and the fishing-lantern slung from the mast swings to and fro, so that all the shadows sway. One of the deck-hands winds in the warp on the steam capstan, and the rest of us are at work on the nets, hauling them in over a roller on the side of the ship. If the catch is a good one, the nets will come up laden with fish hanging in the meshes – a mass of glistening, quivering silver. If we are lucky, too, we may see the nets as they leave the water ablaze with green fire – the phosphorescence of the sea. Once aboard, each is shaken so that the meshes are pulled open, and the herrings either fly upward or fall down from the net according to the way they are facing. There is the rich smell of sodden netting; and as each net is shaken, the air is filled with fluttering silver scales, glittering in the lamp-light like a shower of tinsel. The deck is piled with fish, which from time to time are shovelled through the circular openings to the holds below . . . So the fishing has been going on through the summer and autumn since it started in June in Shetland. So, too, it has been going on, year after year, for centuries. There has been no real change in the actual method of fishing; old drawings of the sixteenth century show the Dutch using just these same kind of drift-nets and floats; steam and diesel engines now carry the drifters more quickly to and from the fishing-grounds, but that is the only difference.'

In the above extract Sir Alister Hardy was recalling his impressions of fishing trips in the North Sea made around the time of the Second World War. The

23

passage conveys much – the fecundity of the sea, the rugged, romantically-hued life of the fishermen who prowled its surface, the drama of net-hauling, the ancient lineage of the fishing, and the beauty of the quarry.

The herring once swam the whole North Atlantic in gigantic shoals which used to number millions and stretch as much as nine miles long and two or three miles broad, moving towards the shore when ready to spawn. At different times of the year they could be caught round the coastline of the British Isles making the herring fishing potentially unrestricted by season. Around Britain alone there are several different races.

The spawning is unique. Most fish's eggs float on the sea's surface, moving with wind and tide. The herring's eggs (10,000–30,000 per fish) are heavier than seawater. They descend to the seabed plastering the subjects on which they fall, some drifting to crevices between stones which afford protection. It has been suggested that this indicates the herring was originally anadromous, breeding in rivers, where eggs need to lodge amongst stones, and fattening at sea. The herring's diet is principally a minute crustacean related to the shrimp and mainly found in the surface layers of the sea. It sieves them through its gill-rakers. Accordingly, the herring is a surface, or pelagic, fish, making its capture simpler. When young herring tend to seek shallow water it is then that they are caught, along with sprats, as whitebait. Herring join the main shoals as adults from age three, by which time they are ten inches long, three quarters of their final size.

During the odyssey of growth the herring faces many adversaries; almost everything of same size or bigger is a potential predator. The eggs are eaten by haddock, and surface-patrolling terns and gulls scoop off the fry. Adult herring pay for their surface-feeding habit to the benefit of larger gulls, gannets and cormorants. Porpoises follow their shoals, and herring is also an important food for tuna, dogfish, hake and coalfish. The old fishermen used to say that when the herring spawned everything else thrived.

The herring has had a commercial and dietary importance for man since long ago. It is believed that herrings were fished from Yarmouth in England shortly after the landing of Cedric the Saxon in 495 AD. By 670 the fishermen of Lowestoft and Yarmouth were paying tax on their herring fishery to the Abbey of Barking. The earliest export of salted fish, most likely to have been herring, was from Scotland to Holland in the ninth century. In 1088 the Duke of Normandy allowed the Abbey of Sainte Trinité at Fécamp to hold a herring fair. By the middle of the twelfth century herrings had become a major commodity. A.M. Samuel, talking of England, goes as far as to say, '. . . that national system of economics which was for the most part represented by the words wool and herring.' Market manipulation was an old thing in fish: a statute was passed in 1357 to nail herring spivs and break up a market-makers 'corner' in herrings in Yarmouth. Philippe de Maziere in 1382 states that there were many thousands of boats fishing herring between Denmark and Norway, in addition to 500 vessels for gutting and packing. The herring began to figure as one of the staple traded items in the consortium of north European towns which formed a trading syndicate known as the Hanseatic League. For two hundred years herrings caught in the Baltic by Danes were traded over Europe for commodities like wine and wool. Herring had become the very currency of trade.

The herring then did something which it has, elsewhere, done since, to the supreme inconvenience of those who wish to harvest it: it disappeared from its traditional grounds, in this case to the south of Sweden. The herring waned and with it the trading domains of the Hanseatic cities. Shortly afterwards another herring stock, as it subsequently transpired of a different race, materialised in the North Sea, and it began to be harvested by Holland.

From 1397 a Dutchman, William Buekel, discovered how to cure, salt and pack herrings, starting a fishery which lasted for four centuries. From their two-masted, square-sailed herring 'busses', sturdy, slow, but roomy boats of up to a hundred tons, the Dutch fished from Shetland down the east coast of Britain as far as Essex. With small catcher boats called 'doggers', crewed by twelve men and two boys, filling up the busses, and processing and barreling of the herrings being done onboard as fishing continued, the Dutch built a prosperous industry employing in all about a million citizens. The fleet burgeoned to the extent that in the seventeenth century it was said that four out of every five merchant ships were Dutch. The prohibition on eating meat for two days a week, and for forty days up to Lent, increased the market for fish throughout Christendom. The Dutch business was said to be worth £3 million a year. In 1679 they had 4,000 vessels engaging 200,000 sailors, a figure that had fallen to 3,000 vessels and 40,000 men by 1747. Even on this reduced scale the fishery is reported as worth £5 million. No other fishing nation of the period could compare with it. During this time there were frequent conflicts and sea-battles over herrings. Even the seminal question of whether the sea was a hunting-ground open to all was aired. Herrings were important to every nation's power standing. A.M. Samuel writes that the soldiers at the Battle of Agincourt in the fifteenth century, and the sailors who destroyed the Spanish Armada a century later, lived off dried and smoked fish. Cured fish was the food which sustained travellers and wayfarers.

The grip of Holland's mercantile supremacy was first shaken by Oliver Cromwell, following a dispute about the herring fishery which had started earlier with the Stuart monarchy. An Act passed in 1651 forbade any transportation of herrings in or out of England or Ireland except by own-nation ships, thus torpedoeing the Dutch near-monopolistic freight trade. However, it was not until the early nineteenth century that Britain initiated the escalating herring fishery which saw and stimulated so many changes of technology. It was an extraordinary point in world fisheries, reached just before the First World War, when England, Scotland and Norway, and to a lesser extent Holland, collectively extracted over a million tons of herring a year from local waters. The rise of the herring fishery is a phenomenon of interest to historians, demographers, economists and fishermen alike. Nowhere did it develop faster than in Scotland, a narrow country with an indented coastline favourable to spawning herring, nicely situated in the heartland of the southward-moving migration.

Scotland had a long history of industrial herring fishing, the earliest tax on herring catches being four pennies on every ten thousand fish, introduced in 1240; by 1424 this had increased to a penny for every thousand fish for every Scottish fishing village. The Crown's realisation of the commercial importance of herring is signalled in the 1488 decree that only royal boroughs could export them. From 1540 herring barrels were branded, and unbranded ones were subject to confiscation. Quality controls had started. By 1808 barrel contents had been regularised: each barrel was to hold 200–224 pounds of herrings excluding the salt and brine which acted as preservative. To ensure that dirty tricks were ferreted out, all young fishery officers were trainee coopers. It is interesting to note how this sophisticated branding system stood the test of time: the last year barrels were branded in the traditional way was 1939.

Scotland's world-leading herring fishery was to centre on the unlikeliest of places, a straggling row of humble buildings on the far north of the east coast with a population in the 1790s of under a thousand. Despite lying just off the great herring migration, Wick, destined to be the herring capital of the world, did not even have a harbour. Cod was the desired fish until the 1780s. Herring started to be landed and in 1790 Wick had a total of thirty-two boats lying on the shore, or the 'bound', where they were not infrequently battered by storms. These boats were twenty feet long, had a crew of four, and used eight to ten nets, working over a season of sixteen weeks.

Displaced agricultural tenants were pouring coastwards from the Highland glens and by 1835, 830 boats (thirty feet long by now) were fishing from Wick. By 1862 there were over 11,000 boats, fishermen from all over Britain joining the onslaught on the 'silver darlings'. What was behind the transformation of the little borough of Wick? The critical event was the 1786 decision by the government to introduce two bounties.

The details of this were discussed lengthily, but by 1808 an Act had been passed including in the preamble the statement: 'The improvement of the British herring fishery is an object of most essential importance to the wealth and commercial prosperity, as well as to the naval strength, of this kingdom.' A bounty (£3 per ton) was paid on all vessels fitted out for herring fishing of between sixty and a hundred tons; and a further bounty (two shillings a barrel) was paid for all herrings cured and packed according to the Act's provisions. This has been described by Philip Street, in *Beyond the Tides*, as 'probably the most important single enactment in the history of British fisheries.' It transformed the economics of fishing, because fishermen could then be advanced credit against the value of the herrings they hoped to land. Herring fishing became highly profitable. A man could reasonably hope to emerge from a season owning his own boat with cash to spare, having embarked on it without even a net.

The next turning-point was the identification of Wick, by the engineer Thomas Telford, as the locus for a major harbour for the expanding herring fishery. The British Fisheries Society, a joint stock company created by the government and financed by local landowners to stimulate fishing enterprise in the Scottish Highlands and Islands, was granted £7,500 by Parliament in 1806 with which to build it. In the new harbour settlement of Pultneytown it was decreed that no fisherman or cooper could own more than his own allotted space. This encouraged the creation of a new class of professional fishermen. Up to this time there had been no professional fishing class, fishermen being agricultural tenants on smallholdings (crofters) who practised seasonal fishing.

The explosion of life in Wick is best demonstrated by photographs. Wick was at its zenith in 1864, four years after the introduction of cotton nets had trebled the catching capacity of hemp. Photographs of that time show a forest of masts enclosed by Telford's superb stone jetties. The naturalist Frank Buckland, appointed to the Herring Commission of Enquiry in 1877, wrote of a visit to Wick that the boats were so crammed together you could scarcely see water between them. On the quay tens of thousands of barrels, neatly stacked in pyramids, stretched the whole length of the immense stoneworks. Standing in front of trays of herrings, their hands bandaged against nicks, oatmeal stuffed into the sores caused by the salt, were the legendary fisher-girls, who often worked till midnight, whose gutting knives flashed so fast that observers were dazzled; to gut sixty fish a minute was considered a good rate. It is a scene of astonishing endeavour and labour, an engine of mass food production, based on a fertile sea, and made possible by clever governance.

Not all was pain, sweat and tribulation. In 1847 it is recorded that 800 gallons of whisky were being drunk each week. This rose to 500 gallons a day at the season's crescendo. A distillery and a brewery were constructed. Public houses were not hard to find: for every hundred citizens of Wick and Pultneytown there was a drinking house. A contemporary commentator remarked: 'During the fishing season enormous potations were indulged in.' In response the Abstinence Society 'organised furiously' in the words of Frank Foden, and a Temperance Hall was built in 1841. In 1844, before the herring peak, a residential populace of 3,200 swelled to 14,500 in the summer. 3,500 were fisher-girls, the same number active fishermen, and 400 were coopers constructing wooden barrels. By 1861 the number of fishermen had risen to 5,000. North Highlanders would walk 130 miles to find work in Wick in the herring season. The largest kippering kiln in Wick held 90,000 herrings and at one time fifty companies were producing kippers there. The record herring catch for the port was 50,160 crans landed in two days. A cran was approximately 1,000–2,000 herrings so this 48-hour catch consisted of 50–100 million fish!

Some of the fishery's characteristics were richly idiosyncratic. In the fishing museum at Anstruther on Scotland's east coast is an egg-shaped lump of lead attached to stiff wire. This fish locator was lowered from the boat and held taut. When fish hit it a skilled listener could detect a herring shoal passing beneath.

More macabre was the source of the buoys which held up the herring nets. They were made from the skins of dogs, treated with alum to keep them flexible. There were special dog farms and different sizes of dog made different buoys. Bladders of oxen were also used. History relates that the man whose job was killing the dogs had an unhappy time walking the streets of Wick; local dogs, sensing the menace, snarled and barked at him.

Naturally such a vigorous industry as the herring fishery, acting as an entre-pôt for legion different people, did not always proceed smoothly. One year the militia was despatched all the way from Edinburgh to separate local men from fights with Highlanders. The herring trade changed the population of Caithness and the county's roll-call rose from 19,000 in 1800 to 43,000 in 1863. In 1843, in the midst of the heady years, a three-mile fishing exclusion limit was declared for Wick boats which were thenceforth to be numbered. A year later fishery officers reported that during the season they had branded, or authenti-cated, 183,000 barrels of herrings. Such was Wick's pride in its efflorescence onto the national stage that it prompted a hyperbolic reference in the local newspaper to 'Our Northern Athens'.

Today Scotland exports the cream of its fish. The same happened in the first half of the nineteenth century to the herrings caught off Caithness. Early in the period the herrings were shipped principally to Ireland, to the colonies in the West Indies, and to the industrial households in England and Scotland. As the regularisation of the industry strengthened, and the 'Crown Brand' burned on the lid of the barrel established a good name for conformity and quality, a better market into Europe was consolidated. Dutch herrings clinched the top quality marque, and the Scandinavians sold lower-grade herrings. Meanwhile the Scottish cured herring established a virtual monopoly with the important German Customs Union or *Zollverein*. Barreled herrings left Wick for Russia, Poland, and the Baltic seaports, and export traffic peaked when fifteen cargo ships left the little Highland town every week. Spanish, Belgian and French markets were more awkward because of high import duties; and Russia allowed in Norwegian herrings at a much lower tariff than Scottish ones. Prices and supplies were relatively steady, although a series of cholera outbreaks in 1833 kept fishermen away for three years.

It is not only the volume of the Wick fishery which was remarkable. Wick's herring days say something about the nature of fisheries. Scotland, after all, had many towns which grew exponentially during the Industrial Revolution, some, like Wick, staffed by homeless Highlanders. But Wick's munificence – and fortunes were made – relied not on industrial or technological change, but on the removal of a natural resource, a migrating fish. Herrings were a self-replenishing gold rush. Unlike in the Firth of Forth, where an earlier herring bonanza had occurred, the Caithness fishery did not fade out. The fish kept faith with their routing. As the south Swedish coastline fishery had shown cen-turies earlier, this fidelity to traditional spawning grounds could not be relied upon. Undersea eruptions, climatic change, and a variety of imponderables could alter fish passage. With today's knowledge of herrings and migrations it is likely that even with over a thousand half-decked fishing boats sailing from Wick the human catch was an infinitesimal part of the whole herring shoaling. Many developments and phases in fishing were to pass before the herring was finally, about 140 years later, to meet that fate which must have seemed inconceivable in the 1830s in Wick – to be almost fished out.

The Industrial Revolution was slow to affect fishing, for wind and tide came free of charge. George Stephenson had made the first steam train in 1814 but it was not until 1881, preceded briefly by the use of steam and paddle tugs to tow traditional sailing smacks, that the first fully-fledged steam trawlers were launched in England. Their catch rates were four times those of sail smacks. The industrialisation of fisheries had commenced.

Fisheries might be thought to have come far, but had they? The types of fishing gear – nets and hooks – had been in use for thousands of years. Fishing rights in the world's major fishing grounds were in many cases unresolved, even after nations declared 200-mile limits in the 1960s and 1970s. The earliest throwers of nets and jiggers with hooks had comprehended the importance of these rights long before. By and large the particular fish eaten, and considered most toothsome, were the same in ancient civilisations as today.

The middens of shells beside the ruins of settlements in northern Europe, such as Skara Brae in Orkney, testify to man's use of the harvest from the sea from thousands of years before Christ. The theory that we are best known by our rubbish is supported by examinations of middens; the presence of bones from the deepwater fish, hake, at a midden at Cnip on the Isle of Lewis proves that in 500 AD Lewismen possessed nautical skills and vessels to match. How much earlier than that they had fished deep waters is conjectural. Spearing fish in the shallows dates further back, requiring simpler procedures. The discovery on Easter Island of a hook made from human bone gives an idea of localised priorities. Which implements were used for inland lake and river fisheries, and which for sea fishing is not always clear, but it is known that the ancient Egyptians caught red tunny in fixed nets and small mackerel-like fish with a type of seine (a circular net drawn tight as fish amass). Also documented was the practice of Egyptians to post look-outs on cliffs to signal incoming fish; Scottish salmon netters, amongst others, sometimes still do this.

The Romans were more fish-orientated, and seafish-orientated, even than the Greeks. They used horsehair lines and sometimes laced their baits with scented hooks. The use of wicker creels, baited traps, seines fished from the beach, and sailboats in which to go to sea, were all standard kit. Oppian records the use of dolphins to drive fish towards the shore where they were perforated with tridents, either from land or boat, a curious early example of mammals being used to catch fish. Famously, the Chinese used tame cormorants, with neck-rings to prevent fish slipping down the gullet, to fish for them. Oppian lists a dazzling array of fishy edibles from the sea including whales, sea urchins, and razor shells. As the whale dived the fishermen released ballooning bladders on the line; when these resurfaced the whale was tiring. The Romans were epicures: in Diocletian's edict of 301 AD the best seafish were deemed twice as desirable as freshwater fish. Sturgeon, mullet, and turbot ranked highest, pride of place going to the red mullet. There was open access to the sea in Roman times but foreshore dwellers could stake off a cove and take fish in tidal traps. Dolphins were mostly sacred, but tuna and swordfish were widely eaten. Using flickering feathers dancing on the surface, pulled by fast ten-man boats, Aeolian describes tuna-catching methods which in essence resemble American sea anglers fly-fishing techniques today. Aristotle reports the use of kites to catch fish, by holding hooks near the surface, also an American sport fisherman's tactic coming into vogue in the late twentieth century. The ichthyophagi from the Arabian Gulf are described as being solely fish eaters, dressing in fish-skins, and inhabiting houses made of whale bones. Their nets were immense and made of palm leaves.

Taken as a whole it is interesting, given the permanence and universality of sea fishing, that in the voluminous texts from classical civilisation fishing receives so few mentions. The activity of fishing never impinged, it seemed, on national affairs. Perhaps one reason is that fishermen generally held low social rank. When Christ recruited his disciples from fishermen the symbolic point was that his emissaries, coming from the bottom, implicated all of society. The second point was that fishermen, though as the Bible says, simple folk, had discernment and integrity. If they were persuaded of Christ's authenticity, so might others be. Also fishermen had developed social cultures rich in superstition and perennial celebration and, working in the capricious elements, were inclined to piety.

Many of the features of the fisherman have survived the industrialisation of fishing, and even its computerisation, and persist in fishing communities today. Had fishermen through history had higher social status there would have been a different evolution of national fishing rights. The stalwart reputation of fishermen, not untouched by romance, is partly why fishing disputes in modern times so readily focus the sympathies of sovereign states. A surrender of fishing rights takes part of a nation's soul.

Medieval history shows why. The search for fishing-grounds, and battles over them, developed and delineated the key nations in Europe's history. The Portuguese, French and English adventurers who made expeditions across the Atlantic to the New World sought gold, booty, and trading links, and found instead fish. They found fish in concentrations hitherto unknown. From the accounts sent home, rowing through a shoal of cod on the Grand Banks was something like paddling across a sea cage on a modern fish farm. The weight of fish held back the oars.

The discovery of the Grand Banks started the first distantwater fishery. Herring were caught commercially, on a local basis, and exported from an early time, but the cod fishery was quite different. For the first four hundred years of the fishery, up until this century, the deep-swimming cod were caught on hook and line, not in nets. The whole Atlantic had to be crossed in fleets of boats to reach the fishing-ground. Provisioning the ships, requisitioning the men (sometimes a brutish activity), and even making a semi-accurate landfall, were major undertakings. These voyages changed not only commercial history, but determined the early history of the Americas. The Norsemen may have reached America 500 years earlier, but the Europeans staked it out. Distantwater fishing was to last until the 1960s when countries began to define and ring-fence their marine territories. Re-drawing the fishing map occasioned mini wars, such as

Britain's cod wars with Iceland between 1958 and 1961, and 1972 and 1973. Today distant waters refer to only the big oceanic expanses, and even here catches are slowly becoming subject to regulation. The Grand Banks cod fishery did more than transfer a food resource from the Americas to Europe; it took a part in defining several European nations, and it was the largest factor in the development of eastern Canada. Of all the participants, Britain was the one to go on and dominate the western world at sea and become the pre-eminent fishing power. The experiences gained in the prosecution of the New World cod fishery, crossing the Atlantic twice a year during the season of equinoctial gales, in vessels of only fifty tons burden, contributed not a little to the formulation of a navy which was eventually to ring a great colonial empire. In the process cod was to become a familiar taste to dwellers throughout the northern hemisphere.

In 1871, L'Abbé Ferland wrote in his journal on a voyage up the Gaspé coast of present day Quebec, 'It is the land of the codfish! Your eyes and nose, your tongue and throat and ears as well, soon make you realise that in the peninsula of Gaspé the codfish forms the basis alike of food and amusements, of business and general talk, of regret, hope, good luck, everyday life – I would almost be ready to say of existence itself.' What was the basis for this? As in all fishing conglomerations it was a mixture of vital elements, geography underlying all.

In the Americas the cod range from Greenland and Baffin Island in the north, southward along the continental shelf of Labrador, Newfoundland and the Gulf of St Lawrence, and down the coast to Maine. Depending on water temperatures cod can even be found as far south as North Carolina. The main populations of 'Northern' cod flourish on the shelf which stretches east and south from Newfoundland, extending over 200 miles into the Atlantic. Termed a 'submarine delta' by the French writer Marcel Herubel, the Grand Banks are a moving tableland of broken shells, sand, and pebbles, mostly undulating in form with long spits of sand, and with some deeper holes caused by the eddies and backwashes of the Labrador Current meeting the Gulf Stream. In pursuit of capelin, a small herring-like fish, cod move onto the shallow shelf, which is only 150–360 feet down, in spring, when they shoal in huge agglomerations for spawning. Cod sometimes even ground themselves onshore in pursuit of capelin.

Western Atlantic cod need water between 0 and 10 degrees centigrade, but the preferred temperature for spawning is 3–4 degrees. This is found in 900–1,200 feet of water on the edge of the continental shelf. Cod eggs initially float to the surface, but as they grow (the rate dependent on temperature) they sink to the bottom. Cod are fecund, long-lived, and capable of attaining huge size. Conventionally described as 'groundfish', demersal or bottom-dwelling, cod in fact are not confined to any particular depth. The capelin hunt is near the surface, and although in some fisheries they feed near the bottom in shallow water, they will also move in summer off New England down to 1,500 feet seeking cooler water.

The continuation of the presence of cod may well be highly important; within their habitat they are most active performers. Their perpetual hunger and capacity to go on growing leads them into a huge diet range. They eat the principal crustaceans – clams, crabs, mussels, lobsters, cockles; they will engorge sea cucumbers, starfish and sea squirts; even spiny sea urchins are fair game. They consume sculpins and flounders, and hunt through seaweed for rock eels. Molluscs are swallowed whole, and the empty shells are then neatly stacked in the fish's stomach. All the finfish of their own ilk which can be swallowed are potential partners with those already mentioned in the stomach of a cod.

They can lay six million eggs, live for as long as twenty-six years, and have the potential, rarely accomplished, to top 200 pounds. Being also firm, white-fleshed, and very good to eat, it is logical that the cod fishery once assumed such global significance.

In 1497 John Cabot sailed westwards out of Bristol, England, in a small boat, with only an eighteen-man crew, to find Cathay or northern China. Born of humble stock in Genoa, Cabot rose to distinction first in the fleets of Spain and then of England. Searching for sponsorship for his enterprise, Cabot was written of by two Spanish ambassadors as 'poor, visionary, skilful, daring, persevering and boastful'. His famous account, 'The sea, so full of fish, which are not only taken in a net, but also in a basket; a stone being fastened to it to keep it in the water', was no boast. On a later Cabot voyage in 1508 an observer reports watching the bears rushing into the sea to try and claw the cod towards land. Before stern trawlers arrived on the Banks in the 1950s children in Newfoundland were still catching fish in dip-baskets. Cabot did not tarry long on land, but one other thing he records is of particular note: he found a net-making needle. The native Indians of Cape Breton, for that is most likely where he had hit land, were fishermen.

This was probably inevitable. The traveller Perret said of Newfoundland, just to the north: 'In Newfoundland as nowhere else can one be made to feel the contrast between a land that is infinitely silent, motionless, poor in vegetation, above all poor in its variety of living creatures, and a sea which harbours every form of life.' Although the French and Portuguese had arrived on the Grand Banks by 1504, cod capture preceded them. England's appearance on such a scene of plenteous opportunity was only delayed because of the vigorous cod fishery being prosecuted off Iceland.

By 1578 England had become the dominant fishing nation on the Grand Banks with fifty boats. By 1600 one hundred and fifty boats were fishing there, following the introduction by the Danes of fishing licences for visitors to Iceland – surely a very early example of a licensed fishery – and the same number of boats hailed from France. By 1597, Dutch, French and Irish boats were calling at Plymouth, England, to buy fish from the incoming Newfoundland fleet. Dried Newfoundland cod were being shipped from London to the Mediterranean. It is astonishing how quickly, given the rudimentary state of navigation and sailing, an international trade in cod had been built up. The fish themselves were splendid specimens; cod that were five or six feet long were recorded.

David Cushing in *The Provident Sea* depicts the typical fishing operation: 'A vessel of 100 tons carried 40 men (24 fishermen, 7 headers and splitters, 2 boys to lay fish on the table, 3 to salt fish and 3 to pitch salt on land and to wash and dry the fish). Such a vessel would carry nets, leads, hooks, lines, bread, beer, beef and pork, which supported employment of many bakers, brewers, coopers, carpenters, smiths, net makers, rope makers, line makers, hook makers and pulley makers in the West Country. Salt was bought from Spain, Portugal and France.' The season's catch, he continues, would have been 500 tons, plus purified cod liver oil (contained in hogsheads), and the fishing season in Newfoundland would have been from May to the end of August.

In the latter half of the seventeenth century the English fishery off Newfoundland suffered competition from France and New England, and from conflicts with natives. Prior to that it is said that 300–400 English vessels visited Newfoundland each year. Portugal and France, both largely Catholic and weak in agricultural development, were present in numbers too. It is one of the mournful facts of this fishery that a tradition started by the omnivorous French, of catching and eating, and salting for bait, great auks, mostly from the Funk Islands, must have contributed heavily to that romantic bird's final extinction in 1844. For nearly 350 years this flightless and hapless bird succoured the sailors in Europe's cod fishery.

The cod remained super-abundant. One Captain Cartwright, in his Labrador journal, reports holding up capelin and cod snatching them from his fingers. Along with cod guts, or pieces of herring, capelin served as bait. So thick were the capelin shoals that they could be driven ashore, where they were loaded onto carts. The seas of the Americas were indeed a bountiful place prior to the incursions of the Old World.

It is interesting to note the early operation of the gastronomy factor in the market of France. The French, pushed by the English to the outlying parts of the fishery, found a place which fished well earlier – from January – and where they claimed the cod were fatter and sweeter. The fish were salted and packed in layers for three or four days, re-salted, and made ready for the journey. Even if only half full, by Lent these ships would sail for home to take advantage of the Paris market when best prices were paid for the first cod. The same ship might return twice to the Banks in the year. The catch was practically all for Parisian consumption. As each one hundred ton ship would bring back 20,000–25,000 fish the Parisians must have been hearty cod-lovers. When not cod-fishing the French sailors hunted whales and walruses.

Today the same principles determine fish imports into France. The best Scottish fish or crustaceans go direct to France, or to Spain. In Tokyo, where the highest prices in the world are paid for premium quality fish, products are jet-freighted alive from enormous distances to satisfy the tastes of the modern fish gourmands. A discriminating palate in fish dates in Japan from the sixth century when the country became Buddhist and eating four-footed animals was discouraged.

From the early days of distantwater fishing the catching of fish served a wider political agenda. When Portuguese and Spanish fortunes waned, England opened markets which the French could not reach. The demand for cod translated into a need for ships and men. In the words of Harold Innis: 'Cod from Newfoundland was the lever by which she (England) wrested her share of the riches of the New World from Spain.' In this context England had the major advantage of being closer to Newfoundland. Fish contracted for in England were sold in ports like Marseilles, Nantes, Bordeaux, and Bayonne. From the 1670s when the slave trade began to bulk up, low grade fish from the New England fishery went to the West Indies, to feed sugar plantation workers; when slaves were emancipated in the nineteenth century the structure of the British trade was undermined, as was that of the Spanish dried and pickled fish trade when her colonies abandoned slavery later on in 1880.

From the point of view of investment and income the trips to Newfoundland paid excellent dividends. Unlike off Iceland, where the season was a narrow window of opportunity offering only one trip a year, Newfoundland had a spring season ending in June, then an August to November one. Captain John Smith commented that the cost of outfitting a one hundred ton ship and a two hundred ton ship would be cleared in a single trip if the catch was freighted straight to Spain. As for the sailors, their pay – calculated on numbers of cod tongues – was treble that on offer in any other fishery. Fishing was a rising profession. In 1680 William Petty wrote: 'In the highest place in the scale of labour is the seaman.' As the seventeenth century advanced, the cod fishery off New England offered developments in parallel, through employment to onshore locals. C.L. Woodbury, writing in 1880, remarked, 'It was the discovery of the winter fishery on its shores that led New England to civilisation.' In Newfoundland this process was clinched when in the early nineteenth century changing European clothing fashions led to a demand for sealskins, forming the basis for springtime employment. Sealing had started in 1799. The sea, when so well endowed, was continuing to have direct political, economic and social importance.

In fishing terms little, fundamentally, was to change in the Newfoundland cod fishery in its first four hundred years. The first custom-built cod boat, developed by Newfies (Newfoundlanders) in the early eighteenth century, was the shallop. It was around thirty feet long, decked fore and aft affording protection from weather, and five men would remain on the fishing-grounds in their shallops for up to six days. In that time they might have caught 44 tons of cod. Ponder on the fact that the big cod were at least five feet long, and that they were being hauled from some 500 feet down, writhing and straining against the weight of water, and one imagines that after six days the shallop men sorely needed a break.

The French, traditionally fishing from larger boats and supplying salted cod for home consumption in France, employed fishing methods which, in the drawings of the precise eighteenth-century graphic artist Durhamel du

Monceau, look extremely bizarre. The fishermen lined the side of the top deck of triple-decked boats, each one ensconced in a sort of confessional-style cubicle, which was constructed en route and fronted with straw. They stood for protection in large hooped barrels lashed to the deck, and for further protection they wore leather aprons up to the chin. Each man had eight to twelve baited lines dropping down about 500 feet. Unlike the English ships which worked at anchor, the French drifted as they fished. Behind the fishermen laboured the de-headers and salters. The fish were salted for one or two days. Cod roes were a perquisite of the crew, air bladders were turned into glue, and the liver oil was used for lighting lamps. One can imagine it must have been a queasy business, rolling side-on to the swell, standing all day hooking cod. This might explain why in 1743 strong liquor was banned because of the incidence of drunken mutineering.

Harold Innis says that in the fifty years following 1783 the French fishery was less efficient than the English. French boats were less seaworthy and French seamanship less expert. Wages and provisions were worse, and more ships ended on the seabed. Structurally, the French fishery had always differed. It was conducted from scattered seaports in France rather than from one area. The English fishery was centred on a few ports in the West Country, a disproportionate number of boats hailing from Bideford. The French served a domestic market whereas England's, and later Britain's, boats used the fishery to enlarge its trading empire. In Adam Smith's words in *A Wealth of Nations*: 'To increase the shipping and naval power of Great Britain by the extension of the fisheries of our colonies is an object which the legislature seems to have had almost constantly in view.' The cod fishery was used by England to better advantage and during the eighteenth century French participation gradually decreased. The jostling for position by these two nations was replaced by increasing friction between England and its American colony; which resulted ultimately in the American Revolution. Access to fisheries is referred to in the culminating independence treaty.

What complicated the picture over the course of the eighteenth century was the emergence, into the fray between European nations, of a native Newfie fleet. Beginning the century with 674 boats, by the 1790's Newfoundland was sending 1,259 boats to sea. The nineteenth century fishery off Newfoundland was characterised by an increasing number of British fishermen becoming resident in Newfoundland in order to fish from closer to home, a resurgence of the French presence after the Napoleonic Wars shadowed by a decline of the British one, more and more American fishing boats, and catches, which mounted, if not evenly, then inexorably. As the century advanced the fishery extended northwards and was conducted off Labrador as well. Towards the end of the century Great Britain was finding increasing difficulty in protecting her Canadian maritime colonies from encroaching American vessels. From the time of American independence Nova Scotia took over the provisioning of the West Indies and became the focal point of resistance to American incursions into Canada's fishery.

Meanwhile the fishery was widening. Jigging for mackerel, a pelagic species, was introduced in 1804; by 1831 an astonishing 450,000 barrels of mackerel were landed by a fleet of 900 vessels. The invention of a trap for cod late in the nineteenth century meant that fishermen could accrue fish while they were asleep. The discovery of cod on one of America's prime fishing-grounds, Georges Bank off Massachusetts, was made surprisingly late, in 1821. French extractions from the Newfoundland fishery rose through the last half of the nineteenth century and then Norway also entered the fray. The Norwegian fleet, sailing from the sub-Arctic Lofoten islands, with a workforce of 21,000, sent dried cod to Spain. The Norwegian catch rocketed from 118,000 tons around 1846, to 308,000 tons in 1877.

The figures were daunting and there was talk of the grounds being exhausted. This is interesting. The anxiety seems to have originated from the fishermen themselves; some scientists were slower to come on board with the concept of stock exhaustion. Newfoundland produced its own legislation to curtail the fishery in 1858 with limits on the size of net meshes. These were designed to let smaller fish escape. The appearance of seine nets had caused consternation, and local hook and line fishermen argued for their suspension. There were worries about the amount of capelin (used as bait for cod) being deployed as agricultural fertiliser, and about the quantity of another baitfish, herring, being caught on its own account. The build-up of pressure in the cod fishery, with England trying to fight off an increasing number of predators on her Newfoundland colony, was partially diverted by fishing for salmon, and by the lucrative seal trade. Steam ships took part in sealing for the first time in 1863.

An array of technical advances and improvements changed the cod fishery in the twentieth century. The advent of refrigeration overturned the basis of the salt cod business, eradicating as well the demand for dried and pickled fish, and vastly increased the ends to which cod could be put. Improved communications – telegraph and telephone – more streamlined facilities for handling fresh fish, and bigger, faster ships powered by steam, modernised the industry. Filleting processes arrived in 1921, and were followed by the rapid freezing pioneered and immortalised by Clarence Birdseye. After about 1928 diesel trawlers (aptly called 'draggers' in Canada) considerably increased catches and by 1931 over half the cod landed were scooped from their saltwater domains by diesel-powered nets.

The day of reckoning for the Northern cod came extraordinarily late; and after a 450-year term of relative stability. It is a noteworthy fact that in the 1950s the same nations – Portugal, Spain, France, Britain, plus the Canadian province of Newfoundland – were fishing the Grand Banks as in the 1550s. Probably the underlying reason remained that Newfoundland's cod market was principally in faraway hot countries of Catholic denomination, with low standards of living and purchasing power. At the same time Newfoundland was competing with a closer, more mainstream country, also with good cod stocks, which had its own newly-industrialised fishery. The costs of the goods Newfoundland needed were also determined by America, and that country's industrialised agricultural production in commodities like flour, pork, and beef had increased in price more rapidly than cod. So the faraway destinations for Canadian cod never altered, nor did the national fleets that caught them.

It was the appearance of factory freezer stern trawlers that set the Newfoundland cod fishery on the fateful course that was to decimate the stock. Dr Ran Myers, of the Department of Fisheries and Oceans in St John's, Newfoundland, has called their introduction 'The single most important event in the five hundred year history of the Northern cod fishery.' The first of these vast fish extractors, or 'vacuum cleaners of the sea' as small boat fishermen called them (they weighed up to 8,000 tons), was British, and hove into view in 1954; two Soviet and a German one rapidly followed.

These stern trawlers revolutionised both catching and storing fish. They out-performed their predecessors, the side trawlers, in two regards: they could keep hauling nets in heavy weather because instead of having to sit dead, drifting in rolling seas while the catch was pulled in portions over the side, stern trawlers could move, while hauling, into the wind; secondly fishing time was increased because setting and hauling nets was quicker from the stern. Next, the new models had gadgets called multiplate freezers which parcelled up the product, refrigerated and hydraulically pressed it. Next, the fish were filleted by automation. Lastly, there was an on-board fish-meal plant, complete with dehydrating presses, grinding mills and drying ovens, with which to process the two thirds of the cod carcass that was not fillet. There was one unanticipated problem, however, for the first British factory freezer stern trawler, a ship commissioned by the innovative ex-whaling firm, Christian Salvesen: the boat caught too many fish. The netfuls of cod inching over the stern weighed over twenty tons. Tows were shortened to as little as twenty minutes to reduce the catches. Nonetheless, some of the results were remarkable – 650 tons of cod fillets from 37 days fishing – and fishermen today reflect back on them with wonderment. Extraction resource history was being made.

For the first time offshore catches exceeded those caught close to land. The vessels of twenty countries hauled Newfoundland's cod aboard. Catches from the Grand Banks almost tripled. Meanwhile the consequences of such bonanzas were multiplying too. Doom was close.

The Newfoundland cod in the 1990s was the first major commercial fish species to collapse while supposedly under fully-modernised scientific management. The North Sea herring had gone in the 1970s, but it had not been monitored with all the present-day aids and in four years the moratorium was lifted. Cod, in a sequence like a slow-motion car accident, was fished to death while the talking trailed on. Such a thing had happened before, not to fish, but to a member of an order of sea mammals, an order which included a creature larger than any dinosaur, the blue whale. The once-and-for-all process of thinning down the world's stock of blue whales had occurred half a century earlier.

Chambers Encyclopaedia of 1888 finishes the entry on whaling by saying: '. . . it may seriously be doubted whether the results of the fishery continue to justify the persecution of the unfortunate whales.' In modern fishing terms this constitutes an early admission from a scrupulously detached source, that the continuation of whaling might be in doubt. Despite whaling's extraordinarily ancient origins – from early times American Indians, Japanese, Tartars, Eskimos and Norsemen all practised it, Norman and Anglo Saxon literature refers to it – there is the simple fact that to kill, land and dismember an animal of such immensity, from a realm so limitless and mysterious, disturbed many of the participants in the hunt and commentators through the ages.

The instinct being stirred was connected with the fact that whales are mammals, like ourselves: their young are born alive and suck milk; their blood is warm and they have a four-cavitied heart; and their bones, muscles and nervous system in structure resemble those of other mammals. They are, in evolutionary terms, huge four-footed animals that have taken to seawater and adapted to it. Their fins are attenuated limbs, and in some whale species, where the hind legs should be, small bone remnants are buried beneath the skin. Whales communicate over very long distances by sound and are, probably, like bats, capable of echo-location. In their ability to sieve krill from seawater, the largest animal in creation, as Buckland observed, preying on the smallest, their swimming and diving abilities, their capacity to thrive in extremes of climate and temperature, in their intelligence and social organisation and migratory achievements – some whales of Antarctica push their great bulk a quarter way round the world each season – whales are simply remarkable animals which inspire awe. Sheer size is an emotive factor. Jonathan Swift touches on the impalpable emotions with which we confront sentient creatures of great size with Gulliver and the Lilliputians.

Whilst it is true that in the heyday of whaling in the early twentieth century there was extreme competition to enlist on a whaler, and the money was excellent, it is also true that whalers were usually working to an end, an end situated onshore. Sir Alister Hardy in his book *Great Waters* set in Antarctica reports that most of the numerous Norwegian whalers he met (Norwegians being the pre-eminent whaling race) were set on making enough money to go home and farm. Down in the whaling station of Grytviken in South Georgia their talk was of farms at the other side of the world. Despite their conflicting feelings, however, whalers have been whaling, in different parts of the world independently of each other, for over a millennium.

The Basques pursued black right whales, which hove to their shores, as far back as the twelfth century. No less than six towns in the Bay of Biscay area have a whale in their coat of arms. Even the word 'harpoon' is allegedly derived from a Basque word 'arpoi', meaning to take quickly. Lookouts signalled approaching whales by drum, fire, and bell-ringing, and boats ran the whales ashore. This fishery lasted until the seventeenth century and latterly up to sixteen sailing ships from Biscay whaled off Iceland, a distantwater fishery that started as early as the fourteenth century, and others hunted whales between Newfoundland and Labrador. The value of a single whale was said to be worth two large galleons. Today, the Biscayan whale is itself verging on extinction, being represented by a few individuals.

From 1611, following the early explorers, a whale fishery started in the extreme north Atlantic. Dutch, Scandinavian and English ships enjoined a whaling operation which by 1618 had thirteen English and twenty-three Dutch vessels working off Spitsbergen. They depended on Biscayans as harpooners and also as fat cutters. The targets were right whales (from the 'right' ones to get), slow, docile whales which had the important requisite of remaining buoyant when dead. The product sought was whale-oil, and at one point enough whale-oil was being landed in Europe by Dutch ships to upset the international price of the competitor, seed-oil. It is arresting how fast shortages and over-kill destroyed even this relatively primitive fishery: by 1645 a shore station built by the Dutch on Amsterdam Island off Spitsbergen had folded. The packed whales boldly coasted into the bays and along the shoreline and were summarily exterminated.

Considering the size of right whales (sixty feet long, up to ninety six tons in weight) the statistics are astonishing. Dutch whaling ships accounted for 58,490 Greenland right whales, or bowheads, off Spitsbergen over 110 years; a further 6,986 were caught in the Davis Strait. Many whales unaccounted for will have been fatally harpooned and escaped. From these whales the harvest was 363,466 tons of oil. 623 whaling ships sank to the bottom, and nearly 20,000 men, their crews, were either drowned or, a small proportion, rescued by other ships. These are startling figures portraying an extraordinarily strenuous exploitation of a wild resource at appalling human cost, a type of operation the world had never seen before. From the start of the detailed records in the fishery, lasting from 1660 until nearly 1800, there is a continuous though uneven decline in catches and vessels and, of course, by extension, in whale numbers. By around 1760 the Dutch whalers were being outperformed by British boats.

The British whaling fishery was stimulated by the need for oil in cleaning wool, Britain's principal eighteenth-century export, for dressing jute, which then assumes a silky lustre for conversion into carpets, curtains and ladies' chignon, and also for lighting. By the 1740s oil from whales lit London's street lamps, and spermaceti oil, from the cavernous heads of sperm whales, in the shape of candles, was the de luxe form of domestic lighting. More famously, whalebone or baleen, which is actually soft, pliable, and, like fingernails, made of keratin, was used for umbrella spokes and ladies bodices, and radiating stays in the fashion for hooped skirts. So Britain exploited whales of both main types, the toothed whales, of which the sperm whale is the largest, and baleen whales, being the group which for its sustenance filters seawater through screens of felted fibres made of, and suspended from, whalebone.

The British whaling fleet off Spitsbergen and in the Davis Strait consisted, according to the whaling writer and latterly Christian minister, William Scoresby, of 250 ships. They were square-sailed, three-masted, weighed 400 tons, and were double and triple hulled to withstand ice. Small catcher boats, from which the action took place, were carried amidships. They left north Scotland in March or April, February for those bound for the far-off Davis

Strait, and spent four to five months accumulating the valuable cargo. The advent of coal gas lighting, and oil made from rapeseed, hit whale-oil prices, and furthermore whales started to get scarce. G. Jackson in *The British Whaling Trade* presents a statistical table 1815–1842 showing the total number of British whaling vessels in the North Atlantic declining from 146 to 18. In the best year, 1823, around 2,000 whales were caught; by 1842 the number was 54. Many fewer boats sank than in the Dutch fishery in the previous century, and, with a total of 21,548, many fewer whales were caught. A whaling fishery directed from Scotland trickled on in a minor key till 1904, but by this time the main harvest was of seals.

David Gray, captain of a steam whaling ship from Peterhead, who fished in the aftermath of the boom years, is interesting both on how whales feed and the complex filtering of small crustacea from seawater in the whales mouth, and on whale-catching. He commends whales memories, observing that if a whale when first harpooned makes the customary dive, and yet still manages to escape, it never repeats the mistake, but accelerates away on the water surface, ripping the harpoon from its blubber. A contemporary of Gray, a Professor Owen, pointed out that it was the absence of valves in the veins that caused speared whales to haemorrhage so badly when they dived, which action increased the pressure on their hearts, making the capture of such huge and powerful animals possible with relatively flimsy gear. David Gray notes that the Greenland whales' habit of generally heading in the same direction when they surface after feeding on mid-water shoals of krill (small crustacea) is important to whale fishers because the harpooner has to approach from tail-on. Where they will surface to breathe in, and complete their swallowing, can be comfortably predicted. He reports that three days before and after both the new and full moon were times when whales could be most easily found.

Different species of whales have been fished by different people at different times. Some small, localised populations – of Canadian Indians, Eskimos (Inuits), Faeroes islanders – have taken small numbers of seasonally local whales for many centuries, and in the debates on quotas in the International Whaling Commission there is considerable sympathy for their plea to be allowed to continue their token whale hunt for traditional, symbolic or ritual reasons. But whaling has a notoriety in the history of man's exploitation of the sea for a separate reason.

The great whaling epochs ended with a severe diminution of the original stock of whales. These creatures, which were routinely obliged to surface for air, could be hunted down to numbers so low that even when whale products were selling for sums that were phenomenal (whalebone or baleen fetched £3,000 a ton in the late nineteenth century, and the capture of three or four fin whales would pay for one trip) it finally became uneconomical to continue their pursuit. Thus species of whales, despite their often huge range, were the first sea creatures which could be virtually eradicated by fishermen. The ocean was, to the surprise of many, a place where there was nowhere to hide.

The other feature of whaling which grimly enforced the power and finality of man's hunt and destroy mission was the disturbing fact that some populations of whales will never recover. Barring a few species, such as the minke, grey and bowhead whales, which may be benefiting from the surplus krill no longer consumed by other whales, most whales are down to low populations, even a long time after they ceased to be hunted. For example, the stock of the blue whale has remained obstinately low. Whale watchers tentatively believe this figure might at last be rising. The Greenland and right whales are no longer found in the seas where they once provided such a bountiful harvest. In the same seas the original population of fin whales has been cut from an estimated 400,000 to 2,000. The troubling thought arises that for blue whales to find mates, no matter how remarkable whale-song and long-distance whale communication, must now be a touch-and-go affair. With industrial whaling, man made a permanent alteration to the biological balance in the oceans.

Ambivalent attitudes to whaling are comprised of this almost universally felt guilt, and also recognition that in whaling there emerged some of the more extraordinary tales of human daring, resourcefulness and sheer toughness in marine exploration. The excitement and drama of the chase easily is captured by observers with lively sensibilities who were fresh to the scene. Read again Sir Alister Hardy writing in *Great Waters* of his visit to Antartica aboard a Norwegian whaler.

> 'The whale rises repeatedly in front of us, throwing up his puffs of steam. We are gradually catching up. Each time you feel sure the captain will strike, but no, he withholds his fire. He's not satisfied – he will get still closer. The gunners never fire at a range of more than fifty yards and usually much less. The whale, although large, only exposes a small part of himself above the water at once. He rises arching his body in a curve; first the snout and blow-hole show, and then dip forward as the back comes up to be again followed by the dorsal fin. His action has been aptly likened to a great wheel turning with only a fraction of the rim showing above the surface. He presents a small target for a gun which is continually jumping this way and that as the ship pitches and rolls. The gunner – our captain – is on his toes, his knees are bent and he looks like a boxer sparring as he rises and falls with the ship, swinging the gun barrel to keep it facing into the water just ahead of us . . .'

Whaling had a symbolism in the lists of human endeavour before Herman Melville wrote *Moby Dick*, but his mighty and enduring book did more than underline this: it used the metaphor of whaling to depict something self-destructive, life-destructive, fatally buried, in the soul of man. Whaling, post-*Moby Dick*, became a sort of mirror activity for the operation of society's relationship with nature itself. The sperm whale fishery which Melville depicted, in which the harvest was purely oil, or spermaceti, the sperm whale having no baleen, was for various reasons one of the strangest and most bizarre exploits in human history.

Consider the sperm whale itself, otherwise, and more attractively known as the cachalot. The cachalot has an extraordinary, un-whalelike appearance. Its head is appropriately called the melon and is a huge blunt organ containing, as in a vat, the prized spermaceti wax. The purpose of the cachalot's physique is not precisely known but some biologists believe that a sound boom is generated inside the melon in a series of high-pressure sonic pulses, helping to stun and immobilise the deepwater squids which are a staple of its diet. The diving ability of cachalots, which can take them 8,000 feet into the abyss, is legendary, and it remains a mystery how even cachalots which have lost their sight can zero in on the equally mysterious squids, which have the capacity to move fast in any direction including backwards. What is certain is that cachalots, through operations within their melons, have refined the art of catching deepwater squid: over 18,000 squid beaks were taken from the inside of a single whale. It is squid beaks immured in fatty substances in the whale's digestive tract which made the prized lumps known as ambergris; this material had the signal property, in a malodorous age, of being able to intensify the power of perfumes.

Cachalots are the whales which stimulated the early stories of malign and terrible avengers of the deeps. They have turned on boats, smacked into them and sunk them. Moby Dick was a cachalot. The males, which are either solitary or in bachelor pods, are far bigger, weighing up to 64,000 kg, in comparison to the female's maximum of 15,000 kg. It is only at age fifteen years that a young male cachalot will have acquired the physical bulk and social status to challenge dominant males for the females. The female matriarch governs the familial pod. Some of the more pathetic accounts relate how the females, indifferent to the danger to themselves, will protect their calves. Indeed, whalers learned to hit the calf first, to ensnare the mothers.

One of the classics of sperm whale hunting, applauded for its literary merits by Rudyard Kipling, is *The Cruise of the Cachalot*, the story of a three year round-the-world whaling trip in the eyes of an eighteen year old English boy aboard an American whaler sailing from the Massachusetts whaling capital of New Bedford. Set in the late nineteenth century, the sailor/writer Frank Bullen details the oil-gathering epic in luminous detail. He is fascinated by cachalots, and his awe is set firmly in place when a captured whale regurgitated, as they are prone to do, its last dainty. This consisted of a squid limb as wide as a girthy man!

His description of operations inside the blubber-room is riveting. Squeezing between the greasy masses of fat the men, 'became perfectly saturated with oil, as if they had taken a bath in a tank of it; for as the vessel rolled it was impossible to maintain a footing, and every fall was upon blubber running with oil.' The difficulties of slicing mountains of blubber into thin strips with slippery hands in a tomb-like, rocking hold must have been awful. Bullen mentions *en passant* that the ship needed no lookout guarding against collisions, for the 'blazing cresset', being the flames from the 'try-works' where the oil is extracted from the blubber, could have been seen for many miles.

The most dreaded task was moving the huge casks of oil which when full contained 350 gallons. Shifting over the rolling deck coated in grease, when only four men could get a grip on the cask, 'with the knowledge that one stumbling man would mean the sudden slide of the ton and a half weight, and a little heap of mangled corpses somewhere in the lee scuppers', he modestly states was, 'the one thing dangerous about the whole business.'

Bullen's book is fascinating for its portrayal of life aboard the American whaler with its ployglot crew traversing the world's oceans whilst the holds slowly filled with 'the clear and sweet oil, which after three years in cask is landed from a south-seaman as inoffensive in smell and flavour as the day it was shipped.' Many things about his own compatriots struck him: their good nature, their complete lack of curiosity about all the maritime wonders which stirred his own active imagination, the immunity of all sea creatures other than cachalots – and sharks – from wanton violence from the whalemen. Sharks, because of the threat they posed to a man overboard, were regarded differently. A practice existed of disabling captured sharks by driving a sharpened stake through both upper and lower jaws preventing their mouths from operating, then returning them to the water to the mercy of their fellows. Bullen articulates too the condition of ennui stretched to agonising tension from being marooned on a still sea: 'The ceaseless motion of the vessel rocking at the centre of a circular space of blue, with a perfectly symmetrical dome of azure enclosing her above, unflecked by a single cloud, becomes at last almost unbearable for its changeless sameness of environment . . . some of the crew must become idiotic, or, in sheer rage . . . commit mutiny.' Such accounts, aside from their literary value, became part of the whole culture of whaling which merged with the culture of the time. Whaling played a part in man's relation to the sea; it is in some ways the reference point of all fisheries.

The sperm whale fishery was mainly an American affair. In 1785 the American government introduced a bounty on white spermaceti of £5 a ton. Spurred by this, American whaling ships had roamed down to Peru and the Galapagos Islands by 1800; two years later they were in the Moluccas and the China Seas, their sailors eagerly scanning the horizon for the cachalot's idiomatic blowing, which spouts forward and to the left of centre. The waters off Japan, New Zealand, Australia and the Seychelles were also visited. The fact that sperm whale oil lasted for so long meant that vessels had no reason to hurry home. The apex of oil production, and cachalot destruction, was 1857; over five million gallons of sperm oil were brought back by a total of over 500 whaling ships. The number of cachalots killed per vessel had started to decline early on in the 1830s, but vessel numbers, and fishing effort, continued to escalate. It was not until the 1850s that the world's stock of this unique mammal began to fall off. By the time the sperm whale fishery faded out, in the first years of the twentieth century, very few whales were being found (one commentator said they had become 'shy'), and fossil oils were in use.

Spermaceti remained in circulation in America until the 1970s discovery that the jojoba bean, commonly used in soaps, could also take the place of spermaceti in jet turbine lubrication. The same plant was then unavailable to the Soviet Union which continued sperm whaling for lubricant uses in its space programme until the 1980s, by which time jojoba bean production had been established in the plains of central Asia.

The right whales and the cachalots started to be hunted because the opportunity arose. Specific parts of their anatomies served specific functions. The time was to come when every part of the whale was to be utilised and when all whale species could be turned to some purpose. As ship engines got faster, facilitating the capture of quicker whales, as the on-board processing of whales became mechanised, as the reach of ships extended further, whaling became universalised. The old days of the thin-planked catcher-boats with rope and harpoon, and the occupants of small, flimsy craft hanging on for dear life as whales surged down or away from them, gave way to industrialised whaling, often in the most inhospitable parts of the world. (It is an interesting aside into the mentality of whalers that some whaling communities, for example the Azoreans, refused to adopt modern methods. Despite the continuing loss of life and harrowing experiences they insisted on deploying hand-held missiles from their long wooden rowing boats, hunting for cachalots as they liked it, 'face to face', a practice clung to right up until the cessation of whaling in the Azores in the 1980s). However, by the 1950s eighty per cent of the world's whale oil came from the southern oceans and little-known Antarctica, by far the most prominent and most historic whaling nation being Norway, followed by Britain. Whaling was always a take by northern hemisphere nations from the world's far south.

The physical dangers of whaling were still great. This was because, firstly, in the Antarctic, gales could blow up in seconds of a force which is unequalled elsewhere, in temperatures unequalled elsewhere. But principally the physical horrors of whaling – and all observers have referred to the smells, the grease, the quantity of offal etc. – are to do with the size of the beasts involved. Yards and tons were terms eschewed in the descriptions of Dr Robert Blackwood, who sailed in 1950 on one of the biggest Antarctic whaling trips of that season and wrote of his experiences in *Of Whales and Men*.

Of one average-size blue whale they caught, 'fat-nice' in the skipper's words, Dr Robertson reports that its length was that of a railway carriage, its fins were the size of a large dining-table, its flukes could have been fighter aircraft wings, its tongue would have overloaded a fair-sized truck, it would have taken six very strong men to lift its heart, its blood would fill seven thousand milk bottles, every rib was bigger than a man, and so on. (I have seen a blue whale idling on the sea's surface at close quarters, from a low-slung inflatable; it resembled nothing so much as an aircraft carrier). The whole of this immense creature had to be cut, chopped, and sawed into bits and processed in suitable ways. The only refuse items pushed overboard were the innards – gobbled up by Cape pigeons and killer whales – and once-prized baleen, the usefulness of which had been by-passed by time. It was in the complex dismemberment that the dangers lay, and the ship's bosun told Robertson that during the operation of swinging the immense mass of baleen by winch over the ship's side alone, he had seen six men killed. If the ship rolled inopportunely, or the head flenser's (blubber cutter) shout of warning came late, or the winchman's timing with the brake was out, 'the jagged lump of bony whale' could swing into a deckhand with fatal results. Whaling was neither for the lily-livered, nor for those too tenacious of their well-being.

Robert Robertson witnessed whaling in action, talked to hundreds of whalers, and became so fascinated with those who chose to practise this egregious career that, spurred by his interest in the mental as well as physical side of doctoring, he developed a theory on the whaleman type. To an extent his theorem has a bearing on all fishermen as a class, and it certainly has relevance to fishermen's status with the land-based, usually urban, societies which, one way or another, they have chosen to circumvent.

He labelled them 'psychopaths', hastening to add that he meant nothing derogatory. 'I use it rather as a term of superiority. The psychopath – "the man with the suffering mind", to analyse the word etymologically – is a type I have spent my life studying, and my conclusion long ago was that his mind is healthy – too healthy to be acceptable to, or to accept, the civilisation into which he was born, and therefore doomed to alienate itself from that civilisation in some way . . .' He concludes that such people are 'forced to make an actual material getaway to some of the few remaining parts of the world where they will not encounter, and so will not clash with, their orthodox, average, and usually intolerant fellow humans.' He talks of them getting out of 'the human jail', and goes on to instance figures like Galileo, Newton, Darwin, Dostoevsky, Thoreau, Socrates, Marco Polo, Ernest Shackleton, and Columbus. Many of society's oddballs, he believes, left civilisation 'in the fo'c'sles of ships, and especially whaling ships'. Curtly dismissing the 'Viennese psychological hypotheses' he remarked that they bore no relevance to the minds and spirits which so interested him on whaling ships. Summarising his thoughts on whalers Robertson describes, most amusingly, the Christmas celebrations in Antarctica, starting with restrained and elegant conversation – 'An observer might have thought we were the elders of a Presbyterian kirk in session' – set in a tinsel and flower-decked smoke-room, all surrounded by ice, the rough celebrants incongrously in collars and ties, and proceeding, inevitably, to a brawl, broken up by the giant bosun from the Shetland Islands.

The Antarctic whale fishery was, looking back at it in the context of the present dearth of whales, a gruesome affair. Of humpback, blue, sei and finwhales only vestigial populations remain in the Antarctic ocean. As one species became scarce whalers turned to the next. From right whales they turned to the docile, coast-hugging humpbacks. When humpbacks became scarce economics dictated that the largest whales should be pursued, exploding harpoons being as effective in big whales as small. Accordingly around the Antarctic Peninsula and South Georgia the huge fin and blue whale were thinned down. As had happened all along technology rode to the rescue. Just at the moment when all whales close to shore had been hunted down a Norwegian whaling ship learnt how to process whales moored to pack (drifting) ice. Whales could now be taken far out beyond any nation's territorial waters. The untouched Ross Sea was penetrated. In the early months of 1924 in the Ross Sea the largest blue whale ever measured was killed. It was female and 31.4 metres long. It has been surmised that this whale could have been the largest animal on earth at that time.

Concern about whale numbers led the International Council for the Exploration of the Sea (ICES) to set up its first whaling committee in 1926. ICES scientists in the 1930s confirmed that whales reproduced slowly, that they had short lives, and that they were being fished hard, but could see no way to guide the situation past the practical difficulty that contemporary whaling was carried out in free international waters, beyond control, surveillance or regulation.

The hunger for whale oil in the inter-war years became extreme. In the summer of 1930 over 40,000 whales were killed in Antarctica, finally surfeiting the market. Whale oil was now being used in soap-making, and margarine manufacture while war-torn countries could not afford butter. Beefing up for another military conflict, both Japan and Germany sent whaling fleets to Antarctica; by 1939 there were thirty different whaling fleets operating there. The whales' respite during the Second World War was short; in shattered post-war Japan whale meat formed nearly half the meat diet. This concession by the Allies was balanced by the condition that all the oil from these whales be transferred to American and European ships.

The whaling fleets of the 1950s and 1960s are described by David G. Campbell in *The Crystal Desert* as 'Floating armadas, pelagic industrial sites into whose stern slipways were winched most of the remaining blue and fin whales, and many of the sei and minke whales.' He describes a typical new whaling ship, the *British Balaena*, 15,000 tons in weight, with a fleet of ten catcher boats which resembled powerful trawlers with a harpoon gun equipped with whale-seeking radar mounted at the prow, and supported by two amphibious planes. Fitted out with steam saws, grinding machines, boilers, storage tanks etc., these ships could process twelve whales every twenty-four hours. The crew was of about 400. By this time electric harpoons, which stunned the great mammals instantly, had started to replace the exploding harpoons, which had done their gory work by detonating the creature's insides using a time-fuse buried in the harpoon-head.

Blue whales were already fished right down, but catches of other baleen whales reached an unprecedented peak between 1950 and 1960. The sperm whale catch, stimulated by the demand for jet turbine oil, peaked a decade later. The harvest of the world's whales can be given perspective by considering the 1951 total whale oil production of just over half a million tons. The oil production peak had been 600,000 tons in 1930–31. Whales were certainly paying their dues to mankind.

The International Whaling Commission (IWC), formed in 1949, was stuck with problems about how to prove that whale stocks were being seriously affected. Various stock analysis methods were tried and found faulty. This predicament presaged many similar in other fisheries. Scientific proofs to justify action were refined too late, and long after a general recognition, including in this case amongst whale fishermen, of the need for regulatory decisions. The

IWC meantime contented itself with creating a closed season for whaling and trying in a general way to limit catch numbers. By the time the IWC, a body without statutory teeth, began setting detailed quotas for maximum kills in each species in the early 1960s, the big industrial whaling countries, with the exception of Japan, were already extricating themselves from the southern oceans.

The failing economics of whaling halted the hunt, not politicians. For instance, the obvious quota for blue whales, the most valuable resource under the IWC's jurisdiction, was zero. By the 1950s a third of those caught were immature. However the zero quota was not effective until 1967. The demands of the whale oil interests had exploited the weak structure of the commission and the lack of agreement amongst scientists to good commercial effect, but at a price to the balance in ocean systems which it is too early to quantify.

Today there is a moratorium on whaling as recommended by the IWC. A few nations, such as Norway and Japan, still take token catches, described as for scientific purposes. Norway, which abandoned commercial whaling in 1986, took 425 whales in 1996. But the Norwegian whaling interests are overhung by the sort of ironies which could only arise from the politics of an intense international conservation drive pitted against isolated nations refusing completely to surrender industries which, in their day, gave those nations definition. The blubber, once so prized for its oil, has no market in Norway. Norwegians enjoy whale meat – braised in red wine – but not blubber, which constitutes a considerable part of the animal. The Japanese, conversely, love blubber, as well as whale meat. They will eat it raw, as sashimi, or cook it in strips like bacon, and pay handsomely to get it. However, like elephant ivory, whale meat is prohibited from being traded between nations. So, Norway, whose environmental legislation forbids the dumping of blubber, is sitting on a rising and costly pile of this now redundant natural product held in cold storage. The wheel, one could say, has come full circle.

Because whale meat is not a staple food for most peoples, and because whale populations were easier to reduce with comparatively simple technology, whaling was the first of the great international fisheries to founder.

Whaling is not dead, as explained. Nor, curiously, is it dying. Circumstances in the world change. Gradually we draw more distant from our human past. The whaling of some whale species remains inconceivable, but not all whales are rare, and in contradistinction to the long epoch of whaling just described, we are in the phase of management of stocks. For some human communities whales remain a need. For others whaling is logical as part of the order of things in fisheries, where the sea is regarded as a resource in its totality. In 1996 when I asked Thorsteinn Pálsson, the Icelandic Minister of Fisheries, about whaling his answer was immediate: 'It is necessary for Iceland to resume whaling. It is dangerous to overprotect some stocks. There is a need to balance the ecosystem.' He went on to refer to the political problems, saying Iceland was undecided yet when to resume, before fleshing out the ecosystem picture: 'Whales eat four million tonnes of biomass in our waters. One and a half million tonnes is the average catch of Icelandic fishermen. It is obvious we cannot indefinitely overprotect whales. Their proportion of consumption in the sea is rising annually.'

Norway, Iceland and Japan are strong voices in fisheries. They all manage their marine resource profitably and sustainably, a rare combination. Iceland last caught whales for commercial purposes in 1983, and in the harbour at Rejkjavik, side by side, sit four old steam whaling ships, privately-owned, kept spick and span, as they said on the dock 'ready to go'. We are entering the phase of undertaking to manage all components of ecosystems. As I will argue later, the most numerous whales need to be drawn into the overall management strategies for the oceans. The fact that all talk about whale killing is emotion-alised is right; it shows an awareness of the past, and of the particular sensitiv-ity which arises when dealing with such wondrous mammals. The cessation of whaling was perhaps too recent for any imminent reversals. However, because the subject of whales is so touchy, does not mean it should be left untouched.

Chapter 2

Herring Gone, Cod Gone

'If I'm not a fisherman, what am I?' (Newfie fisherman)

'Collective fishing, like collective farming, results in collective misery.' (Julian Morris, letter to the *Daily Telegraph*)

The herring is fated to be always at the forefront of fishing history. Its phenomenal reproductive powers, if nothing else, ensure it. Several factors make it an indicator fishery in modern times, not least its particular suitability for stock assessment, and moderately accurate management, in theoretical terms, by fishery scientists.

We left the herring fishery in Britain as steam drifters entered the scene, pumping their black smoke skywards, leaving port with mizzen masts set, in small fleets, the year around 1800. The target stock was the North Sea herring, a fish 25–30 centimetres long which matures at two or three years, and which spawns close to British shores. By 1900 many of the larger ports had fleets of herring drifters, so called because they set their nets and then drifted through the night. Night-time was when the herrings were prone to shoal and feed near the surface.

Locating herrings in the days of the steam drifter involved skilled use of instinct as well as the logical brain. Herring was always a fish that could behave in curious ways. Although herrings could be located somewhere round the coast at any time of year – December in the English Channel, October to December off East Anglia, June to September between the coasts of Aberdeen and Orkney, and May to September off the Outer Hebrides – it was a fish with foibles. Herrings disliked certain diatoms; they would swim away from these miniature creatures whose huge populations could spread over several thousand square miles. This could deflect a migration. Peculiarly though, herrings thrive around the perimeter of the diatom-cloud. The diatom interface may have been the factor capable of permanently shifting a herring stock's migratory route. At one time so many herrings annually swam into Loch Broom in western Scotland that a fishing station was constructed there. Then, around the 1850s, this herring population totally vanished. Environmental changes may have been behind the sudden change in movement pattern. No one knows for sure. What is definite is that such erratic shifts still occur, frequently bedevilling the modern systems of division of the resource, sector by sector, and state by state.

Some skippers of drifters sniffed the air and reckoned they knew when to pay out the net. Others looked for darker patches on the water which they said indicated herrings below. The nets were cast, many sections around 34 yards long and 14 yards deep being attached to each other, and they hung vertically in the water, forming a curtain hanging just below the surface. As the boat drifted on the tide herrings were caught by their gills and held in the net during the night. The men whiled away the time in their bunks reading, sleeping, or smoking their notorious black shag tobacco.

At dawn, as the fish began falling lower in the water column with the brightening sky, the nets were hauled slowly in by steam-driven winch, and the fish shaken off into trays. Dogfish and conger eels gathered for the kill, even taking herrings as the net was drawn. Skippers had to watch the weather; one wave slopping over into the hold could wreck the catch.

Drift-netting was tough on the men, making physical demands of a sort which has work on a modern trawler looking a doddle. The foot-rope could be winched in by automation, but the top of the net had to be hauled hand over hand, inch by inch, hour after hour. These basic mechanics of drift-netting remained the same until the late 1960s. To haul the drift net empty, on a barren drift, took two and a half hours. This time was spent bending over the gunwale, heaving the sodden material. If the net was full – and frequently almost every window in a drift-net had a herring twisting in it – the operation could last twelve hours. That is, twelve hours without a break. Drift-net fishermen consequenty became famously muscled, with massive chests and shoulders. It was said you could tell them at a glance, and at a distance. In the Scottish Fisheries Museum at Anstruther visitors can watch an old pre-1914 film taken aboard a herring drifter. The sheer physical demands on the fishermen, the excitement and drama of a big catch, are wonderfully communicated in the grainy film through the flickering, jumping picture. From one point of view, steam drifters, when they first appeared, offered considerable physical advantages to fishermen. Where before, in the 'Skaffies', 'Fifies', and 'Zulus', and other evolutions of the open-decked sailing boats, the herring fishermen had to sit through freezing winter nights on the boat's unprotected thwart, with only their three pairs of trousers and three 'gansey' jerseys for protection (some were chilled to immobility and had to be lifted out of their vessels on return to port), now they could eye the weather from a covered wheelhouse. Catches in the steam drifters may have been heavier, but there was rudimentary comfort and warmth in which to steel themselves for the next hauling.

The steam drifters ousted sail. To start with many fishing families owned a steam drifter for the herring season, based in a big herring port, keeping the sail-boat for inshore work with white fish (cod and haddock) in the meantime. As the profits from herring fishing became more conspicuous the fishing family tended to concentrate on herring, and settle down in the larger fishing port. Thus many small fishing villages emptied. Their beautiful and enduring stone-built quays and harbour-stairways testify to a life out-run by time.

From 1900 steam drifters also started to change the financial structure of fishing, moving the industry to a more modernist, capital dependency. The steam vessels cost three to five times what a sail-boat cost. With the advent of year-round herring catching, moving to wherever the fish had migrated, some of the finance could be raised from operating profits.

Inevitably the running costs, on a ship which burned sixteen tons of coal in a month, were of a larger order. The tradition of settling accounts with the chandler 'at the Back of the Fishing', or season's finish, was overtaken by the more modern concept of higher overheads, hopefully justified by higher returns, paid relatively quickly. Speculators and outside finance entered the scene. People other than those dependent on plentiful herrings became involved. Entire fleets started to be accumulated by companies or even by individuals. Iain Sutherland, in his detailed book *From Herring To Seine Net Fishing*, records that a Mr Coupar from the small Sutherland fishing village of Helmsdale, who made his money from a monopoly on curing salt, had accumulated ownership of nine drifters by 1908. The path was laid for a heavily capitalised fishing industry. This, in time, would see the costs of a fishing-boat rise to a million pounds sterling, then ten million, with the consequent necessity to catch, process and sell huge quantities of fresh fish in as fast a time as possible.

So far, the herring resource seemed inexhaustible. Confidence was such that by 1911 there were almost 800 steam drifters in Scotland – always the herring heartland – and Iain Sutherland's figure for the modern-day capitalisation costs is £300 million.

The first half of the twentieth century was a period of inventions, and their rapid modification and polishing. Oil replaced coal in drifters, for example, within the space of seventeen years. Semi-diesel engines were refined into full diesel engines in the 1930s, a process that reduced the incidence of on-board fires caused by the blowlamps needed initially to heat the engine to turn it over. Just before the First World War radio started to be deployed on fishing-boats, and radio telephones were being installed by 1930. Echometers replaced piano wires swinging lead weights.

The British fishing industry played its part in the First World War in many ways, some remembered, some forgotten. One of the saddest debts of war was that paid by the fishermen who lost their lives in the aftermath, blown up by mines laid in the North Sea and never swept, whilst trying to supply a nation short of food.

After the First World War the fishing industry was again involved in a novel form of financing, once more a government-inspired stimulus to what was seen as a key industry. The government commissioned the building of fishing boats to replace war losses; these could be bought on hire purchase. In detailing these developments Iain Sutherland mentions an interesting point: fishermen had unusually good access to banking facilities because a higher percentage owned their own houses, which could be advanced as security, than in any other sector of the working class. As the bulk of the Scottish fishermen of whom he is writing were herring fishermen this indicates the herring fishery's high profitability. The fact that from Scotland alone 1,200 fishing-boats were either volunteered or requisitioned at the start of the War, demonstrates that the aims of the politicians who in the early nineteenth century kickstarted the herring fishery with bounties, expressly mentioning the benefits to naval strength, were still bearing fruit. From the herrings that swam the coasts, and Britain's exploitation of them, was developed a maritime culture that played a substantial part in world history.

Although drift-net fishing for herrings was to continue until the 1960s, the inter-war period saw the introduction of new means of catching fish. Steam-driven trawlers began to drag the floor of the sea for cod and in the process, according to herring fishers, interfere with herrings. To add salt to the wound the trawlers which were working both inside and outside the British territorial fishing limit of the time (three or six miles from shore) were mostly those irrepressible herring fishers, the Dutch. After the First World War white fish began to be worth more than herring, the international market for cured herring having suffered from the chaos following the Russian Revolution in 1917, and the tottering economy of post-war Germany being afflicted by widespread starvation.

The second new fishing instrument was the Danish seine, a bag net with a long rope attached to each side of the mouth, operated by three or four men. The bag was winched slowly against the tide, the pressure keeping the mouth agape. The Danish seine, which was swept in a circle off an anchored buoy, was smaller (202 feet long) and lighter (made of cotton) than a trawl, and did not need the side-boards which kept a trawl-mouth open. With an operating time of one hour Danish seines had huge advantages over trawls which needed five hour interludes between hauls. Against the greater speed and manoeuvrability of seines, which could shift to a new area if a blank was drawn, drift-nets offered the hope to the crew that, should they intercept a large shoal, they could, in Iain Sutherland's phrase, 'gross the kind of money early seine net skippers could only dream about.' The further shortcomings of drifters were the larger crew needed – eight or nine – and their vulnerability to net damage in bad weather owing to the time needed to get the gear aboard.

These developments took place against a dark background. The herring price was lower than it was pre-war; markets had shrunk; the cost of the new boats precluded many traditional fishing families entering the business; and other species, like plaice, were gaining recognition. The inter-war period was one of uncertainty for herring fishers. For the first time different fishing gears were in conflict because of damage caused by one to the other. Seiners fishing in a circular sweep could collide with each other, which prohibited night fishing.

Another feature of the unstable 1920s was the greater variety of fish hauled ashore (nets were sometimes pulled singly to the beach for sorting) by the seines. Fishermen now would look with admiration and amazement at the size and condition of some of these fish, taken from stocks which by today's standards were untapped. Yet of the flatfish landed sometimes only lemon soles and plaice commanded a market at all. Witches, dabs, megrims, and brill were, records Iain Sutherland, 'shovelled over the side'. Monkfish, catfish, whiting, and haddock were often unwanted too. Little effort was made to secure every fish in the haul under these circumstances, and many slipped out of the net. A wider, more volume-demanding market was eventually established in Britain through the medium of fish and chip shops. The fact that echo-sounders with which to find fish were not adopted until the 1930s, although fully developed in wartime, is eloquent of the richness of stocks. Having to hunt down a particular shoal was, so far, unnecessary.

None of this, however, was especially pertinent to the herring fishery, starved of the huge market in Russia (one million barrels a year before the War), and in Germany and the Baltic States (a further 1.4 million barrels). Cheated by bad debts with her herring trading partners, and a political battle of wills which prevented the British government acting in her fishermen's interests to restore the market in Russia, British exports of herring had fallen to 733,000 barrels by 1933. Dumping occurred, and in 1934 1,800 crans were jettisoned in Wick, 1,600 in Lossiemouth. The herring price was a pitiful ten shillings a cran, equivalent at today's prices, assuming the cran is 1,320 herrings, to 26 herrings for one penny. The government responded by passing the Herring Industry Act of 1935 which appointed a Herring Industry Board and funded it with £600,000 with which to modernise the herring fleet, refitting newer boats, and scrapping old ones. The Herring Board set and sold licences, not only to fishermen but to curers, kipperers, and their salesmen too. Fishing areas and close seasons were described, and imports of herring from Norway regulated. The board was empowered to manage the whole herring fishery, financing included.

The traditional way of fishing for herring came to an end after the Second World War. A succession of government acts in the 1940s had attempted to refloat the herring fleet, more modern boats being fitted out with novel equipment such as radar which, twinned with electronic navigation, saved many boats from calamities in fog or from getting lost, and with improved radio and echo-metering. Financial assistance was provided for forming co-operatives and assisting the industry's self-management. In modern equivalents capital grants of £25 million and loans of £40 million were paid out. Meanwhile, as the seas were patrolled by U-boats and ships of war, herring had multiplied. Govey Cargill, a line fisherman from the Scottish port of Gourdon, recalls the flukes of their boat's anchor when pulled in had herring spawn hanging from them. British fishermen took full advantage of the bonanza; but it was a decade later that the herring fishery really was to boom.

The mid-1950s has been referred to as 'the second golden age'. The once-bountiful inner fishing grounds, like the Moray Firth on Scotland's east coast, had been exhausted. The new gadgetry could track and ensnare fish comprehensively. Boats were much bigger, and they stayed out until the holds were full, instead of sleeping in home port. Polypropylene twine and rope had replaced hemp, and latterly sisal. This permitted much bigger nets, in turn requiring stronger and faster engines to tow them. Herring fishermen from Scotland began ranging further out, to the northern isles, then past them to the Norwegian and Danish coasts. To quote Iain Sutherland's flavoursome remark, 'There was hardly a fisherman worthy of his name who was not prosperous'.

The 1950s and 1960s were boom years. Herring, of some species or other (there are over sixty) were important catches to many countries – Norway, Japan, Canada, Iceland, the USA, Germany, Russia, Holland, France, Sweden – which assisted in maintaining the internationalism of their trade. The fishing industries of Norway and Holland were particularly herring-dependent. A table of 1951, grouping herring along with other small pelagic fish, such as anchovies found in the warm temperate waters of both the Atlantic and Pacific, pilchards (the young are sardines) from the Pacific and the Mediterranean, and also the alewives and menhaden important to the USA, shows Norway as the one nation catching over one million tons. Japan and America are next largest, with 852,000 and 740,000 tons respectively, followed by West Germany (317,000 tons), Canada (272,000 tons), and Britain with 180,000 tons. Two decades later the anchovy fishery alone, prosecuted off Peru, was to produce a harvest of over ten million tons.

Britain is reported in the 1950s to have had a particularly high rate of production per fisherman, some 3,000 men landing 35 tons of fish each, a rate exceeding that of both Norway and the USA. This is attributed to the generally advanced fishing methods deployed by British fishermen, and the high proportion of large boats. The herring, referred to by Walter Wood as occupying an 'unassailable position as the lord of fishes', had become part of British culture, inseparable, in the intricacies of history, from Britain's still formidable naval strength and maritime identity.

The history of North Sea herring in the 1960s and early 1970s has an atmosphere reminiscent of a gold rush. There was an international free-for-all. Britain at that time had only a three or six mile territorial limit. The main participant in the 1960s was a Norwegian fleet of purse seiners. As the name implied, a purse was a deep circular net which was tightened when the sweep was completed. Iceland and eastern Europe gorged on the North Sea herring too. When the Russians appeared on the scene their fleet consisted of 90 purse seiners which offloaded onto factory ships holding by. Bulgarians, Poles and East Germans raked the same waters with factory trawlers. The factory ships processed the herring and dumped anything else they caught. The British continued to fish for herring, 95 per cent of the catch being off Scotland.

No one has collated figures for all these catches. Jim Slater, chairman of the Pelagic Fishermen's Association for Scotland, says it was reckoned the east European fleets alone, which salted most of the herring for human consumption, took a million tons in one year. He says it was always inevitable that such pressures on stocks could not last; it was covertly acknowledged that the fishery would at some point be forced to close. No conservation policy was practised because without consensus there was no point. No policy of any sort was pursued, or even talked about. The pressures on herring were fatalistically stepped up, no one daring to think of the morrow.

Politics took a hand first. The year 1973 marked the entry into the Common Market of Britain and Ireland. Their fishing waters became Europe's, but more to the point the Common Market claimed by 1977 to control a 200-mile territorial fishing limit stretching out from all Europe's extremities. The fish-rich north Atlantic was carved into fishing zones by the Common Market bloc and also the other relevant fishing states, Norway, Iceland, Denmark and Russia. The Baltic was separate and managed communally by its own coastal states from 1977. Only a small part of the east Atlantic fell outwith the embrace of these new fishing territories and as the pressure on fish stocks was to continue mounting, that area too, in the north-east, was to become the subject of dispute and political wrangling. In this way the North Sea herring resource, in part of its range, became subject to control by an organised international body. All the distantwater fleets which had been delving into it were expelled, with the exception of Norway, which was to continue its traditional herring fishery on a percentage basis of the scientifically-calculated stock.

It is in the nature of collaboratively managed stocks that often for political, economic and social reasons fishing cannot be wound down quickly enough for the fish to recover. So it was with herring. Had the free-for-all continued the closure would have come earlier, and with the herring stock in an even more parlous condition. Pulverising the stock in the way that had been done was inevitably going to be costly. In 1977 this great, prolific, historic, fishery came to a forced closure for the first time in hundreds of years of yielding up its bounties. Modern technology, not matched by modern wisdom, had pushed the natural resource too far. In 1978 the ailing, smaller west Scottish herring fleet closed too. Things were never going to be the same again for North Sea herring. In 1981 the fishery was re-opened; there had been a partial recovery.

The critical event for the North Sea herring's future was the agreement between European Community members to a Common Fisheries Policy (CFP) in 1983. This policy covered a vast area of fish-thronged seawater, over one million square miles excluding the Mediterranean. The CFP was the result of exhaustive, detailed and intricate negotiation and the policy that resulted was complex, catches for each fish stock needing to be agreed annually by the European Commission. At the same time member states agreed there should be a conservation policy for fish, an idea that was alien to most states whose principal interest was in organising a high percentage of the catch, and building structures in which to market it. Only Britain had a long-established history of fisheries conservation, having introduced a minimum mesh size regulation as far back as 1937, and having been involved in cutting-edge fisheries research for many years. The conservation policy is the part of the CFP which has generated by far the most heated debate.

Herring was unique in the allocation of what were called Total Allowable Catches, or TACs. All other fish fell into one of three different groups, separating out through the importance of species. Herring was alone in its own fourth group, because the stocks had so utterly collapsed.

The 1977 ban had actually prohibited catching small or immature herring for processing into oil or fish-meal. But as there were so few adult herring which would have been suitable for human consumption, all herring fishing had by default become illegal. When the CFP was being formulated in 1983 the stock was painfully climbing back. However, unlike most other fish stocks herring negotiations had to involve a non-member state, Norway. Norway had to be handled with kid gloves, having plentiful fish stocks, and being a potential EC member. Such was the pressure to re-open some sort of commercial fishery for herring that by the time it was done the allocation to Norway had still not been settled. This figure was not reached until 1986, and Norway managed to secure an advantageous arrangement whereby the higher the assessed stock the higher Norway's share of it. Herring had in the meantime shown legislators the difficulties of attempting to manage a free-swimming fish in a large sea, which shifts its migration routes. The largest herring catches were being taken further north all the time, closer to Norway's own territorial waters.

The CFP has been called one of the worst management regimes ever devised, where political forces repeatedly over-rule scientific recommendations, where the science is flawed and fails to incorporate some of the important data, and where, anyway, too many fishing boats are chasing not enough fish. There are truisms in this, and later on the CFP is examined in more detail. But undeniably the CFP has had its successes, meagre though they are. Given the awkward remit, its managers have achieved a degree of stability in fisheries which contrasts with a trend towards dangerous instability before it. North Sea herring is one of the successes, up to a point, and for a period.

Catches between 1982 and 1990 increased in most of the seven European Union (EU) fishing areas, and doubled over the whole area. Catches measured against fishing effort, a standard indicator of fish abundance, also doubled. The total TAC for the EU's herring area never fell below 400,000 tonnes, a volume maybe not to excite the dreams of avarice, but sufficient to afford quite a few fishermen a livelihood. In any case the problem for fishermen today, as Jim Slater maintains, is not how many herring, but how much money can be generated from the catch. If there is a shortage of herring in the world market fishermen may still make more money than if herring are being caught in super-abundance.

The difficulties with the present-day marketing of herring originate in the 1977–1981 closure. The processors of herring sold their machinery and lost their skills. Housewives, or fish consumers, learnt to substitute other products in their diet for traditional herring fried in oatmeal, kippers, or salted herring in brine. When herring appeared on the fishmongers slabs again, eating fashions had moved on. Younger housewives did not know what to do with this small, rather bony fish. Blander tastes were coming into fashion, and foods needing fast preparation. Herring were never a gimmicky or foodies' fish, and what clung to them was an air of being old-fashioned, a fish eaten in times of need, not in the era of food excess. Herring languished on the slab as shoppers gazed at these unfamiliar creatures, once Britain's, and many countries', staple fish.

The solution was found in a phenomenon with the correct gold rush overtones – klondiking. This was the tradition whereby Russian or east European factory ships appeared off west Scotland in the herring season and bought the catches straight off fishing boats. The ports like Ullapool were over-run – in the latter days when they were actually permitted off the ship – by Russian seamen trying, often successfully, to swap goods onshore, like cassettes and pocket radios, for vodka. From the hills above Ullapool the fleet of klondikers seemed at night like a glittering floating city, far bigger in extent than the little west Highland coastal port. The chains anchoring these huge processing ships were stretched in such a cobweb that local boats complained they could not reach the quay. It was, in reality, a variation on a traditional trade, Scottish waters providing Russians with one of their favourite foods, salt herring. Only a proportion of the Scottish catch went for freezing.

In 1996 this scene was changing. The North Sea herring population was once again under pressure. The 1996 TAC for all EU nations fell well below the 400,000 tonnes benchmark, to only 313,000 tonnes. Norway, drawing on the EU's increasing desperation to get her and her fish into the Community, negotiated for herself a higher than usual proportion of this. The situation has been confused and exacerbated by some fishermen taking fish illegally from the North Sea and then claiming they were caught off western Scotland, in a different fishing area, from a different stock which is not under pressure. Despite this the scientific evidence, which has a better record for accuracy with herring than in most stocks, points unmistakably to the need for caution. Mindful of the herring closure, branded on living memories, and of the increasing concern from the media and conservation bodies about overfishing, fishery ministers seemed to act in a cautious and conservative manner.

In the meantime methods of removing herrings from the sea have once again moved on. Up until 1996 most herrings were being caught by purse seiners. The modern seine net is 600 metres long, 200 metres deep, and represents an investment of £200,000. It has 14 tonnes of lead weighing it down to keep the netting taut. The sections are put together by hand. There is a recent trend away from pursers into trawls. The nets are less deep, and costs range from £20,000 for small trawlers to £100,000 of net on the larger vessels. Bigger winches – new developments in winches occur regularly – and greater horsepower on the boats, enable skippers rapidly to move up and down the water column. 'Midwater' trawling, introduced by the Germans in 1969, has the advantage, particularly with herring, of being so precise as to almost eliminate by-catch. What is gained on contemporary trawlers in manoeuvrability is lost in quality of catch. Seined catches are not damaged and bruised by fast towing. This is relevant in a fish market that is rapidly becoming more quality-conscious.

Two recent controversies concerning herring illustrate two persistent dimensions in fisheries, the cultural and the political. One concerns the taking of herring in full roe purely for the eggs. These were flown by jet straight to Japan. The concept of taking fish solely for their roe, in an era of greater transparency over resource exploitation, and of antagonistic and well-staged public protests about by-catch and waste, was vulnerable. The fact that scientists said a roe fishery was supportable was irrelevant: it looked obscene. The situation defused itself. The pernicketiness of the Japanese market about in what state the roe had to be stripped, how many days before natural expulsion stripping was performed, and so on (the Japanese insist herring roe should be harvested only during a three-day period, about two days before natural expulsion), meant British fishermen were obliged to abandon roe fishing.

The second controversy was principally one between two EU member states, Britain and Denmark. It concerns fishing herring as an 'industrial' species. Industrial species are a CFP category of fish in which there are no quotas, like sandeel and Norway pout. Giving member states part of an 'industrial' catch is used as a sop if higher-profile fish are being denied. Industrial fish are political sweeties, to be handed out to the disgruntled. The Danish fishery in the Skagerrak, which commenced just before the last war, for what were ironically called 'waste' fish, is thought to be supplied by western North Sea fish blown to the eastern side. The Danes took these 'sprats', as they were loosely described, from just beyond their harbour walls, small boats going out, filling up, and returning to sea again two or three times a day, and the fish were processed into meal for fish farms. The fishery grew and grew till recently it reached 200,000 tonnes, Danish fishermen finding it most convenient. A catch this size of a pre-adult stock in any fishery would be a lot; in a localised stock it is patently absurd, and W.C. Hodgson, writing in *The Herring and Its Fishery* in 1957, remarked on the stupidity of ransacking the nursery ground of the fishery which has served human use so well for centuries.

True to form it took EU legislators awhile to realise the significance of the fact that Hodgson had guessed, that this sprat fishery was of immature herrings, the future stock of an already over-harvested mainline commercial species. Finally, long after the first major herring stock collapse, a rumpus has broken out. Sprats are a quite different fish to young herrings being only six inches long, and greenish rather than blue. Sprats should be the cherished restaurant starter dish, whitebait, partly so good because of the huge oil content, which is lacking in young herring.

Herring being caught as sprats has caused considerable angst amongst herring fishers. One herring collapse has occurred already. Scottish pelagic fishermen's leader, Jim Slater, maintains if the Danes could be financed out of the industry the North Sea herring stock could rise again to the healthy stock level of one million tonnes. At time of writing the dispute is ongoing; the latest development has been an EU Commission limit on young herring catches, a recognition, maybe, that an industrial fishery is a normal fishery suffering the curse of misunderstanding.

Despite the 1996 TAC cut Jim Slater is relatively bullish about the future of herring fishing. His son, sidestepping good academic qualifications, is now skippering a pelagic boat. Asked about the future Slater's reply instantly focuses on the 500-year North Sea herring history. He remarked with approval on the divisions of the profit, decided in 1605 when Lord Fraser launched four vessels from the port of Fraserburgh, between boat owner and crew, remaining recognisably the same today. In the early 1600s the Scottish east coast boat owner took 25 per cent of the value of the catch; today, with the huge capital investment in high-tech million-pound boats, it is up to 50 or 60 per cent. The historicity of fishing here is interesting: how many industrialists asked about the future in the third millennium would start off discussing financial structures of the seventeenth century? Slater goes on to point to the reassuring fact that the North Sea has always been an internationally exploited resource; Scotland and Norway are embedded in it. Care, of course, is needed not to over-reach into the core of the stock. As his association's members are highly aware, herring spawn better the older they get. The machinations of the CFP have, by happenstance, favoured Scottish pelagic fishermen. But on the verge of the twenty-first century the British pelagic fishermen's most senior representative feels a future can be anticipated with confidence.

One might have said the same about the prospects for cod – if one were writing in 1986 and not 1997. In the early 1990s cod, in its premier grounds, disappeared. Although several cod species occurred off Newfoundland the main one, representing 70 per cent of the whole, was the aforementioned Northern cod. As said, it was the first time a major stock under scientific management had vanished. Its collapse caused immense grief and soul-searching in Newfoundland and Canada's maritime provinces, and immediately put 40,000 people out of work in Newfoundland alone. Sociologists, economists, politicians, fishermen and fishery biologists have agonised long and hard over what went wrong, and how the blame should be apportioned. Any study of modern fisheries must address Canada's cod collapse on the Grand Banks because confidence in fishery management has never been the same since. Until the sorts of disciplines absent in Canada's east coast fisheries in the early 1990s are installed, there could be other parallel catastrophes, on other great fishing grounds. There are many who can see them coming now, for example in the North Sea. Canada's cod collapse demands close consideration.

The Canadians, it could be argued, should have taken on board the recent lessons. The North Sea herring had been overfished; haddock catches on America's Georges Bank had fallen from 155,000 tonnes to 5,000 tonnes in only a few years. Moreover the Georges Bank collapse had happened after the mould-breaking Magnusson Act of 1976, which extended US jurisdiction from twelve to 200 miles offcoast. After the foreign vessels were expelled more efficient American boats, using every modern fishing aid, had doubled the catch. All this occurred under the new regional fishery management councils, in the full public eye. Staring everyone in the face still was the haunting example of the Atlantic halibut, a magnificent fish which never returned in numbers after target fishing for them ceased a hundred years earlier. Surely the Grand Banks cod was just as vulnerable?

The answer is it was, and it was known to be for a long time. To understand the Newfoundland cod fishery it is necessary to know something of the background. Newfoundland, from where most of the fishery was prosecuted, is a rocky, rough, inhospitable place. Poor people from Ireland, an old fishing nation, were the colonists. The east coast where the fishing communities grew up has few natural harbours, few trees, and an unrelenting wind. The wooden houses cling like limpets to the rocks. The climate is cold as well as rough; icebergs and ice-mush float from the north damaging nets and fishing gear. No one, in short, would contemplate living in such a place unless the fishing was good, very good. In fact Newfoundland's cod fishery was once the largest in the world.

The contemporary communities of fishermen fully understood their reliance on cod and the need to protect the resource. David Ralph Matthews of McMaster University, in his book *Controlling Common Property*, shows how well-organised and complex were the self-imposed regulations of the fishery developed by the communities exploiting it, all designed to prevent overfishing. He describes how in the 1960s these durable arrangements began to be compromised by state-wide regulations which aimed to tidy up the fisheries sector and provide a core of financially-stable fishing families using bigger boats. He explains how, in his view, efforts to rationalise the fishery imposed by theoreticians from outside, discriminating against the small-scale inshore fishermen in favour of fewer, bigger boats, ended up breaking it.

The isolated communities in eastern Newfoundland had varying fishing traditions. But in Bonavista, once one of the largest inshore fishing communities in the world, the traditional fishing methods were the handline (naked hooks jigged up and down to attract the fish), the trawl (in Canada, trawls were lines of baited hooks travelling over cod 'runs'), and the cod-trap, which was capable of catching, and keeping alive, several hundred fish at once. Different methods were used according to season. Rules governed the procedures of fishing, and attempted to soften conflicts between fishermen using different gears.

In the 1970s the provincial government introduced gill-net fishing, prosecuted from boats called longliners, to increase fishermen's incomes. There was immediate protest. Handliners, whose boats drifted over the shoals, hooked up on the gill-nets. Gill-nets took fish from the traditional netsmen and also obstructed access to their prime fishing spots. It was thought they took the very large, deepwater 'mother' cod, which had the capacity for enormous reproductive success and were not caught by other inshore fishing gears. More hurtfully, traditional fishermen said gill-netters wasted the catch, letting it rot in its nylon entanglement for days, fish's eyes being eaten out, the fish body gathering scum. They said gill-netters themselves wanting a cod to eat ate a hooked or trapped one for preference.

The final affront for traditional inshore fishermen was the introduction of drag nets on the offshore grounds. The iron bars which held the nets to the bottom damaged the seabed and its denizens, said the inshore men. They took fish indiscriminately and destroyed marine plant life on which cod fed. Dragging, said one, 'takes the flowers off the rocks.'

The introduction of heavier gears, with large boats now competing with the traditional small boats (less than 35 feet long, deployed only within sight of land), the consequent movement further out to the offshore grounds, and the fact that many skippers of the new, more expensive vessels had dug themselves far deeper into debt than was customary in these societies, took place against two other relevant socio-economic changes in Newfoundland. In the 1970s the logging industry sharply contracted, and in the 1980s, as a result of the EU ban on seal product imports, sealing also ceased to be an alternative part-time activity for fishermen. Pressure mounted on the cod resource to outperform itself. Already, by 1980, the numbers of inshore fishermen had grown from 12,800 in 1974 to 35,000.

Before plunging into the crisis of the 1990s it is worth noting that out of the many studies which the cod collapse gave rise to, several have challenged the easy assumption that up until the 1960s the 500-year-old inshore fishery had been conducted on a sustainable basis. Some argue that the inshore fishery was being overharvested as long ago as the early eighteenth century. A Newfoundland provincial government report of 1877 worried over the fishery's decline. In the nineteenth century the Newfoundland fishery moved northwards to coastal Labrador; this is advanced as possible proof that Newfoundland's own shores were becoming depleted. The Labrador-based fleet burgeoned quickly, rising from a few vessels at the start of the nineteenth century to 1,400 schooners a century later. Certainly it appears that between 1857 and 1911 Newfoundland's catch rates dropped, and dropped sharply, by sixty per cent. This would have been enough to turn adventurous fishermen's attentions to Labrador.

Two considerations worth noting in these hypotheses are that scientists broadly agree that the inshore fishery has access to around a quarter of the whole harvestable Northern cod stock, the rest swimming further out on the Banks beyond the reach of small boats; secondly, it is now known that Northern cod only mature after seven or eight years, i.e. very slowly. By contrast North Sea cod mature at three. This was to become a critical factor when the gross overfishing got underway in the 1980s.

We left the cod fishery in Chapter One with the onslaught on Canada's cod stocks of the factory freezer stern trawlers. The first of these ravenous giants appeared in 1954. The Northern cod catch reached a historic peak of 810,000 tonnes in 1968, eleven nations' fleets gorging on the heart of the population. Various different gears took advantage of the unregulated fishery. Spain favoured pair trawlers, being two boats sweeping a net between; Portugal used otter trawlers; and Canada itself used gill-net longliners. Most of the fishing took place between June and September. Ten years later the catch had fallen to 139,000 tonnes. These catch figures are now known to be opaque. In the heyday of the early bonanzas, and increasingly as catches fell back, in order to evade the strictures of the quotas introduced for the first time in 1972, which were based on historic catches, there was substantial misreporting.

Very soon after the foreign stern trawlers hove into view, offshore catches for the first time exceeded those of the traditional inshore Newfie fleet. Off Labrador the inshore fleet's catch dropped by two thirds between the advent of the stern trawlers (1954) and 1977. The inshore fishermen responded by spreading their season, or fishing all year.

Canada was becoming increasingly concerned. Human populations in the fishing provinces, like Newfoundland, were growing fast. The natural resource was being ransacked, without benefit to the coastal state. In 1977 the Canadian government declared a 200-mile territorial limit. A critical point about this, to become significant a few years later, was that the 200-mile limit did not cover the whole of the Banks. The furthest extremity of the Grand Banks, called the Tail of the Banks, is one of the few places in the world – the north-eastern half of the Bering Sea, and the Patagonian Shelf are others – where a coastal shelf extends beyond 200 miles from shore. As it happens, the Tail is where parts of the chilly Labrador Current clash with warmer 'slope-water' currents. It is a water-mixing ground particularly suitable for fish: redfish, pollack, sole, flounders, dabs, and, yes, cod too, notably young cod.

A second point on which to focus in the 1977 declaration is that it came, as mentioned, late in the day. From around 1968 the otter trawl catches of France, the USSR, East and West Germany, Poland and Britain, had plummeted. The period between the peak catch of 1968 and the 1977 declaration was a grisly one for Canadian cod.

One of the reasons that fate overtook the cod so precipitately is that the years of decline, starting in the 1960s, were also an era of rapid and radical technical improvement. This pattern is discernible in so many fishery stories of the late twentieth century. The fishing boat in the 1950s was a technical dinosaur by the late 1970s. Far more trawls and nets could be set by fishing crews operating the new machinery. As in the herring fishery, radar and, most pertinently, echo-sounders, entered the attack against fish in the 1960s; and navigation systems jumped forward in the 1980s and are still leaping on in the 1990s. Japanese trap designs modified the age-old Canadian cod-trap reducing escapement, and also making it possible to set the traps on hitherto unviable seabeds. Fishermen by the 1980s knew precisely where they were in a featureless sea, where their nets hung, where the fish were, and where everyone else in the fishery was. As the Canadian fisheries scientist, Dr Ran Myers, put it: 'The drastic decline in inshore catches throughout the 60s and 70s was largely a consequence of the unregulated factory trawler fishery. The inclusion of trawlers in any sustainable fishery is of widespread concern . . .' Newfies today, waiting for the cod to reappear, often say that the 'seek and destroy' technology must itself be banished, and that fishermen should return to the old jigging and trapping methods which sustained the fishery and its people for hundreds of years.

The continuing collapse of the fishery when under home state management after 1977 still needs explaining. After all, most coastal nations have now declared 200-mile limits. Canada is a fair test-case.

Having evicted the foreign vessels, Canadians thought the once-great fishery would become great again. They had demonised the foreign fleets for so long; now that they were gone everyone expected the dream to come true. The history of fisheries, however, demonstrates that self-determination is not always a panacea. There was now a dangerous thing in Canadian fisheries – an abundance of optimism, except, signally, among the thoughtful fishermen themselves. The old timers who found they could no longer go out and jig a cod were pensive, and the fish being landed were seven or eight pounds, not thirteen or fourteen. The fishermen realised something the buoyant politicians did not, or did not admit they realised: eight-pound cod were too young. Instead of harvesting the surplus from the stock biomass, they were still harvesting the core stock itself, as the foreigners had.

There was political pressure for the cod fishery to succeed. Newfies were crying out for employment; the fecundity of the cod grounds was legendary. Everyone focused their hopes on it. Article 62 of the United Nations Law of the Sea Convention says that coastal states are obliged to let other countries harvest surplus fish stocks if the home state is unable to. Accordingly the government initiated an energetic boat-building programme and Canadian boats replaced the British, Spanish and Soviet ones. There was no point in ring-fencing the fishery if you just sat back and sold fishing licences. Federal and provincial government expenditures on the industry soared. Not only were fishing fleets built up; the processing sector was too. The government itself bought out seven insolvent fish firms at a cost of over Can $100 million. A fisheries expenditure figure of Can $22 million in 1973 had transmogrified into Can $125 million by 1981. Canadians really cared about the cod.

It was against this background that Canadian scientists had to set a TAC. They realised full well that stocks needed rebuilding. TACs were lowered from yesteryear, when no one agreed them anyway, and in broad terms the new limits were adhered to by Canadian fishermen. The harvest rate was set at eighteen per cent of the stock.

How was this figure arrived at? Canadian scientists took data from two sources: from their own research vessels, and from commercial fishermen. However, their respective findings differed. The research vessels, fishing randomly, kept coming up with figures suggesting stocks were low and getting lower. Commercial fishing boats, fishing where they knew best to fish, revelling in all the new fish-finding gadgetry, continued to fish with moderate success, even sometimes with bigger catches. From commercial fishing vessel catches it appeared that, based on catch measured against fishing effort, the stock between 1978 and 1985 trebled. Research vessels could not corroborate this.

The folly of using only the commercial fishing boats catch rate data as the basis for estimating the stock is now known. Amazingly enough though, when the commercial vessels data suggested stocks were higher than even when the historic catches of 1968 were clocked up, TACs were raised. The scientists' own data was disregarded. On top of the stock miscalculations there were over-optimistic calculations about the growth rates of cod. During the crucial period after 1977, and into the 1990s, the sea was abnormally cold, chillier by about three degrees. Fish growth slowed. The logical step would have been to ease fishing pressure and let stocks adequately recover. No account in the commercial vessels catch rate was taken of the mounting fish-catching and fish-finding power of the new technology.

The use of commercial fishing boat data had another inherent flaw. It was at the time not known that cod shoal when in small numbers. It is an established protective biological reaction to low population stresses in many species, not only of fish. Haddock also do it. When the commercial vessels deploying their echo-sounders found a big shoal on their screens they thought, of course, 'Good, another big shoal. Stocks may be spread out, but all is fine.' It never occurred to them that these were the last shoals of a fish stock fighting for its survival. Around them was empty seawater. It did not occur to anyone.

Another question, more fundamental than any of the rest, has arisen. Even if scientists correctly estimated the numbers of fish on the Grand Banks, were the systems and formulae they were using capable of adequately predicting what stock levels would be next year? Several top scientists have said the scientific modelling of stocks in play at the time was simplistic and bound to lead to overfishing. Two assumptions made by stock assessment scientists are that natural mortality is constant, and that it is independent of age. Neither may be true.

Another scientific assumption on which there is disagreement is that 'recruitment', or the number of fish which survive to a catchable size in the year, can remain stable regardless of the size of the stock. The scientific explanation, at first glance puzzling to the layman, is that fish like cod lay so many millions of eggs that a small change in their percentage survival rate makes a huge difference to recruitment, far more than how many parents there are. It sounds like gobbledegook – or like a licence to issue munificent TACs. Common sense says that in a depleted stock what is needed is more adults. The importance of the spawning stock itself is only starting to be realised. So many cases have now piled up where overfishing of the spawning stock has resulted in a collapse that scientists are having to review both their habitual nostrums about stock assessment and their understanding of the mechanisms behind recruitment. In the Grand Banks case it is thought that the scientific miscalculations resulted in a take not of eighteen per cent of the stock, but something nearer sixty per cent. Naturally, such a rate is devastating.

The reasons for the cod collapse are still argued over. The official line, from some fishermen's organisations, is that cod were not fished out, they migrated off the fishing-grounds. Ran Myers, an energetic, free-speaking and eloquent scientist, advances the overfishing theory, whilst convincingly dismissing the rival cold water cycles theory. Another scientist, Tony Dickinson, Professor of Fisheries at Memorial University, St Johns, Newfoundland, said of Myers, 'He is a very good scientist. But he upsets people. He tells it the way it is.' The way it is in Myers view is that Northern cod were on the verge of commercial extinction in 1977. He believes the declining proportion of older, ten to fourteen year old females, was critical. In 1962 the old 'mothers' produced half the Northern cod eggs; by 1991 their contribution had dropped to a tenth. The age structure of the stock was ripped apart. Myers thinks the signs of over-exploitation date to the mid-nineteenth century. He argues persuasively that catch rate data from fixed gear, instead of seek and destroy trawls, would be more useful. Ultimately he believes fishery managers must recognise the 'tremendous uncertainty' of the stock assessment methods presently in use.

Teasing one possibility after another from the web of possible causes for the cod collapse, Myers ends up pointing the finger at overfishing. Many of those responsible for doing the damage would agree. As a Newfie remarked once to a British scientist, 'The Newfoundland cod migrated all right – into the fishermen's holds.'

Dr Dickinson himself is a marine mammologist. Seals come into his brief. A favourite explanation among fishermen for the cod collapse was the over-abundance of harp seals following the 1982 ban on seal imports. Off Canada's east coast the harp seal numbers reached five million. This constitutes a vast number of big carnivores. As Dr Dickinson said, 'They don't eat turnips.'

Seals are seldom honestly spoken of because they have become a totem of animal rights activists. Anyone in favour of restricting the growth of seal populations is immediately considered anti-conservationist. The British government, for one, has demonstrated its extreme reluctance to even consider reinstating a cull of grey seals, although numbers are now at record levels and their damage to fish stocks on a prodigious scale is beyond doubt.

In Dr Dickinson's view seals 'had a role to play' in the cod collapse. But the exact component of cod in harp seals diet is unknown. It is thought to be around five per cent. The analysis of seal diet is complicated by the fact that they often eat chunks of the flesh of fish and not the durable bones which can be counted in seal droppings. Because seals are opportunistic feeders the constituents of the diet vary. The collapse of the market for seal products affected Newfies profoundly, beyond its significance as a seasonal employer. 'Sealing' as Dr Dickinson put it, 'was part of the Newfoundland psyche.' Seal flipper pie was an important traditional dish. A Newfie fisherman told me that in his youth everyone knew the taste of seal. Such is the strength of local feeling about seals that in 1996, despite the howls of protest from European greens and threats of counter-measures, the east Canadian seal cull was re-opened. Although the numbers killed were around a quarter of a million, it is reckoned that the reduction will still only succeed in reducing the population's rate of increase; the core stock will remain untouched. The policy for protection of the northern oceans most numerous and voracious large carnivore, whilst fish stocks are under pressure and fishermen are losing their livelihoods, will need to be addressed in a rational and unemotional way. As a Newfie bitterly remarked, 'Seals: they only look pretty for two weeks of their lives.'

Those arguing for a cocktail of circumstances for the cod collapse include the vivacious and formidably eloquent Icelandic scientist, Ragnar Arnason. Acknowledging the role of overfishing which he calls 'a disgraceful management failure', he thinks too that recent evidence points to the contributing factor of widespread failures of cod over the whole north Atlantic. Alerted by the finding that cod traditionally thought of as Icelandic stocks were in fact being found in the early 1990s mixing with Barents Sea cod, and that cod stocks suddenly materialised as a novelty off the Faeroe Islands, he believes now that cod stocks are not discrete as formerly thought. They can interbreed, and he thinks they do. There is a good biological rationale for different but related populations to intermingle and breed together. It spreads the risk of climatic, and so feeding, failure, in one part of the cod range. He believes the idea of separate stocks is misleading. Cod, he remarks, can swim thirty miles in one day. He thinks, like some of the Canadian fishermen, that part at any rate, of the Newfoundland Northern cod stock swam away; on top of which, something climatic injured breeding – over the whole spectrum of North Atlantic cod, not just Newfoundland's. The reason for the degree of uncertainty, he says, is hardly surprising. When talking of the relationship between one population of fish in the sea and another, between cod and their staple food capelin, for example, it is well to remember that 'Scientists know almost nothing about the ecology of the oceans.'

There are still management aspects of the cod collapse which need explaining. Why, for example, when in 1990 scientists made it clear that catastrophe was looming, did the authorities fail to cut the TAC? It was not until 1992 that Brian Tobin, the Newfoundland Premier, banned further fishing, even subsistence fishing. The fishing science has been looked at; so have some predator effects; so has the evolution of fishing methods; Ran Myers examined the possibility of some form of natural mortality and drew a blank. The last question – why the prevarication? – is in many ways the most important of all. Overarching the fishery itself were the economics and politics of Newfoundland. They determined what happened, and they explain why the scientists were overruled, in the same way that the scientists of the EU's CFP are routinely overruled in another great northern hemisphere fishery. In his book *Fisheries Management: The Case of the North Atlantic Cod*, Rögnvaldur Hannesson, of the Norwegian School of Economics and Business Administration in Bergen, puts the cod fishery under the economist's microscope.

Using comparisons with Iceland, for volumes of fish caught, Hannesson reckons the Newfoundland fishery was vastly overcapitalised. Quoting the work of Professor William Schrank of Memorial University, he says the value of the trawler fleet grew from $73 million in 1976 to $131 million in 1980. There were too many boats and too many fishermen. Thirty thousand Canadian fishermen, of whom about half were part-time, landed the same amount of fish as six thousand Icelanders did. Because official reports embraced in their considerations Newfoundland's employment needs, recommendations for TACs far exceeded those of the scientists. The fishery was being used as an employment safety valve. Remember, the scientific recommendations were themselves based on a false premise. Thus it was that in 1990 the Newfoundland government was advised to authorise a TAC of 190,000 tons when the scientists had recommended only 125,000 tons. The discrepancy of 65,000 tons represents a lot of breeding cod.

Behind this apparent lunacy were the peculiarities of the benefits system. By working for ten weeks a year and fishing for at least six, fishermen could claim unemployment benefits for 27 weeks at over half their 'insured' earnings. Politicians wanted to avoid the spectacle of starving, or at any rate shouting, fishermen. Fishermen needed the opportunity to work. The consequence of this was that, as income from fishing dropped, unemployment benefits increased as a proportion of income. By 1990 self-employed fishermen in Newfoundland relied more on benefits than on earnings. The magnitude of government support became staggering. Between 1980 and 1990 it is reckoned at $3 billion; over a similar period the gross value of the whole fishing catch was actually less, at $2 billion! From a giant of production the fishery had become a ravening Moloch of resources. Brian Tobin said much later in a BBC television programme on fisheries: 'If the fishery becomes the employer of last resort, you break the back of the fishery. It's impossible for the stocks to bear the burden.'

It is a politician's view, and as the Canadian politician who has done more than any other to address the fishing issue, Mr Tobin deserves respect. But an outsider, or an economist, might say that the welfare system should never have developed a relationship of interdependence with fisheries to this degree. In a country where tougher treatment of socio-economic issues was normal the Newfies should have been told to move to alternative employment, if necessary outside Newfoundland, rather than continue to attach themselves to the carcass of a beast in its final convulsions. As Mr Hannesson, a non-Canadian, remarks, 'Newfoundland provides a vivid illustration of a population which has outgrown the resources available to support it.' The fishery was mismanaged, he writes 'because the management process is driven by politics and not by profitability.'

William Schrank, who is a Canadian, has studied the unemployment and benefits system in detail. He highlights its absurdity in a paper written in 1995, 'Instead of falling incomes triggering benefits, the usual case, fishermen's benefits actually fall as their incomes fall.' Deploring the 'pothole theory of government' which leaves 'the existing system intact and fills the potholes with money', Professor Schrank advocates the linkage between benefits and the fishery being broken. The Department of Fisheries and Oceans, he writes, 'has suffered from a policy of schizophrenia, never being able to determine whether its chief goal is to set and implement policy for the fishery as a viable industry or whether it is to maximise employment and save non-viable rural communities.' 'The social goal of the fishery' he continues, 'has consistently taken precedence over the goal of economic viability.' Such questions are important because, to an extent, they apply to all fisheries. There are coastal communities all over the world clinging to dependency on ailing fisheries. With different shifts of emphasis these are the problems which Europeans are witnessing, in the prolonged death-throes of their CFP.

One of the themes chorused by Canadian scientists as the results of the numerous studies and researches into the cod collapse were synthesised was that the Canadian stock of Northern cod could not be adequately managed if the moment it nosed outside the 200-mile exclusive economic zone it was swept up by predatory vessels sitting on the boundaries. For reasons mentioned, the Tail of the Banks was just such a place, where fishermen from many countries tried their luck in international waters. What especially got up the noses of scientists like Ran Myers was that the cod stocks which tended to drift out to the edge of the Grand Banks shelf were the young stock. 'There has to be some control' he argued, 'You can't have a stock that's killed freely beyond 200 miles. And they are taking the babies.'

For those unfamiliar with the trials and tribulations of eastern Canada's cod fishery the course of events in March 1995 was astonishing. The remnant cod stocks, now part of a 'sentinel' fishery where the only catches were caught up by government scientists for research, upset in their normal inshore migration patterns, moved into deeper waters, and onto the grounds straddling the EEZ limit. There the Spanish fleet, described in the international press as 'The most ruthlessly efficient hunters in the world, notorious for regulations abuse and with a population at home attuned to the flavour of immature fish', was waiting for them.

Canada, the sleeping giant of the world's foremost seven industrial nations, stirred to action. Warnings to Spain were fired off, and ignored. Now fishing Greenland halibut (known in Canada as Newfoundland turbot) rather than cod, the Spanish fishing boats, numbering nineteen, were amazed when a Canadian fisheries vessel appeared and tried to arrest one of their number. The Canadian vessel fired warning shots. The Spanish trawler *Estai* was eventually brought to port and examined. To defeat mesh restrictions the Spaniards had been using a 'blinder' net, or one net inside another, thereby closing mesh holes. Sixty-nine per cent of the Spanish-caught Greenland halibut were smaller than the palm of the Canadian Fishery Minister's hand. The hand belonged to Brian Tobin, the same redoubtable Newfoundlander who had halted the cod fishery in 1992.

Brian Tobin has emerged as the strongman of fisheries. He is from a small 'outport' in Newfoundland and, it is said, 'talks like a fisherman'. Speaking of Newfoundland, he said 'The price for conservation must be paid. It is not negotiable.' His action was wildly popular in Canada, and also in all the fishing nations in Europe which felt that at last Spain, the fishing bully in the EU, had been seen off. Canadian flags flew in fishing ports all over Scotland, Ireland, England and Wales. The EU Fisheries Commissioner condemned the Canadian action, but amongst fishing nations sympathy lay firmly with the country trying to protect the last of its lost cod fishery.

The Tobin action, though, had a much more lasting importance. The Canadians had acted blatantly outside the law. They had fired on the fishing boat of another nation in international waters. Defensible and understandable in terms of stock conservation, their action was against international law. Canada was not a country reputed for this sort of behaviour. It raised the issue which will now never go away: should nations manage their own fishery stocks, or only arbitrarily-defined sectors of ocean? Can you have a free-for-all in international waters? Tobin spoke of a 240-mile limit. Should the management area be defined by the number of miles, or by the range of the stock? What if, as in Newfoundland, climate change, or any other influence, pushes the stock off its customary range? At what point in a permanent shift does it then become someone else's stock?

In theory, these waters had been under management at the time of the incident, a quota allocation on the halibut having been agreed through the Northwest Atlantic Fisheries Organisation (NAFO). But the effectiveness of such organisations, relating to waters which are seldom, if ever, policed, is gravely in doubt. Policing is by coastal states, but only of their own boats. Fishing participants whose country has provided no police are at a major advantage. Member states anyway have a habit of declaring unilateral quotas, and behaving much as they think they can get away with. There can be few parallels with fisheries in any other international agreements.

The question is already topical: how are straddling stocks to be managed? Whose are they? How is their exploitation to be policed? Brian Tobin's action, widely admired in the fishing world, gave definition to Canada's conservation effort, split the EU, and exposed chasms of disagreement over the CFP. It also unleashed some dangerously prickly issues.

At the time of writing, a surprising postscript to the Newfoundland fishery has emerged. At first, indeed, the news from Newfoundland seems like a fairy tale: the fishery is now grossing more money than in the heyday of cod without any cod, except from sportfishing in small boats in some places, and a few from the sentinel fishery, being caught. Shellfish, such as snow crab and northern shrimp, have become the number one earner for Canadian fishermen. The shellfish resource is being protected and nurtured with a care never seen before. Seasons are limited and quotas are limited. Most fishermen are not full-time. The Canadians have gone through a painful self-examination. The cod humbled them, but with the resourcefulness for which they have always been admired, they have come back from the brink and thought out new strategies.

The regimes of management are complex, but the principal changes in the fishery are as follows. Firstly, there is a commitment to rebuild the groundfish stocks which appears to have universal support amongst fishermen. Conservation as a *sine qua non* of the fishery's rebirth is accepted. In the fisheries which have replaced cod there are quotas for each participant and they are transferable from one fisherman to another. These are therefore the controversial Individual Transferable Quotas (ITQs). The Department of Fisheries and Oceans (DFO) is reviewing the licensing system. Its commitment to straightening out the causes of the cod collapse is apparent in the scientific team convened to study it: the team is of 180!

The policing has been beefed up to a level non-existent in comparable fisheries. The DFO operates fifty-five patrol boats, and aeroplanes and even submarines. Satellite surveillance is being looked into. Landings are closely monitored. Fines for malefactors are heavy. Above all, the atmosphere is different. The illegal landing of fish which took place before in the cod fishery, which made scientific projections of stocks and TACs even more impossible, is now regarded as an abomination which could harm everybody. Fishermen paying money for quotas do not want to see the resource in which they have invested illegally meddled with. The 'black fish' problem is small. Most importantly for some fishermen the new shellfish business is providing handsome returns. It is not all 'jam tomorrow'.

Another development has been creeping up. Furthermore it has crept forward in an area most people had put out of their minds for the foreseeable future. The sentinel fishery monitoring the performance of the Grand Banks cod is finding more of them. Not lots, but more. This moved Fred Mifflin, Canada's Minister of Fisheries, to re-open the Newfoundland and Labrador cod grounds to small boats for six days in September 1996. Handlines and sportfishing rods were used. Amazingly, 1,400 boats launched into the waters they had only been looking at for nearly five years. The cod are slowly getting bigger. None were allowed to be sold. Mr Mifflin threatened that the slightest abuse of the rules would result in the fishery being closed again. It would have been interesting to have been there. The 1997 fishery opened up a little more. The inshore fishery was allowed 6,000 tonnes. Only long lines could be used, numbers of hooks were limited, so was the length of participating boats – to 65 feet. The whole thing was monitored inside out. Again, any infringements would have risked closure of the fishery.

Fisheries are usually not completely wiped out. The target fish can move territory, indeed frequently do. Even after the most rapacious overfishing somewhere, tucked away, survives a nucleus which fishing vessels never found. It is uneconomic, after all, to take the last fish; it would need too much finding.

Communities in Newfoundland meanwhile have changed. This is what happens if a fishery is run for social objectives: the fish defeat the aims. The Grand Banks were managed for the fishermen, not the fish. Although many Newfies cling to the windswept rocks and gaze nostalgically to sea, others have upped and out. Some migration has occurred to the big cities, of Canada and even beyond. Some of the younger and more enterprising fishermen have gone west and practised the only art they know. Off British Columbia they catch prawns, rockfish, or cod, or halibut, or whatever is most profitable. Their boats are often their homes. The traditional high-value British Columbian fishery, all allocated out and carved out, is in salmon. The roving Newfies and Nova Scotians get licences to catch as many species in as many different fishing zones as they can afford. Locals say they fish too hard and too aggressively. The incomers refute this. The sea-change in attitudes was highlighted in a remark made by a Newfie to a newspaper about the west coast fishery. He said there could be a fishery forever on the west – if they adopted the tight regulations now in use on the east!

Meantime as Doug Letto, a local politician, said, by the time the Northern cod is back in commercial operation many of those with the right skills will be too old to fish. Much local folklore will be gone. It is a consequence of human mismanagement on a large scale. Brian Tobin has warned against a new race for the resuscitated cod. He has a vision of a fishery in which the fishermen are professionals, 'like doctors and dentists'. 'Until the inshore fishery is producing 100,000 tons we would never re-start the dragger fleet. We're talking about a completely different kind of fishery.' He has said the bankers to whom boat owners are indebted should not be allowed to run the fishery. 'If they do' he warned, 'we will get a marine desert for a helluva long time.'

Chapter 3

Iceland and the Falkland Islands

Iceland – 'The first and the most vigorous of the good stewards.' (William Warner)

'It is in our blood to respect the sea.' (Icelander, Orri Vigfússon)

'50 years back most Icelanders worked in summer in agriculture, fisheries, or fish processing. Now even the processors have never done this. People today are unaware of the chain of production.' (Jon Reynir Magnusson, former President of the International Fishmeal and Oil Manufacturers Association)

The common perception is that fisheries are being wiped out and man's greed at sea is unbridled and ungovernable. In the nature of things it is bad stories – and the Canadian cod collapse is a very bad story indeed – that get highlighted. Rampant violations against nature or natural resources affords some people a satisfying *schadenfreude*.

Fisheries are not inherently unmanageable. Some countries have been managing fisheries with general success, although there are always dips and cycles, for a long time. The more important the fisheries are to those countries, and the greater the absolute imperative of getting it right, the better they are run. Fish-dependent countries put their top scientists into fishery research, pull in their best economists, and the business of catching and processing the fish, in the country's pre-eminent industry, attracts the highest-calibre skippers and manufacturers.

Just as there is no fishing state which is irredeemably damaged, so there is none which is perfect. The Falkland Islands, however, a dependency of Great Britain lying 8,000 miles from the mother country in an isolated position in the south Atlantic, has as good a chance as any place of professionalising the operation of its fishery. It is probably unique in having a combination of powerful attributes: it has fish thronging its waters; it has the muscle (two fast vessels and two aeroplanes) to make a fair fist of protecting them; and it has no fishermen. Management of the Falkland Islands fishery involves measuring the stock, assessing a TAC, and then selling licences to catch it. In contrast with other countries, Falkland Islands fisheries take place in a pleasant, intellectual, almost abstract, vacuum. It might be said that the energy with which some other fisheries, for example the EU, are trying to do away with their fishermen suggests that politicians have spotted the attractions of just selling licences, and are trying to disband their own fleet as quickly as possible.

No one interested in the resources of the Falklands goes far without referring

71

to the British Government's second study of the islands compressed into a report chaired by Lord Shackleton and published just after the British–Argentinian war in 1982. It studied the economy of the islands in their entirety. Lord Shackleton's conclusions on the utilisation of the Falklands fisheries have been proved wrong; little did he suspect that by 1995 the Department of Fisheries in the Falklands Government would write of 'the almost total dependence of the Falkland's economy on fishing licence revenue.' The Falklands in Shackleton's day was a curious place: one seventh of the whole gross national product (GNP) was generated from selling Falkland Island postage stamps to philatelists. In 1982 this contributed over £600,000 to island coffers!

Shackleton did not neglect fisheries. He investigated them thoroughly. However, the desperate rush for fish taking place in the last years of the second millenium was not foreseeable then. Fish was still a low-value food in Europe in 1982. No one had identified the full range of dietary benefits afforded by fish. Industrial fisheries, turning low-grade fish into meal and oil for poultry, sheep and pig farmers, were smaller, and the demand for fishmeal for salmon farms was in its infancy. The different perspective about marine resources is expressed when the report addresses shellfish fisheries: '. . . it is only a high-value product such as shellfish that is likely, in the near future, to overcome the high processing and transport costs to northern markets.' Yet today, Falklands' squid is principally fished by Japan, Korean, and Taiwanese boats. Peruvian and Chilean horse mackerel and anchovies are converted to meal and oil and sold to China to feed farmed fish; another huge market for South American industrial species is Germany. The Germans find a margin in re-processing it and exporting it again. Fish and fish products are a global market. They are in the mainstream of the phenomenon of globalisation. How the world has changed since 1982!

Shackleton's report estimated some potential for shellfish, and also for salmon ranching. Neither have materialised. The currents around the Falklands sweep away anadromous salmon into inimical warmer water; there is nothing for them to ride back on. Shackleton mentions the potential for blue whiting, a low-grade fish converted into meal, and finfish in general. Neither have taken off particularly, the catches within the Falklands Conservation Zone in 1995 being around 20,000 tons for both. The fish identified as of possible future value was hake, although the fact that hake spoils faster than cod, and the consequent difficulties of marketing it, were noted. Hake fishing has not prospered in the Falklands and catches have fallen from 50,000 tons in 1988 to a paltry 1,700 tons in 1995. The hake resource is, unfortunately for the Falklands, shared with Argentina, and Shackleton noted throughout his report the difficulties of getting investment companies to research the commercial possibilities whilst the Falklands and Argentina were at loggerheads. The Fisheries Department's present view of the Falklands hake is that 1988 was a one-off hake influx year, and that no permanent stock of any size swims in home waters.

One critical factor in Shackleton's report was that it was written when Britain only had a three-mile territorial sea limit round the Falklands. The report advocates establishing the 200-mile limit which most countries had done by this time, neighbouring South American countries being the earliest of any to declare EEZs. It went on to suggest that Polish fishing boats which were already taking hake in these waters, could be an inexpensive means of gathering more information about the potential for offshore fishing. Without the 200-mile limit the licence-based fishery today prosecuted by the Falkland Islands would not have been easy to predict. The Falklands then had 1,800 inhabitants instead of over 2,700, and lacked the confidence, exuberance and commercial keenness which developed out of a victorious war, vigorously establishing the islanders sovereignty as British subjects, and territorial security. Lord Shackleton, certainly, could not easily have envisaged a future situation where licences for fishing illex squid were worth five sixths of the fishery, with rental values of over £20 million a year.

The Falklands shelf, properly called the Patagonian Shelf, which roughly corresponds to the reaches of the 200-mile limit, is phenomenally productive in two valuable species, both of them squid. One is the illex squid, a large member of the family measuring about a foot, the other is the loligo squid, half the size, principally fished by Spain, and eaten, like calamaris, in rings. Both these species feed on zooplankton and lobster krill (different from Antarctic krill) and are cannibal. They are problematical to manage in that they only live one year. Their populations are inherently volatile and they can move a long way. Flourishing squid fisheries have already been exhausted in the North Pacific and the North Atlantic. The South Atlantic's is now the largest in the world, accounting for ten per cent of the global catch. Around a hundred large fishing boats are involved and, taken in all, some 10,000 workers. The most acute management problem for the Falklands is the illex squid because it occupies a huge area in the south-west Atlantic, is totally absent from the Falklands EEZ for some months (e.g. February), and the other main nation involved in its exploitation is Argentina.

Most people know that Argentina has laid claim to the Falkland Islands and that in Argentina politicians who make bellicose noises about recapturing Las Malvinas, as they call them, can expect a cheer. The Argentine has always been South America's bully-boy. Independent constitutionalists see no basis in the claim: the truth is that the Falklands is proving a useful piece of territory rich in oil and fish, a stepping-stone to the minerals in the Antarctic, and Argentina is its closest mainland neighbour. Since international recognition of 200-mile EEZs the Falklands, once identified as the bleak haven only of penguins, have taken on a quite different lustre.

From a fishery point of view the Argentines sharing the illex stock is more than an inconvenience. It is the opinion of the Falklands Director of Fisheries, John Barton, that their own catch having fallen from 223,000 tons in 1989 to 64,000 tons in 1995 is almost entirely due to Argentinian overfishing. In the Falklands it is believed that Argentina is deliberately overfishing stocks to apply economic pressure on the islanders; they have already been offered cash payments on a *per capita* basis in return for surrendering their independence.

The Argentines in the first half of the 1990s set themselves TACs, without troubling to provide scientific substantiations, which they then failed to catch. This was a poor look-out for neighbours given that fishing licences cost less than in the Falklands and therefore attracted roving fishing vessels, and that there was no shortage of fishing effort. Some of the illex squid move outside both Argentine and Falkland waters as far as 400 miles to the north of the Falklands, and here they are in a high seas fishery open to all. Fishery patrol officers in the Falklands say that Argentinian vessels are a continual problem, hovering around the territorial limits and then popping in and out in a two-step of perpetual provocation.

At a scientific level relationships are better. British and Argentinian scientists work alongside on stock assessments for illex, and also for blue whiting. Joint research cruises are enjoined, in an Argentinian vessel, in both fishing zones, funded by the Falkland Islands. Neither Argentinian scientists nor fishermen are allowed to land in the Falklands, a sore point with Argentina. An 'early warning system' exists by which either side can request the fishery stopped if the illex is being overfished. This was invoked by Britain in both 1994 and 1995. Argentina complied, but tardily.

It is vital that illex squid, and for that matter loligo too, are effort-controlled. Licensed boats go and catch as much as they can as fast as they can. Estimations on stocks are done by seeing how much is caught in every two hours of fishing. If the fishery is suddenly closed the Falklands Islands Government makes some small percentage reimbursement to licence holders.

If the Falklands only had illex squid they could lay no special claims for their management. Spending a disproportionate amount of money on research and conservation, as they do, is not enough. The loligo fishery by contrast is entirely in Falklands waters, and bears every sign of being astutely managed. Since 1987, when the Falklands' licensed fishery got properly underway, loligo catches have been very stable. In the following nine seasons it averaged 76,000 tons. Recent years are showing maintained catches, with fewer boats.

The Falklands Islands pride themselves on their management and plainly the effort is highly-motivated. The Fisheries Department in the capital, Stanley, is led by young, keen people who enjoy their work. They pride themselves in particular on daily catch reporting. They say fishing skippers do not especially object, and it enables them to monitor the stock using the formulae of catches per hour. Seemingly daily catch reporting is unique to the Falklands, although naturally it would have much less relevance in a TAC fishery, or one in which the target fish were long-lived. With squid being so short-lived the time-scale is tight and the room for overharvesting large.

As trawling for loligo squid with small-mesh nets still produces almost nothing but loligos in the catch, and jigging for illex (it is done with very bright lights glaring into the waters at night) seldom catches non-target species either, there is minimal by-catch. Without TACs there is no reason for skippers to mis-report what they have caught. Similarly 'highgrading', or throwing back small fish to make room for big ones, especially in a fishery where there are no catch limits, is not a problem. Finally, the Falklands fisheries people think the fact that all fisheries management takes place in one department, including policing, helps to keep things simple. The Falklands is a relatively small place in a big sea, 'a stone sitting in the middle of the river' as Dr Conor Nolan, the senior fisheries scientist, put it, which helps team spirit.

This is particularly important in policing because the waters filled with squid are sailed by some of the most polyglot fishing crews anywhere. The skippers can be Korean, Lithuanian, Peruvian, Chilean, 'whoever is cheapest' I was told. The crews are almost invariably Oriental, Chinamen and Vietnamese being favourites as they are willing to work for the least. The squid is principally fished by Spanish, Japanese and Korean boats. Eleven large stern trawlers are registered in the Falklands themselves, usually in joint ventures, mostly either Portuguese or Spanish. For some time the Taiwanese also came into Falklands waters, but their tricks and duplicity were legendary. Several boats would fish under the same name. When fishery protection officers boarded these boats they found as many as four different combination names on stencils. Even the life-raft lettering and ships documents had been tampered with. In the view of one protection officer it was more a matter of the companies not policing their own people than scallywaggery institutionalised at the top. Nonetheless Taiwan provides one of the Fisheries Department's favourite boasts: a fisheries protection vessel once chased a Taiwanese malefactor to South Africa, logging 4,364 miles before giving up. When wrongdoers are actually frog-marched into the courtroom in Stanley the rough ride is not over. A Belize-registered vessel, with a crew of five nationalities, owned by Spanish fishing interests, was found guilty of two illegal fishing charges and fined £1.8 million in spring 1996. If the Falklands was found to be a soft touch its waters would rapidly become crowded, and empty.

The fishery protection force also patrols off South Georgia, the furthest point of which is 1,000 miles away, whose fishery is run from Stanley. The ecology off South Georgia is quite different, the two largest species being toothfish, caught on long lines, and Antarctic krill caught in very fine nets. The protection force goes to South Georgia four times annually.

It is something of an irony that the Falkland Islands is a place surrounded by a great biomass of fish, occupied by a population which does not much eat fish, nor feel inclined to procure any. Flying from place to place there it is a remarkable thing looking down at the indented coastline that in none of the sheltered inlets are there any fishing boats. The eighteenth-century British settlers seem to have arrived in the Falklands, fixated on their sheep breeding, and looked inland. The weather may be windy but in most other islands with a temperate climate resourceful settlers would have come to grips with at least some of the prawns, crawfish and crabs that frequent the coast. Whatever the reason it is a fact that Falklanders do not fish and, although their capital town is on occasion the busiest fishing port in the southern hemisphere, with huge 'reefers' or refrigerated cargo ships lining the docks whilst taking on board the catches of the fishing fleets, and although the Falklands fishery averages around 300,000 tons, and despite the fact that the islanders economy has received a tremendous boost from the scientific management of fisheries, there is little chance the Falkland Islands will ever be saddled with the greatest inconvenience of all for a fishing nation – native fishermen.

The ultimate in a fishery devoid of fishermen is South Georgia, the mountainous island group near Antarctica favoured by some of the worst weather in the world. Only a few researchers, a small military garrison, and a single fisheries officer and his wife, perch precariously there, and for short spells. Given that his father, Ernest Shackleton, in one of human history's most epic travel adventures, actually managed with a small group of men to climb over South Georgia's hideously rough peaks, when the men were half-starved, dressed in rags, and had sailed in a small boat from Antarctica, it is understandable that Lord Shackleton's report depicts a future for South Georgia which to date has proved over-optimistic. South Georgia and the nearby Sandwich islands are British territories and he recommended applying a 200-mile limit in the seas, a move achieved in 1993. One reason was his team's estimate of the formidable quantity of Antarctic krill, a small shrimp, off-lying the islands. The potential sustainable yield he put at 50–150 million tons a year, making it by far the largest fishing biomass to be found on earth. When Shackleton was writing, Russian, Japanese, and Polish boats were fishing 'a few hundreds of thousands of tons a year'. Whilst acknowledging that the processing of krill into animal feeds, and ways of using krill for human consumption, needed a lot more research, and noting that because of their very cold habitat harvesting krill poses problems of spoiling once they are landed, he predicted that in the long term the krill harvest could become a major industry, and recommended researches to prepare for this.

By 1996 this prediction had come no closer. Although precautionary catch limits of 1.5 million tons of Antarctic krill have been set by the Convention on the Conservation of Antarctic Marine Living Resources (CCAMLR) (composed of the countries closest to Antarctica including Britain) the actual catch in 1995, in the South Georgia and Sandwich Islands Maritime Zone, was only 48,000 tons. For the whole CCAMLR area it was 94,000 tons. Only five ships were involved, either from Japan, Poland, or Ukraine, and the krill was used either as sportfishing bait or as feed in aquaculture.

The absence of native fishermen, happily for fisheries round the world, is not a precondition for successful management. The precondition appears to be that fisheries are considered to really matter. The bulk of the world's commercial fish species live in the northern half of the globe and here, where the fishing-fleet is home-bred, fisheries management has been taken to its furthest stages of sophistication. Iceland is a country which, in the form of its northerly island of Grimsey, touches the Arctic Circle. Its seas are some of the best fishing-grounds in the world; its people eat more fish even than the Japanese, and Icelanders will enter the next millenium enjoying one of the best standards of living in the western world. A few decades ago Iceland was a backward country farming sheep on the periphery of the icefield which forms the interior; its transformation to an affluent, industrialised, market-orientated society, is due to fisheries.

Apart from the geothermal energy from hot springs, which is used for nearly all the country's heating and hot water, Iceland has few natural advantages. Devoid of building stone, oil, precious metals, special minerals, or timber, the country in its large interior approximates to a sub-arctic desert, devoid of all forms of life in winter except two introduced herds of reindeer and the arctic fox, the only native mammal. The Icelandic economy rests on sea fishing which has been calculated to represent 45 per cent of gross domestic product. The fishing-grounds are favoured by currents which flow clockwise round the island, and vertical mixing of cold and warm water currents producing multitudinous small life forms occurs in two places. Fish can be caught somewhere off Iceland in all the months of the year. At a loose reckoning the Icelandic fishing-grounds are three times as productive in demersal fish as the North Sea, twice that of the Barents Sea, and half as productive as the Grand Banks. It is almost certainly true to say as well that a reason for Iceland's present considered and professional approach to fisheries is that a major stock, herring, has already been fished out, disappearing in the late 1960s and unavailable to fishermen in Icelandic waters again until 1996. Icelanders have looked into the fishless abyss and have been extremely discomfited.

Iceland is an old destination for fishermen. From the British Isles they began their visitations around 1400. Iceland by this time was firmly associated with fish, the export trade having started soon after the union with Norway in 1262. Fishing was conducted in Icelandic waters spring through summer by Dutch fishermen in the seventeenth and eighteenth centuries, and French fishermen in the late eighteenth century and nineteenth century. Cod, once more a critical species, was the most common demersal fish, and Iceland's seas also contained plentiful haddock, saithe, halibut, and redfish. The first Icelanders to climb from rowing boats into decked sailing vessels only did so in the late eighteenth century. Fishing for all this time was by hook or line or, rarely, net.

The fishing by steam trawlers of Icelandic waters started in 1891 with a Grimsby trawler which brought home plentiful plaice and haddock. Next year there were nine British trawlers. In this fishery, and its later developments, the states of Iceland and her southern island neighbour, Britain, are entwined. Britain was always, from the start, the principal foreign nation to fish Icelandic grounds. The Icelandic fisheries historian Jon Thor finds it useful to classify Iceland as a 'fish-surplus' state and Britain as a 'fish-deficient' state. In his definition fish stocks are of vital importance for the fish-surplus state, which is normally a place where the inhabitants depend to a great extent on fish for their livelihood. In the fish-deficient state fish are a minor item in the economy and few depend on them.

Iceland was developed as a destination for British trawlers because they had started to exhaust the North Sea and needed fresh grounds to fish. The British fishermen arrived in ever-larger numbers, envoys from a country which had an empire covering a third of the world, with the world's biggest navy, making visitations on a primitive agricultural community with only about 78,000 inhabitants. The British boats trawled cavalierly through Icelanders' little inshore fishing nets and lines.

Icelanders are resolute folk and were resolute then. Eyeing the future cannily they had a trawler built – in Britain – which came into play in 1905. Ten years later the fleet numbered twenty, and Icelandic skippers thrived, becoming the first market-orientated capitalists in Icelandic society and selling what was not salted and dried for Spanish and Italian markets, to Britain, packed in ice. Not only did the two countries have a trade relationship but many Icelandic trawler-men got their training on British boats.

Exactly how many British trawlers worked Icelandic waters before the First World War is uncertain. The figure of 120–150 is mentioned. Certainly they overshadowed Iceland's own little fleet. The principal targets were plaice and halibut, and British fishermen often gave the cod they did not want to the locals. The proportion of the takings by British ships off Iceland pre-First World War was huge, between 80 and 95 per cent of halibut and plaice, and usually over 60 per cent of all haddock. Jon Thor remarks: 'Huge value had passed out of the country . . . without any measure of remuneration.' The First World War brought all this to a halt and stocks rebuilt while the seas remained unsafe for fishermen.

After the war the British built new, more powerful trawlers. By 1920 there were over 1,500 British distantwater trawlers, now travelling right up to the Arctic and fishing off the Faeroes as well. In the history of fishing this was a formidable fleet. Efforts by Iceland to protect her fish stocks by extending the fishery limits were brushed aside; not until 1952 was Iceland able to extend its fishery limit from three to four miles, a modest move. The 1894 Act forbidding trawling right up to the beaches had always been a bone of contention between Britain and Denmark, at that time Iceland's sovereign overlord.

The inter-war years saw Iceland participating more fully in her own fishery. For most of the period Iceland took more from her waters than Britain; Germany trailed behind as a third fishing force, and the Faeroese came behind Germany. Britain by this time had started taking cod. It had a growing population needing cheap food, and haddock and plaice were not available in sufficient quantities. Britain kept punctilious fishing records. One of the most telling statistics is the time it took a British boat to catch a ton of Icelandic fish. This steadily lengthened till by the start of the Second World War it took twice as long to catch the same amount. This stock decline is reckoned especially to have affected plaice, also haddock, but not cod.

Icelanders were becoming increasingly worried. They had discovered, as pupils of the British, ways to harvest their most important resource, but not the means to protect it. As early as the 1920s it was widely understood that the resource was far from inexhaustible, and Icelanders assayed tirelessly to get more protective fishing limits agreed. Failure to make headway was a contributory factor in the decision to beef up the coastguard and enforce more rigorously the existing regulations for the inshore fishery. This proved a useful measure and Icelandic courts backed up the coastguard with meaty fines. Nonetheless many of the best haddock and plaice grounds lay inside the three-mile limit and both German and British boats were inclined to dip over, if they thought they could get away with it, and fill a last corner of the hold before heading for home.

Iceland achieved full independence from Denmark in 1944 but the year Iceland became her own mistress on the sea was 1975, when the 200-mile limit was declared. It had been a gradual process helped by the drift of history, partly through the founding of the United Nations, with small nations increasingly determining their own future. The post-war fishery off Iceland up to 1975 was marked by several developments. Oil fuel replaced coal; fishing boats were bigger, faster and more powerful, which improved the freshness of fish; most significantly Iceland took an increasing share of the catch. The British distant-water fleet started the period with 22 per cent of the catch on Icelandic grounds and by the time of fisheries independence had only 11 per cent. The German fleet's share fell from 11 per cent to 9 per cent.

Whilst Iceland gradually resumed her fishing heritage, what was happening to the fish? The performance of the fish stocks closely shadows the rolling pro-gramme of legislation putting Iceland in charge of her own affairs. By 1952 the Icelandic fishing limits had been pushed out to four miles, to twelve miles by 1958, to fifty miles in 1972, prior to the standard 200-mile limit in 1975. Plaice illustrates the effect of this well, because plaice is a typical inshore species which favours shallow water and spawns on a sandy bottom. Young fish therefore cleave to the coast. Trawls and seines could wreak major damage to young fish stocks and they were duly banned as the territorial limit was pushed outwards. When in 1958 foreign boats were shoved further out to twelve miles even Icelandic boats were only allowed in certain zones, outside the nursery areas. Predictably, the fish recovered and improved.

Cod was different, living and spawning in deeper water than both haddock and plaice. The cod stock biomass moved unpredictably as it was swollen by stocks immigrating from Greenland, then thinned down again by fishing. With agonising sureness a familiar picture began to take shape – younger fish in the nets and increased effort by fishermen to maintain the same catches, creating a cycle of over-exploitation. There is one particularly eloquent statistic: right after the War the Icelandic cod spawned on average two and a half times in its life; by 1967 it spawned 1.3 times. Between 1970 and 1974 it is reckoned that 70 per cent of the cod stock was being caught annually, as compared to 37 per cent just after the Second World War.

If anyone is in doubt that overfishing was not mainly responsible for declin-ing catches on Icelandic grounds up to 1975 it is only necessary to look at what happened on other fishing-grounds targeted by the British distantwater trawl-ing fleet at the same time. Prompted by the urgent need for protein in a nation which had suffered a marine blockade and where people were undernourished, the British distantwater fleet performed a vital and life-saving function. They motored to grounds in the Barents Sea north-east of Scandinavia, off Bear

Island between Spitzbergen and northern Scandinavia, round the Lofoten islands of Norway, and off the west and south coast of Greenland. These were in addition to the traditional Grand Banks destination.

On these North Atlantic fishing grounds, mostly north of the Arctic Circle, British boats after the War found a veritable cornucopia of fish. Catches were vast and sometimes overfull trawl-nets hauled boats onto their sides and under the freezing waters. On occasion, several fishing boats were lost in a month. The fishermen fished in huge seas and often appalling conditions far from any friendly shore. Distantwater trawling was called Britain's most dangerous job, and in *Lovely She Goes* (1969) the Grimsby fisherman, William Mitford, describes in a matter-of-fact way scenes of heroic combat with towerblock waves fought in temperatures unimaginably far below freezing, all in a light he describes as more a 'perceptible variation in the intensity of darkness.' The mentions of fish themselves in this extraordinarily well-written book are fascinating. He describes the Arctic waters cod as only three and a half feet long at twelve years of age. On the better known and heavily fished grounds the cod weighed between one and forty pounds. But far enough north where fishing boats had never ventured they weighed up to 200 pounds. How old these fish were can only be guessed. He tells of one haul, in a cold where the alcohol in the deck thermometer stood at a record low of minus 82 degrees fahrenheit, wherein, apart from cod weighing 10–200 pounds, there was a 434-pound halibut, a ray with a seven-foot wingspan, and numerous razor-toothed sea catfish (*Anarchichus lupus*), all over four feet long and weighing around 50 pounds each. He records elsewhere the appearance of an Atlantic salmon in the net which measured exactly five feet; it was poached and eaten by the exhausted crew. The Americans have devised a system guesstimating the weight of a salmon from its length. This fish goes off the tables, but on the formula basis that every extra six inches equals 16 pounds weight, it weighed over 80 pounds and may have been the heaviest Atlantic salmon ever caught.

Fishing tales are never straight-line. Returning to our theme, the catches from the Arctic grounds showed the same fairly fast decline as fishing pushed down stocks. Norwegian and Soviet Union fishermen were reducing these antediluvian populations too. The bonanza began to fade from around 1950, only four years after it commenced.

Meanwhile the law of the sea was changing and one of the principal movers, through the United Nations, was Iceland, a state which plainly had great potential gains from territorial expansion over the seawater. In 1949 Iceland recommended to the UN that its Law Commission should examine the subject of territorial waters. Proposals were met by counter-proposals. The British were the most die-hard opponents of expansion beyond the three-mile limit. Almost devoid of allies barring the USA, Britain took up a rearguard position trying to hold the line at six miles. Different nations forwarded differing suggestions depending on their circumstances. The second UN Conference on the Law of the Sea (UNCLOS) convened in 1960. By this time most Latin American states had already claimed sovereignty over fishing limits 200 miles from coastlines, and several states in the congested Old World – Iraq, Libya, Iran, and Iceland – had unilaterally declared 12-mile limits.

The 1958 declaration of a limit of 12 miles by Iceland caused the first 'cod war'. Although Germany immediately recognised the new Icelandic limit, Britain did not. It is possible that the Conservative British Government, under Harold MacMillan, wished to champion the marginal parliamentary seats of Hull and Grimsby. Perhaps more likely is that the British Foreign Office was misinformed, and failed to understand the unanimity within Iceland about justification for the new limit. At any rate, Britain despatched the Royal Navy to defend its fishing fleet from the perky, defiant, civilian attentions of the Icelandic coastguard, which had only seven small vessels. The 'cod war', a term coined by a British journalist, was no war, rather a diplomatic seesaw theatricalised by some minor incidents at sea.

The next phase in Iceland's adoption of its 200-mile limit was prompted by the appearance outside the 12-mile limit of East European fishing boats chasing cod. Intensive concentration on the herring fishery off Iceland's north coast during the 1960s, while cod recovered from low stocks and exceptionally low market prices, culminated in the herring also being fished down. Iceland sorely needed a firmer framework for its fishery; without it no fishery management, other than protection in inshore waters, could be attempted. In 1972 Iceland declared a 50-mile limit, to immediate objections from both Britain and Germany.

The 1972–73 'cod war' ensued, this time involving Germany too. Now Iceland had a larger area to protect against incursions, and so deployed its new invention of the 'trawl wire cutter'. British skippers responded by fishing in pairs, the second boat behind aiming to fend off the net-cutting Icelandic coastguard. The dispute escalated when the British sent in Royal Navy frigates, despite the fact that the vessels were part of NATO forces and Iceland was a NATO member. The second cod war was shorter and more violent than the first. In all probability it was after the intervention of the chief of NATO that Britain's Prime Minister, Edward Heath, agreed to a withdrawal.

The third and last cod war was a consequence of Iceland's declaration, along with Norway, of a 200-mile limit. They were the first European states on the North Atlantic to do this. After another series of rammings at sea between British warships and tugs, and the Icelandic coastguard, a peace was finally convened in 1976. The first recorded cod war had been in 1415 when the Norwegian king tried to intercede between English fishermen and Icelandic traders. 1976 marked the culmination of a total of ten 'wars' fought over cod. There is nothing new about tussles for control of the resources of the sea.

Ultimately in the last cod war a small nation with a desperate need for healthy fish stocks, by successfully jumping the gun and pre-empting the new order by declaring its own sea limits, had defeated an old colonial state, used to stravaiging round the world's oceans at her whim, with no argument to back up her claim except the right of historic use. The British had hung on so tenaciously because distantwater trawling was a major industry. Iceland was closer than the other rich fishing-grounds, and the mix of stocks, including haddock, halibut, lemon sole and plaice, was popular in British markets. The self-same assertion and need to manage the national resources at sea demonstrated by Iceland is altering the behaviour of countries like Morocco now. The EU, the last great expansionist fisheries colonialist, in its relationship with 'third countries' (see Chapter 5), will eventually have to call home its distantwater fleets too, and fish sensibly within its own confines. Iceland, amongst the small nations, was the mould-breaker.

It has been a rule of thumb that nations which finally achieved full control of their coastal waters for a brief moment rested on their laurels, assuming the benefits of the resource would now flow their way. This never happened. Almost invariably – Iceland and eastern Canada are two examples – native fishermen replaced foreign fishermen, plundering the stocks with ever more efficient technology and bigger boats. This new onslaught, pumped up by the greater economies of home fleets which were closer to the grounds and could therefore deploy more time fishing and less travelling, rapidly brought the realisation that by the 1970s most fish populations were already severely depleted. For the free-for-all of the post-war period atonement was now required.

Iceland moved comparatively fast. By 1977 cod fishing was restricted, to be followed by restrictions on the other main demersal species. Rules limiting types of gear and regulating when and where fishermen could work inshore, in the fish nursery areas, were introduced. To allow smaller fish to escape, minimum mesh sizes for nets were introduced. Fishery management measures, hitherto almost pointless, given the openness of access, came thick and fast. The problem for Icelandic fishery managers was the classic one: to produce legislation and regulation that effectively protected the resource whilst still taking enough from it to sustain jobs and support fishermen too.

The history of fishery management in Iceland is the history of an increasingly affluent and modern social democratic state, egalitarian by nature, attempting to regulate its oldest industry, by going down the avenue of privatisation in a way that seemed fair to the bulk of public opinion. Icelanders are hard-working and resourceful: whenever the government tightened the noose around fishermen's freedom of action on the sea the practical men on the water found a loophole way of catching fish which had not been spotted. When long lines were omitted from regulation to protect old-fashioned fishing in isolated communities, many more long lines were deployed. When the government tried to give a little leeway to operators of smaller boats of less than 10 Gross Registered Tonnage (GRT), exempting them from the requirement to have a fishing permit, boats of under 10 GRT multiplied. They lowered the tonnage to six. Boats under six tons then multiplied. As soon as the main commercial stocks were catch-limited, attention was deflected onto lesser species, like shrimps and scallops. Iceland's fishermen have proved worthy advocates of their own cause, skilful lobbyists and artful manoeuvrers through the intricacies of domestic politics, where the political parties are broadly similar. As profits have increased in the fishery sector they have proved well able to defend their corner by arguing that they are the cornerstone of the economy, and everyone else's wellbeing, to be handled with due care.

To understand the management systems that have evolved it is necessary to understand the embrace of Icelandic waters. Happily the most important stocks live inside the 200-mile zone. Capelin and the Atlanto-Scandian herring stock are partially migratory and move outside the EEZ, and the redfish stock on the Reykjanes Ridge to the south stretches way beyond 200 miles, but the bulk of the rest swims in home waters. Capelin and herring, even when the latter is in a state of recovery, only account for less than a tenth in value of all Icelandic fisheries anyway. So with the demersal fishery alone there is plenty to play for.

The privatisation of fishing rights was seen as the answer to the predicament famously described by the Canadian Garret Hardin as 'the tragedy of the commons'. This was the theory that if a natural resource like fish was equally available to everybody, it disappeared under the pressure of harvesting. Much angst was felt about starting down the road of distributing property rights in fisheries because it moved civilisation once and for all away from the pre-industrial age of community living when, in a barter economy, people on the outskirts of the main thoroughfares of the world could subsist on their gleanings from the wild. Privatisation of fisheries addressed the last arena in which this way of life, often idealised and romanticised by city folk, persisted.

It started in Iceland in 1975 in the exhausted herring fishery. Only boats with a historic record of catch in herrings were eligible for a quota. The quota altered annually, depending on scientific assessments of the stock.

The difficulty was that too many boats, or quotas, were now eligible for a stock that had apparently vanished. So in 1979 the quota became transferable. In other words, one boat, to remain economically viable, could accrue several catch quotas. At the same time the quotas were made permanent, although, of course, if the TAC was zero, they could not be fished. This was the first fully-fledged ITQ system to operate in a significant ocean fishery. Some years later, in 1990, ITQ systems were applied throughout Icelandic fisheries.

In the view of Iceland's foremost authority on fisheries management, the youthful, combative economist, Ragnar Arnason: 'Provided such quotas are permanent, transferable, and perfectly divisible they constitute true property rights.' Herring fishermen, whilst the operation of the herring fishery had been in almost total suspension, had landed themselves a big catch. Quota rights in Iceland, and other places, can be used as collateral for raising money, bequeathed to successors, and in all essentials constitute an important capital asset. The fishermen have had to do nothing, having, the rancorous might say, first fished out the stock.

A far more important fishery than herring was the demersal one, whose star performer was cod. In 1977 the government restricted fishing by saying fishermen could only go out on so many days. The number of boats grew, and the number of days was accordingly shrunk. This was obviously wasteful, with so many boats used so seldom. In their muddled progress towards an efficient system the government at one time had an individual quota system operating alongside the days-at-sea option. Fishermen could work under whichever system they pleased. When ITQs were introduced in 1990 the days-at-sea option expired. With the 1990 Fisheries Management Act all fisheries were embraced, including the shrimp, lobster and scallop fisheries developed during the 1960s, and the capelin fishery which had outgrown itself in the 1970s. All these transitions, or transfers of ownership, took place either against a background of financial crises relating to poor markets, or, more frequently, stock collapses. Prior to 1990 it was trouble that initiated change. The fishing industry had promoted and forwarded the quotas concept, but this would not necessarily have progressed without the support of the key Icelandic economists. These pivotal figures argued that if the fishing industry owned the resource it would have a vested interest in good stewardship, to the benefit of all. Hot debate is taking place today in Iceland about whether such a transfer of the nation's resources was right, and if the terms should not be amended.

What has prompted the debate, which has wider implications, because the ownership of resources in a world which can no longer sustain their prodigal misuse has itself come under scrutiny, has been the unexpected reality of the ITQ system in operation. Fishermen, after all, were handed shares in Iceland's main resource, free of charge and free of binding commitments. The industry was state-run; that, and the market-place, were the only bridles on their behaviour. How has the resource fared in which they were given stakeholdings?

It is a sobering fact that in 1997 Icelandic fishermen will enjoy the first rise in the quota for cod since ITQs were introduced in demersal fisheries in 1984. For many years the critically-important stock of Icelandic cod remained suppressed by a combination of too-high TACs, with politicians routinely ignoring scientists' cautionary warnings and recommended catches, of high-grading, where smaller, lower-value fish are chucked overboard in order to load up the catch with bigger, higher-value fish, and of variable cod breeding seasons. The victims of high-grading, or 'discards', are a wasted resource sacrificed on the altar of market convenience. To replace the fishing in home waters Icelanders have done exactly what the British did after fishing down the North Sea: fish in distant waters, where quotas are irrelevant. Icelanders have been doing what they would shudder to contemplate in home waters, committing acts of gross over-exploitation elsewhere and arguing flimsy claims to historic use to support them. Stocks of cod in the Barents Sea, shrimp off Canada's Flemish Cap, and redfish outwith the EEZ on the Reykjanes Ridge, have taken the brunt of Iceland's non-domestic fishing effort. The problem of overcapacity in their fishing fleet has, as it were, been exported. The distantwater fleet accounted in 1996 for about 15 per cent of Iceland's fisheries by value. Fishing out distant waters, as Icelanders appreciate, is unsustainable, both biologically and politically. The rise in the TAC for cod in Icelandic waters in 1997 is therefore, for many reasons, being welcomed with great relief.

The situation with herring echoes the new optimism. The Atlanto-Scandian herring shoals, Iceland's own summer-spawning herring, were fished out of

Icelandic waters in 1968. In 1994 a few herrings appeared right on the edge of the EEZ. By 1996 Iceland's catch had soared to an impressive 190,000 tonnes, most of which was converted to meal. When the stock was last fished down the final fugitive herring stocks were of large fish; much of the optimism about the herring rebound has to do with the present catches being of younger fish, showing successful breeding. One of Iceland's most experienced skippers, who took on board a reporter from the London-based publication *Fishing News*, said that in his 35 years as a skipper, hauls of herring in 1996 were among the biggest catches he could remember; with his holds full he radioed other purse seiners to pump herring from his net into their holds. Needless to say, where there is abundance there is no argument. Conversely, the quartet of traditional vested interests in the Atlanto-Scandian stock, Iceland, the Faeroes, Russia and Norway, in addition to some internecine squabbling, have been angered by the EU, a newcomer, also nosing into the international fishery zone and awarding itself by unilateral declaration a sizeable share in the catch.

On the fate of stocks since ITQs have been introduced there are grounds for cautious optimism. For example, the government decision to aim for a catch of 25 per cent of the cod population looks as if it might be sustainable. One of the principal aims and justifications for the ITQ system was that it would preserve Iceland's key resource, wild fish, in good heart. If there are signs now of modest stock recovery there is ample evidence too that initially, ITQs failed Iceland. Cod in particular was overfished. The discipline of catches aligned to the reproductive capacity of the stocks is inescapable.

It is a fact of modern society that if any business sector does conspicuously well there is a rapid development of interest in curbing it, or redistributing the benefits. In Iceland debate about the justice of ITQs has only arisen as the 'princes of the sea', as they are scornfully called, have done better and better. Critics of ITQs admit that to have asked skippers to pay for their quotas at the outset, when fish stocks were in crisis and fishermen were broke, would have been unfair. It is now, with improving fisheries, that the gift of ITQs look over-generous.

There are various reasons in the detailed operation of ITQs which exacerbate the situation. One infamous quota holder lives abroad and his only economic action, lying, as Icelanders perceive him, in his deckchair in Spain, is to telephone round to find the highest bidder for his quota. Not only is he a 'feudal overlord', as Icelandic newspapers phrase it, but he is absent too, and in a nicer climate.

The shares in the trawler companies are doing spectacularly well. They can treble in a year. One of the reasons for this is that companies with surplus quota, who may be unable to fish it all, can lease it out for a short time. They then get money by trading something abstract, which was a donation in the first place. The costs and inconvenience of actually going to sea have been side-stepped.

Furthermore the number of quota holders is shrinking. They dropped from 535 to 391 between 1984 and 1994. Profiles of the quota holders show that large companies are consolidating by buying smaller ones. Many of these companies are processors who now enjoy the benefits of vertical integration, catching fish, processing and packaging them ready for shipping and putting straight onto faraway supermarket shelves. Some such firms are doing very well. It is reckoned that 70 per cent of the processing plants are in the same ownership as quotas.

The situation is generating considerable heat. Stymir Gunnarson, the editor of Iceland's main daily newspaper, among many, thinks that there should be a resource tax to cream off some of this profit in order to lower other people's taxation. 'It's medieval' he fumed, 'Here are the mighty sealords of the ocean, two per cent of the population. The rest are serfs.' The ITQs have raised the issue of a resource tax as a further philosophical point. There is now talk of charging farmers for their summer grazing of the public lands of the interior, and of charging radio companies for the use of the air waves. However, the fishing quota holders attract the most attention because their wealth, in a country where wealth is never flouted, seems out of proportion.

There is also a distinction in the public eye between fishermen, the men in the oilskins doing dangerous work in a harsh climate, and the fishing industry, perceived as a small circle of affluent individuals, dividing up an increasingly valuable resource between each other, and frightening the government with predictions of widescale economic distress if the main engine of national wealth is interfered with.

The arguments on this side were marshalled by the economist to the Federation of Icelandic Vessel Owners, Sveinn Hjartarson. He said the Icelandic fishing industry was in sharp competition with other heavily subsidised businesses. He said the industry could not bear special taxes, going on unexpectedly to reveal that the mortgage of the Icelandic fishing fleet is presently an enormous £1 billion. How could his members invest and plan long-term investment knowing how politicians could play havoc with the rates of the resource tax to sweeten voters? 'If there is too much argument' he said, 'We could pay higher conventional taxes, on income.' To which his opponents say, somewhat fatalistically, that income levels can always be capitalised and rolled back into the business before they are taxed. Hjartarson believes envy of fishing companies is being whipped up by politicians in a cheap vote-gathering exercise. 'Will Icelanders have the sense to see through this?' he mused moodily.

His anxieties are echoed by Brynjolfur Bjarnason, Managing Director of Grandi, Reykjavik's largest and most modern fishing and fish-processing company. He sees the present debate as an effort to tax a successful business using the resource tax as an excuse. Fiscal stability is important to the fishery business he said: 'We must not let emotion rule reason.' Instead, he recommended effort be directed to sensible harvesting of stocks. 'No stock in history has survived an open fishery', he pointed out. His company being particularly involved in redfish, he is concerned that the international redfish grounds off the Reykjanes Ridge are sensibly exploited. 1996 was the first time the participating fisheries had managed to agree a TAC, a sign of encouragement to him that rationality may prevail.

The economist and fisheries intellectual Ragnar Arnason prefers to cite history to turn the resource tax argument on its head. He claims that post-war Icelandic governments have run an exchange rate policy for high-spending Icelandic consumers. A higher exchange rate for Iceland's krona has been maintained than was justifiable. This, he argues, has hit profitability in the fishing industry which has to export its product. 'These policies' he pursues, 'have operated as a resource tax on the harvesting industry.' The resource tax, another Icelandic economist confirmed 'is truly the nation's most divisive issue'.

I was told about Iceland that everybody had a view on fisheries. 'Everyone has an opinion even if they don't know anything' one scientist quipped. 'A nation of 250,000 fishery scientists', said another. It is not only the national resource, it is the national subject. One aspect which exercises the reasoning, sentiment and conscience of the Icelandic nation is the role in a modern fishery of small boats. Again this illustrates the sorts of moral, social, and economic dilemmas facing a nation which is trying to professionalise utilisation of a resource that cannot be seen, that can move out of Iceland's orbit with a few waves of the tail, which interacts with its neighbours in ways Icelandic scientists are only just starting to comprehend, and which is made even more topical by international shortages and therefore rising values. The small boat conundrum is a classic case of how much the fishery should be run for the fishermen, how much for Iceland's economic wellbeing, and how much for the fish.

From the beginning of Iceland's effort to run a modern, science-guided fishery in the 1970s attempts were made to protect and look after small boat owners, the category which, in the days of international fishing of Icelandic waters, had been harmed by foreigners trawling unceremoniously through the inshore fishing-grounds. The small boat owner is in some ways an Icelandic archetype, an icon of self-sufficiency and enterprise, trimming the fat off the resource.

As recorded, in the early days of quotas small boat owners were excluded from catch limits. As the sizes of the boats brought within fisheries jurisdiction got smaller the number of even smaller boats grew. Finally all boats were embraced. Now, even under quotas, the small boats are still seen as a threat. It is immediately evident why from a glance at a graph of catch allocations. The Icelandic Parliament is predisposed to award a higher percentage of the important catches to the small boats; their share rose from three per cent in 1983 to fourteen per cent in 1990 and has risen more since.

At the quayside the sensitivity of the subject becomes clearer still. The small boats are sturdy, fast craft festooned with every modern ocean-going device. There are large stabilisers in the stern to balance it all up. As my companion remarked, they are either 'Speed boats with fishing gear aboard or mini-trawlers with souped-up motors.' The skipper of one said he could run for home in bad weather at 23 knots. Sometimes they range 30 miles out, further than the nearest trawlers. The boats have either a crew of one or two, and auto-return systems whereby lines reel in when fish strike, enabling several to be fished at once. Up to six or eight power winches can be operated by one man. Everything is designed for small-crew convenience. The laws which regulate small boats measure GRT, not catching capacity. Some of the boats which are classed as small can take 200 tons of cod a year. In addition, small boats can fish a variety of species, exchanging quotas with each other at their convenience. There are 1,600 small boat owners, constituting both a strong voting force and a powerful fish-catching force. The trawlermen accuse them of taking young fish from inshore territories and damaging stocks; the small boat owners say they are the ones who are more selective, and land the catch in better condition than is possible in very large nets.

The views of small boat owners are revealing. They illustrate the historical sense of Icelandic society, the vitality present in the fisheries, and a completely different view of fisheries from that projected by economists and fishery administrators. Arthúr Bogason of the National Association of Small Boat Owners sees his members as ecologically-friendly and responsible practitioners using methods – mainly long lines and jiggers – validated by 1,000 years of history. To him, the ITQ system is a siren voice from big business building up a monopoly in fishing rights, and unsuited to conserving stocks.

His members still have the option to fish under effort limitations. In 1996 they could go to sea for 47 days between 1 February and 31 August. By exercising this option, approximately half the small boat owners have kept a wide berth of the ITQ system. Arthúr Bogason believes that if they had not, the bigger vessels would have bought up all the quota completely, the small operator eventually succumbing to the temptation of profit without effort. 'The ITQ system' he stresses 'was made to kill the small boat fleet, and all individualism too.' He would like to see what he calls 'a gentler system', with the inshore fleet separated absolutely from the trawlers and left to exploit the near-shore resource as they have traditionally done. As to discards, he claims small boat owners bring every fish home. 'There is a market for everything' he avers.

Sveinn Hjartason also has strong feelings – strong feelings are common in Icelandic fisheries – on small boat owners. Whilst accepting that a mixture of boat sizes is important for the national fishing fleet he says the question must be asked: 'What is of most economic benefit to Iceland?' 'On pure mathematics' he continues, 'it's not profitable to build the fishing industry on lots of small vessels. It's like using shovels instead of bulldozers again.' He continued to make several strong points, pinpointing small boats' more circumscribed catching opportunities. 'You're running an industry in the far north, in bad weather and a rough climate, and the markets want fish every day . . . We have to serve the market and make a profit.'

Mr Hjartason thinks Iceland's fishing industry will move towards consolidation anyway. The market will push it that way. 'As an economist I would like to see half as many boats fishing twice as long. TACs will lead to that, the best taking over the debts of the worst. It is the inevitable consequence of the market driving the industry.' Pointing to the movement over the last ten years from the size of the catch to the value of the catch as the critical indicator, he sees the future of Icelandic fishing, now that quotas have been universally applied, in fresher catches, of better quality fish, processed more effectively. Ragnar Arnason agrees, using slightly different phraseology: 'Icelanders have reached the upper limit in their fishery. Catches may rise as stocks fluctuate but we have seen the ceilings. The area for improvement is marketing, and increasing the catch's value.'

The economist's view has always been that fisheries, and Iceland's was originally no exception, are awesomely overcapitalised. Dr Arnason showed this in a dramatic statistic from the early days, charting the development of capital funds in fishing between 1945 and 1984: the figure went up by 1,200 per cent. Over the same time the value of the catch, or the dividend, rose by only 300 per cent. So, capital deployed to catch fish exceeded returns by four times. In this picture, which is indicative of most industrial fishing nations in the late twentieth century, lies one of the key problems for the twenty-first century fishing industry. ITQs were designed to squeeze overcapitalisation out.

If Iceland was only remarkable for its early understanding of the need to control access to the resource, only half the subject would have been addressed. The other half is understanding the resource itself. Iceland may be fortunate in having a comparatively simple fishery with a lower variety of species than in warmers seas, like the English Channel, but prodigious efforts have been made to assess stocks and, more recently, to try and work out their interactions with each other.

Iceland's Marine Research Institute's function is to assess stocks, recommend TACs, and monitor the effects of fishing. It has three research vessels and a staff of about 130. Icelandic scientist Gunnar Stefansson claims the Institute's single species stock assessment is one of the best in the world. He explained that the numbers of young cod coming into the Icelandic fishery each year (the recruitment) can vary from 100 million individuals to 400 million. The average is 200 million. Variability in cod recruitment is much less than for haddock and herring, which can increase or decrease by a factor of a hundred times. In Gunnar Stefansson's view the environment is the main control on cod numbers. Currents must reliably transport cod eggs and larvae to the nursery grounds. Temperature, salinity, and the prevailing wind are important. He admitted that poor recruitment causes were hard to pinpoint.

For the groundfish surveys, Icelandic scientists go out for three weeks in March. Visiting the same locations every year they test-fish 600 different areas with a fine-mesh trawl. The aim is to find one-year-old cod and the tow lasts half an hour. Then they count the catch and measure the fish. So far, close correlations have been established between the groundfish surveys and the rate of fishing catch when the fishermen get onto the grounds to fish in earnest. Gunnar Stefansson believes the groundfish surveys, done by scientists, are vital. He sets the scene by pointing out that in Iceland there are no possibilities for enhancement of wild stocks by human intervention. He is sceptical of the commercial possibilities of sea-ranching cod. Everything, he concludes, is down to measurement, and then management.

As cod is a predator of most other fish in its environment it is important to know what effects it has. Icelandic scientists know, for example, that the more cod there are the higher the mortality of shrimps. Cod eat them. But shrimps are worth a lot more than cod. Where is the most fisheries gain to be found? Presently the strategy is to re-build cod stocks. 1997 TACs show it is working. But at the back of Gunnar Stefansson's mind, at the back of every scientist's mind, is Newfoundland. 'They under-estimated fishing effort, and over-estimated recruitment' said Stefansson, 'It could happen again.'

He believes the quantum leap forward in the future will be good multi-species modelling, twinned with data on migration and predation. However, scientists must always be alert to the significance of detail. Traditionally, slow-growing Icelandic cod matured at seven, some were mature at five, all by the age of nine. Recently they have found cod that matured at four. Is it a warning signal? 'It's a hard place to be, as a fisheries scientist in Iceland' says Stefansson, 'but these sort of problems make life interesting.'

The Marine Research Institute is abetted in managing Iceland's fisheries by a Fisheries Office, which among other things puts observers, often retired captains, on fishing boats; by the Ministry of Fisheries, the government agency which revises legislation and sets the TACs; and lastly by the Coastguard, which made a name for itself in the cod wars. The Coastguard has three patrol boats, a plane and a helicopter. The four institutions collectively employ 350 people on a budget of US $20 million.

How then is the fishery actually managed? Firstly, all catches must be weighed on landing. All vessels are licensed, and quotas logged. There is a day-by-day record of all landings. Fishing over quota is quickly detected. As a double check the outputs from processing plants is also recorded. Skippers who break the rules can have catches impounded, fines imposed, and their valuable fishing licences revoked. Fish buyers tend to know most about the volume of illegal catches: a very large buyer of Icelandic fish in Britain, himself an Icelander, said black catches were no higher than five per cent.

There is a complex system in Iceland for detecting stock emergencies. If the official observers report too many young fish in the catch the fishing area in question can be instantly closed. Most commentators rank this highly as a management tool, constituting effective protection of young stocks. To the same end some types of fishing gear are prohibited in specific areas. For example, purse seines are not allowed in the cod fishery. The Icelandic system of seasonal closures of fisheries is for stock protection, and also to avoid fishing when there is a high likelihood of by-catches. For example, the inshore shrimp fishery is subject to long and variable periods of closure. There are mesh regulations, generally at fish-friendly specifications.

Flexibility in applying controls is one thing, flexibility in harvesting the resource is another. The only item of Icelandic fisheries policy which seems over-generous to fishermen is the rule which allows quota-holders to exceed their quota by up to five per cent in one year, provided they cut it by the same the following year. In addition, as much as twenty per cent of the quota can be harvested a year later than the year it was due. Therefore a catch could legally deviate by twenty-five per cent from the TAC set by the Ministry of Fisheries. It is hardly surprising that scientists find this flexibility excessive. Striving for precision in their stock assessments and TAC recommendations it is irritating to find the resource harvest open to such wide manipulation.

This aside, there is a degree of satisfaction about the way Iceland manages its critical resource. The situation in Iceland, as rehearsed, is unique. The productivity of its fishermen is unrivalled (computed by Ragnar Arnason as seven times that of a Canadian fisherman) and a source of pride. The shake-out in Iceland's fleet – the herring fleet contracted to a seventh of its former size after quotas became transferable – has taken place without causing major unemployment. More importantly, however, public opinion has accepted that in order to save stocks, access to the fishery has to be rigidly controlled. There is pride too in Icelandic fishery science, and in the relatively comfortable relationship between scientists and the fishermen. When scientists admit there are parts of the multi-species assessment that baffle them, there is sympathy rather than criticism.

There is certainly strain between different fishery sectors, but the arguments between trawlermen and small boat owners, who are themselves often part-time trawlermen, is about finding the right balance in the national fleet, not completely supplanting each other. The debate about ITQs is heated, indeed. The beneficiaries cannot be seen to be doing too well, particularly at a time when Iceland is experiencing the stirrings of hitherto unknown unemployment, and when the economy appears to have reached a plateau. The longer heads in fisheries suspect a resource tax, or equivalent, will be imposed before long. The user-pays principle, is, after all, the one applied by coastal states to fleets visiting by invitation. The Japanese, in effect, pay a resource tax to the Falkland Islands Government. Why should nationals not pay likewise to fish home-waters?

The limitations to resource taxes need to be established if the fishing barons are to feel secure enough to invest; and fisheries are a high-investment activity. Other resource users may feel they too could be targeted. The Icelandic farmers, the pre-fishery days mainstay of the economy and now a community languishing in a low-income slough, could, on a similar basis, be asked to pay a resource tax on the Atlantic salmon, the valuable sporting rights for which attach to the surrounding farmland. The howls of protest, from people far from affluent, would thunder through a traditionally quiet society. When the resource tax comes it might be better described as the wealth tax which is what it really is.

Iceland's fishery management enjoys peer approval. Fishery scientists working in the EU, in an impossibly politicised atmosphere stricken with recriminations, look at Iceland's problems with envy. A country in control of its own fishing destiny, where fisheries are taken seriously, is clearly an agreeable spectacle. As Dr John Pope, a senior scientist at Britain's top-class Lowestoft Laboratory, and committee member of the International Council for the Exploration of the Sea (ICES), remarked: 'Yes. To profile Iceland's fishery favourably is certainly fair.'

Chapter 4

Norway

'The better an ecosystem is known, the less likely it will be destroyed.'
(Edward O. Wilson)

Ruminations about fisheries management never go far before the name of Norway arises. Norway like Iceland is a northern hemisphere country with a lot of coast and a shelf forming the basis of excellent fish stocks. Possessing the two northerly groupings of the Jan Mayen Islands and Svalbard (Spitzbergen), Norway also has two huge EEZ marine areas in the fish-richest part of the north-east Atlantic. Its cod catch of 372,000 tons (1995) is a substantial part of the world total, normally 1–2 million tons a year.

Norway is therefore in the senior league of fishing nations, landing around two and a half million tons of fish annually, much of it consisting of high-value species. Blessed with non-membership of the EU (a referendum turned this down in 1994, largely because of the fears in the northern communities that they would surrender their most faithfully self-replenishing resource), Norway too has used to its best advantage the highly skilled pool of fishery scientists. The time-series of some Norwegian fishery studies is over fifty years long, affording opportunities to analyse long-term trends others would envy. Norwegian fisheries, although they do not play the absolutely pivotal part in the national economy that Iceland's do, owing to Norway's possession of immense oil and gas reserves, provide plenty to write home about.

Again like Iceland, Norway is a nation of fishermen. 'Everyone feels them-selves a fisherman' one Fisheries Ministry official told me, 'If they don't fish themselves, they have a grandfather who did.' In a revealing demonstration of conscientiousness Norwegian shrimp fishermen pre-empted their administra-tors by adopting the use of sorting grids, from which smaller shrimps escape, before they became compulsory. The scientific community works alongside the fishermen rather than against them. Norwegian scientists and fishermen enter-ing into discussions with EU representatives over quotas, in the areas where each has access to the other's stocks, were surprised to find the EU fishermen and scientists sitting far apart, seemingly representing different lobbies, whereas their own national team comfortably included the fishing industry and its regulators. Norway is not a state at odds with itself over fish.

In 1930 Norway had only one trawler. Its fishing scene for a very long time persisted as a small boat affair, prosecuted from remote and often isolated settlements along the rugged north coast. Fisheries were regarded as an indus-try which deserved protection because they supported livelihoods in places

92

where no other livings were conceivable. A State Fishery Bank was established as far back as 1921, and in the years immediately preceding the Second World War the majority of trawlers were constructed with government money.

Norwegians not only saw fit to subsidise their fishing industry, but also to protect it. Perhaps befitting a country perched on the top corner of the European landmass, with only one major land-based neighbour (Sweden), with whom relations have always been one of subordination, and therefore lacking natural allies, Norway is touchy about its independence. At the suggestion of their fishermen the Norwegian government prohibited the entry of foreign capital in fisheries, and also prevented people of no experience in fishing from muscling into the industry. The fishermen's faith in their ability to go it alone has proved justified, although the discovery of very large fossil fuel reserves in the 1970s, making Norway instantly a great deal more affluent, has given the fisheries sector both direct supports, and a cushion. This could not have been predicted in 1964 when an agreement signed between the government and the Norwegian Fishermen's Association set the aim of unshackling fisheries from its scaffolding of state support and converting it into a self-supporting, independent industry.

The intentions that fisheries should stand alone were overtaken by events beneath the seawater. The new technologies of purse seining succeeded in fishing down the Atlanto-Scandian herring in the late 1960s. North Sea mackerel was then targeted until it too began to collapse. The drive to catch more and more fish had to turn into one to contain overfishing and restrain catching. 1972 was the first year any Norwegian fishery was licensed and restricted. Purse-seiners bringing in pelagic fish, and shrimp trawlers, were brought under control first, to be followed by the saithe-hunting fleet. Even when the government published its 'Long Term Plan for the Norwegian Fisheries' in 1977, there was no mention of profitability or productivity. Maintenance of secure employment was a leitmotif. At no time did the Norwegian state contemplate the closure of its peripheral communities, made increasingly redundant from an economist's view by the ability of fast modern fishing boats to reach the grounds from further south. When oil profits began to soar in 1975, monetary rectitude became unnecessary. Apart from filling government coffers, the cost of fuel oil for fishermen fell. Nonetheless, by 1981 subsidisation of the industry had reached a new pitch: the Ministry of Finance calculated that subsidisation amounted to nearly 70 per cent of fishermen's incomes. This speaks volumes about the political attitudes to fish, and about the power of the fishing lobby. Professor Rögnvaldur Hannesson, from Bergen's Centre of Fisheries Economics, dryly comments: 'One impeccable if tortuous logic behind the goal of preserving scattered settlements along Norway's rugged and tempestuous coast is that it puts a premium on having people live in places where the weather is bad.'

Norway had chosen a path of its own and different from either Newfoundland or Iceland. Outlying settlements had been supported and economic rationalisation of the industry had been averted. Norway persisted with overmanning and a unit performance by its fishermen of which an Icelander would be ashamed. Overcapacity in the fishing fleet had been pre-served because, for most of the fishery, quotas were not transferable. The policy persists right into 1997. In late 1996 the government announced that trawlers could merge their two unit quotas into one, the eventual aim being to halve the total number of trawlers. But, typically, northern Norway was exempted from the regulation, and trawlers may only merge units within one district. The possibility of merging the quotas of ice-fish trawlers with freezer trawlers, or either with smaller trawlers, has been excluded.

Economists may say the system has been wasteful. However, there is some-thing soul-sapping about a country where the peripheries are left to wither, and Norway is a very tall country in which the blood has to be pumped the whole way to the top. Norway had the money to run a luxurious fisheries policy, and chose to do it. Excess catching capacity in the Norwegian fleet has been main-tained principally because the management of fish stocks, using a similar science to Iceland, has been good, and stock collapses of every sector at once has been avoided.

The idiosyncrasies of the Norwegian system need looking at to make sense of Norway's greater success at conserving stocks. The regulation which is syn-onymous with Norway in fisheries management is the outlawing of by-catch. Fish caught over-quota or undersized must be landed. If quota is so far unful-filled, the by-catch counts against it; so by-catch carries a penalty. There is a continuous recording of catch and if a boat keeps returning with by-catch it is obliged to move somewhere else. Fishermen are compensated, not for the com-mercial value of the by-catch should it be unsellable, but for the cost of catch-ing it. This has encouraged more creative usage of the catch: utilisation of almost all that is caught, on modern Norwegian boats, is almost a hundred per cent. The system is fair and equitable, not punitive. The fact that by-catch cannot be thrown back has altered attitudes. Fishermen in any case hate reject-ing an edible haul. Norway is saving untold numbers of young fish from being wasted. As Dr Reidar Toresen, chairman of the herring group in ICES and senior scientist in Norway's Institute of Marine Research, said: 'Having no by-catch, from a management view, eradicates many unknowns.' Notably, measurement of catch for fishing effort produces a true figure. The Norwegians have a close idea of fish abundance, a cornerstone of stock management. Norwegians have argued their philosophy of no discards with the EU Commission, in whose waters they have various reciprocal access arrange-ments, for years without making headway. Norwegian representatives have repeatedly said they regard young fish thrown back to sea as an environmental crime. As to illegal catches, and some do occur, Norway has guarded against the phenomenon of 'black' fish since long ago. An Act of 1951 made all sales of fish pass through fishermen's own sales organisations. Fishermen themselves have a strong handle on the market.

The likelihood of actually taking by-catch, or undersized fish, is reduced by modern management. Norway more than any other European country has directed considerable resources into the development of fishing gear designed to avoid catching the wrong fish. Rules punish any gear-rigging which harms selectivity. Sorting grids in trawls allow the escapement of smaller fish; they stream out through narrow escape-panels. When trawling for shrimp, the most problematical by-catch fishing operation all over the world, the use of these grids is compulsory. Deploying these grids actually opens up areas for fishing which otherwise would be too sensitive to exploit at all. The Norwegian fishing conglomerate, Aker Resource Group International, one of the most modern fishing companies in the world, claims that even in its shrimp trawl fisheries the by-catch never exceeds two per cent; shrimp trawl by-catches in other places can reach eighty per cent. The critical point about this high degree of regulation in Norwegian waters is that it has led to a higher level of awareness in fishing skippers, and now that they see that it works to enhance and underpin the fishery, they support it.

The Norwegians are sticklers for sensible basic modern management. The Norwegian seas, like those of every modern fishing state, are divided into areas. Before an area is opened for fishing it is test-trawled. After a four-hour trawl if the by-catch component is too high it is kept shut. The same sector is tested again later to see if the species composition has altered. Fishing areas are also gear-controlled. For a long time purse seiners have been considered too destructive on cod grounds and have been prohibited access. The business of converting multispecies assessments into policies has been taken fearlessly to its logical conclusion: the capelin quota, a vast fishery at one million tons in 1983, was zero in 1996. It is only likely to open up again if scientists are reassured that there are enough capelin to feed the high-value cod. The completely unrestricted catch of the North Sea capelin equivalent, sandeels, within the CFP, is regarded by Norway as quite bizarre (at time of writing some curbs on EU sandeel fishing have been agreed, in principle).

It is true that it is easier to run a 'clean' fishery off Norway, where there are a limited number of species, than in the multispecies environment of the North Sea. However, steady common sense has produced its own benefits and good Norwegian stock management has resulted in Norwegian waters having better structured age classes, which itself reduces the non-target catch.

Norway has also taken a different attitude to fishery regulation. In common with any fishing zone with efficient management and sustainable stocks, Norway is tough on policing. Toughness is the official aim. 'If you don't have strong controls, the regulations are worthless', maintains the Fisheries Minister Jan Henry T. Olsen. He regards the virtual absence of controls in EU waters whilst boats are actually on the fishing-grounds as lamentable. EU fishermen in action in Norwegian waters have reacted with anger and bewilderment when subjected to Norwegian-style enforcement. For a start, most fishery surveillance in the EU is at quayside, not at sea where the action is. Norway's fishermen operate under unremitting control and supervision. EU fishermen are used to reins of control so light as to be barely felt.

Unlike in Iceland where the Coastguard is civilian, Norway's is drafted in

6 year terms of service from the Navy. The Coastguard chiefs, situated in the Lofoten Islands or in Bergen, although funded by the Defence Ministry, co-operate closely with the Ministry of Fisheries. The Coastguard has thirteen large vessels and several smaller ones. In addition, Norwegian enforcement has the services of three ships of over 2,500 tons. Trawlers and purse seiners, manned by fishermen, with Coastguard officers aboard, and also planes, are hired by government. Inside the four mile limit, where arguments between Norwegian fishermen focus on differing fishing gears, usually in the north, there are about twelve more enforcement vessels.

The sorts of rules being broken are the usual universals – catches of under-sized fish, under-reporting of real catches, and fishing with illegal net meshes. The name of Spain surfaces quickly when talking about illegal fishing, both in Norway and everywhere else. Non-Norwegian fishermen landing an illegal by-catch may be directed to a fish-meal plant or, depending on where they were caught, treated to a fine. The by-catch cannot be thrown back within Norwegian waters. What gets up the noses of the Norwegian Coastguard is the practice among foreign vessels of not declaring by-catch, steaming out to the international zone and dumping it, legally, there.

The head of enforcement in Norwegian fisheries believes monitoring vessels by satellite is the future way of applying the strict controls to catches that will be necessary. Acknowledging that fishermen do not enjoy feeling spied on, he goes on to point out that satellite monitoring works to a tee. Around thirty shrimp trawlers have been sailing with gadgets called transponders for a year already. He says he wants the law changed to make carrying satellite trans-mitters compulsory, and changed soon. While not telling anyone exactly what is being caught, satellite surveillance can tell exactly where anyone is – tanta-mount in many cases to revealing the former. In an era when European farmers are having the metric areas of their fields checked against their subsidy claims, and when the precise nature of the crop can also be read from satellite images, it seems likely fishermen will be brought within the net soon as well.

There is no talk of resource taxes in Norway. As Johan Williams from the Fisheries Ministry, said: 'The majority feel, as the fishermen do, that the fish are a common heritage.' Although once massively subsidised, Norway's fishery is now self-supporting. The rising value of fish, and the well-tended stocks, have caught up with the seed-funding. The only part of the fisheries' cost paid out of government and not fishery finances, is policing and enforcement. That might be considered not unreasonable as fisheries are now the second biggest export after oil.

The stocks of both Arctic cod, which is managed with Russia, and the spring-spawning herring, are in good shape, and improving. Although, owing to its curving long coastline and rim-position in Scandinavia, there is an obliga-tion to manage straddling stocks co-operatively with other countries, for example in the Barents Sea, and although in return for reciprocal rights in the North Sea Norway has to contend with the presence, in the fishing zones of its outlying islands, of people with a rather more cavalier attitude to stock conservation, the incontrovertible fact is that catches are rising and stocks appear to be secure for the time being. The partial collapse of cod as recently as 1986, due it is thought to environmental reasons and a shortage of capelin, has led to a new application of the theory of multispecies management. Norway's 1995 catch figures for mackerel and saithe were both over 200,000 tons, and the figure for herring neared 700,000 tons. The linch-pin cod catch, like in Iceland representing about a third of the national catch by value, is strong. Scientists' worries about increasing rates of cannibalism amongst cod in the 1980s have been partially allayed by writing in multispecies factors to TACs. The flexibility of the Norwegian management of cod, which brought the 1989 TAC of 500,000 tons crashing down to 172,000 tons the next year, is regarded as crucially important by Norwegian scientists. In most systems the politicisation of fisheries would make such dramatic rescues mere pipe-dreams. Norwegians think it is because of their history of problems with stocks that they are willing to be tough; and it is because they have a low regard for the unbridled licences within the EU that they are determined to stay out of it.

Also the structure of fisheries affairs gravitates towards the acceptance of biology-based management. Major management issues are discussed by a Fisheries Board. This body contains representatives from all possible interests, including a Coastguard member, a Ministry of Fisheries observer, fishermen, scientists, ecologists, and even a Lapp. If the Board gives unanimous advice it is hard to completely override it. Fisheries are not the political pawn they are in some European countries, nor could be. The claim of Norwegian fishery managers to the British House of Lords Select Committee Report that in Norway fish stocks are managed on a biological basis, is credible. Norway has a serious approach – and is rewarded.

Chapter 5

Fated Fish: The Common Fisheries Policy

Question: Why is the CFP not working? Answer: 'Politicians and bureaucrats are trying impossible political solutions where there should be scientific and technical solutions.' (Dave Pessel of the Plymouth Trawler Agents Association)

'The 1970s showed that no fisherman has an incentive to cut back his own activity; on the contrary, it is better for him to have a larger boat, which will reduce his earnings and reduce his tax liability.' (Michael Shackleton, former administrator in the Secretariat of the European Parliament in Luxembourg, *The Politics of Fishing in Britain and France*, 1986)

'However well the state authorities might recognise the scientific case for effort limitation, they could not deny that they were themselves strongly in favour of promoting national production in the face of increasing imports.' (Michael Shackleton, ibid.)

'The Council [of European Fisheries Ministers] makes use of the combination of scientific uncertainty and honesty to postpone difficult decisions . . . we are coming to a situation where the matter is considered by Ministers to be too technical and by the scientists too political.' (Alain Laurec, Director of Directorate C, DG XIV, EU Commission)

'If one chooses the one alternative – a 'common pond' for the whole European Union – one cannot simultaneously choose to have "fishing communities".' (Audun Sandberg, 'Community Fish or Fishing Communities', from *Fisheries Management in Crisis* edited by Kevin Crean and David Symes)

'Everyone wants improvement, and no one wants change.' (Dr David Armstrong, senior scientist in DG XIV, (Directorate General) in conversation)

'At the end of the day there are two rules. Don't catch too many fish, and don't catch young fish. It's an "and" not an "or".' (Dr David Armstrong, in conversation)

'If all the conservation measures were implemented, and the science was right, there would be no decline in stocks.' (Jim Portus of the South Western Fish Producers Association)

There is a perspective on modern fisheries which goes as follows. Fisheries developed on an industrial scale some 200 years ago turning what had been remote outposts of civilisation, or places where boats were pulled up on beaches, into busy, prosperous, productive and thriving sea-ports. The town of Wick, one such place, has already been profiled. Fishing made fortunes, for a few, and lifted the status of some nations, or parts of nations, into the limelight.

As the third millennium approaches we are in the dying days of this epoch of civilisation. The number of people engaged in fisheries has dropped. The status of fisheries and fishermen has dropped. Fishing as a part of the national culture, in nations like Britain and Denmark, has dropped. In Britain, once a world leading fishing nation, catches now contribute only marginally to GNP, by 0.1 per cent in 1993. Some individuals still do well from the world's last major, free-roving resource, but they constitute pockets of economic success rather than whole districts. As numbers of boats and fishermen have fallen so has their clout.

Instead fishing communities are now reverting back to the outposts they once were. The EU is an example of a political bloc in which the fishermen live round the edge, generally in the poorer areas, disadvantaged, within an increasingly centralised monolithic engine of government, by remoteness, small voting power and little influence. In the constitutions of Norway and Iceland there is, in fact, a disproportionate weighting of political representation in the remoter regions, a throwback to the time when farmers and fishermen really did set the national agenda. However, this is already being viewed critically by a technologically-adept, urbanised public, and before long the electoral imbalances will be redressed. The places where the smell of fish pervades the main street are due once again to fade from prominence as societies become centripetal, lifestyles even more absolutely armchair, and as supermarkets' grip on fish marketing tightens.

In this perspective, which is shaped to the northern hemisphere, fishing communities are a thorny problem for administrators. Subsisting on the waxing and waning of natural cycles in populations of fish which are irreducibly wild, and which respond with extreme sensitivity to a wide range of environmental factors over which man has no direct control, fisheries are likely always, for the foreseeable future, to require ordering by governments, research by governments, and a relatively consistent economic background. Fisheries ever since the days of the early pioneers have invariably involved government expenditure.

It is a commonplace that fisheries generally lose money. World fisheries operate on a vast deficit, yet no one is in doubt that sensible fishing policies are an aim worth striving for. After all, if two thirds of the fishing nations merely pulled out of the game, hauled in their lines and nets and burnt their boats, in all likelihood the surviving fishing states would do well, on the scaly back of a resource with remarkable powers of resuscitation.

The nightmare scenario is of nations forced to operate a modern fishery where the political problems appear intractable, and where fish allocated to fishermen to catch become 'paper fish', existing only at the tables of negotiation. Such a body is the EU, whose fishery is frequently described as the worst-managed anywhere, whose fishery scientists are amongst the world leaders, whose waters are potentially fish-prolific, whose political background is bedevilled by a politically expedient treaty containing an idealistic aim which is proving lethal in practice, lastly in which public attitudes have proved unexpectedly strong and passionate.

Fishery scientists from outside the EU look at its difficulties and throw up their hands in horror. Ran Myers thinks it is only the earlier reproductive ability of the North Sea cod, compared with Canadian slow-growing cod, that has saved the cod stock so far. Most outsiders think the EU's Common Fisheries Policy (CFP) is doomed to collapse. Many inside it do too, but have their jobs to protect. Some of these are in awe-struck wonderment that Europe has any fish stocks left at all. Others detect more rapid fish growth and changes in fish behaviour, a sort of survival reaction to the massiveness of the onslaught, a development never documented previously. Despite the scale of the research institutions (Lowestoft's Fisheries Institute alone has 300 scientists) as against the comparative smallness of the CFP's productive fishing zone, with their consequent ability to carry out detailed and large-scale research programmes, it is very hard indeed to find from those within European fishery science anyone who is unequivocally, or even tentatively, upbeat about the future.

The clues to the structural chaos of what was to become the EU's CFP were there at the outset. The six original European Economic Community (EEC) states were quite uninterested in fisheries; fisheries were subsumed in the category of agriculture, in which they were very interested indeed. Accordingly, fisheries principles were drafted, not for a naturally renewable resource, requiring conservation policies, but for an activity whose function was to supply the new market.

Retrospective significance has been accorded to the fact that the day in 1976 on which fisheries got their own department in the EEC was April Fool's Day. It was only when Britain, Ireland, Denmark, and Norway applied to join the Community that the six original states noticed the significance of fisheries. All the four new applicants possessed major fisheries, indeed collectively they controlled (Denmark through the Faeroes) some of the world's finest fishing grounds. Britain owned two thirds of what was to become the common fisheries area; fishing rights were its largest renewable resource. It was literally two hours before the negotiations about these entries started that the wording of the regulations on fishing, by which the new entrants would be bound, was adopted by founder members. These regulations said something absolutely critical: there was to be equal access to each other's fishing waters, with no discrimination between member states.

On this condition a vast amount of destructive fishing has taken place, a juggernaut of 'black' fish catch now rolls forward, discarded dead fish are thrown overboard in hundreds of thousands of tons, stocks have been relentlessly hammered, and politicians have torn their hair out. The Fisheries Ministry in Brussels has become a by-word for hopeless entanglement, political horse-trading, fudging and compromise, and the job of Fisheries Commissioner is recognised as a poisoned chalice. The Commissioner at time of writing makes no bones about her desire to move jobs. When she visited Iceland in 1996,

Emma Bonino told the audience in her first address that she was made Fisheries Commissioner for three reasons: firstly, she was from Italy, a state without serious fishery interests; secondly, because she was a woman; and third because she knew nothing about fisheries. Fisheries have even become such a contentious issue in Europe that they have developed the capacity to break up the EU. Loss of sovereign fisheries is exactly the reason Norway still declines, after several referenda, to join.

Why? How has a minor industry, basically an old-fashioned resource extraction business, employing comparatively few people, acquired the power to break the virtuous circle of jurists, accountants, bankers, businessmen and politicians who are pushing European countries into a closer, tighter, more strictured relationship?

It is perhaps because as countries in Europe lose their identity, with federal laws creeping into every crevice of what was once national life, nation states, many of which only one generation ago were in the deadlock of one of history's ugliest military conflicts against each other, are looking for true nationals. What truer and more emblematic national could there be than the fishermen, redolent of the seafaring adventure which has defined centuries of modern history and limned great expanses of the world map, out on the salty sea in all weathers, bringing home a wild catch from an unforgiving and still mysterious environment?

When the fish-rapacious fleet of Spain was finally brought to a halt by Newfoundlanders with nothing left to lose, British politicians who supported their EU colleagues were shouted into submission. Canadian flags were hoisted near British fishing ports, and the letters columns of newpapers were crammed with diatribes about supporting our kith and kin over the water and bringing to book the rule-defying Hispanics, who anyway, it was pointed out with relish, were only recently emerging from government by a military dictator. There was little doubt in Britain which side the angels hovered on, and politicians scrambled with indecent haste to climb onto the bandwagon. The fact that all three British political parties in their 1996 annual conferences called for some restructuring of the CFP, was in no small part due to the events just outside the Canadian 200-mile limits in 1995.

At the time of writing the drama of the unfolding, or unravelling, CFP has still to be played out. The break year is 2002, when the fisheries policy is due for review, and when the derogation which is presently keeping the Spanish and Portuguese fleets out of the North Sea expires. The Spanish gave warnings of their intentions to get into the remaining fishing-grounds as soon as possible when in 1994 they made their consent to admitting new members to the EU conditional on access to more fishing, not, one might suppose, a germane matter. However, it resulted in the 'Irish Box' agreement, in which Britain and Ireland were hauled over the coals in Brussels, prior to allowing forty Spanish fishing boats into a previously restricted fishing area off southern Ireland. That an EU member could hold Europe to ransom by making agreement to purely political matters dependent on a fishery deal, exposed the reality of decision-making in the Community, a devious process of horse-trading and barter. Such manoeuvrings should never have been allowed. What made matters worse was that Spain had almost no historic fishing rights in the target area. Preserving some respect for nations' historic rights is what had kept the CFP from foundering hitherto. The derogations were proving an unreliable life-raft.

The North Sea is the nub of the whole affair. It contains some 90 per cent of all EU stocks. The Spanish, and their opponents, the breakaway British fishermen's group Save Britain's Fish, say the writing is already on the wall. All member states have equal access at this moment; the treaty says so. Immediate political implementation would have been political dynamite. In order to massage this abstract legal fact into practical fishing the EU adopted a series of derogations stealthily phasing in the huge Spanish fleet, some 20,000 vessels equating to three quarters of the rest of EU fishing effort combined. At the first review of Britain's ten-year derogation in 1982 the sharing-out of stocks between member states was taking place as well. Weakened by its need to get the derogation extended, Britain left the negotiations with only 37 per cent of the catch by volume, which amounted to a mere 12 per cent by value. The country which had once taken the lion's share of commercially useful fish had fared extremely badly in a share-out that was to harden into a key turning-point. One reason was that the division of spoils was based on catches in what were now Community waters, during the 1970s. During this time Britain had been fishing mainly in distantwaters.

Whilst the number of fishing vessels in member states is steadily being reduced, the number of fleets grows. If, as is presently in train, the EU embraces eastern Europe, the common fishing area will also have to accommodate the fleets of countries like Poland and Estonia. These states bring with them either no fish stocks, or virtually none, but plenty of fishing boats. They will bear down on western Europe's depleted but still surviving great fishing-ground. Any attempt to keep them out, unless drafted with Machiavellian ingenuity, would imperil the sacred principle of equal access. It is the unanimous view of everyone in fisheries to whom I have spoken that the North Sea will very rapidly be fished down to uneconomic levels unless the CFP is reviewed and radically changed, and protected from a mêlée. At time of writing there is every indication that powerful influences on fisheries like Spain and France are making sure in plenty of time, while the battle-lines in the debate are drawn up, that any radical overhaul of the policy, in ways inimical to their interests, is prevented.

The upcoming CFP review has had the merit of focusing minds. There is near unanimity amongst coastal states (with the exception of Spain and Portugal which have no proper coastal shelf) that a greater degree of regional control, in fact a restoration to coastal states of control of inshore fisheries where they have been sacrificed to the common pool, is desirable. However, until voting regulations themselves are altered within the EU it is impossible to see how Spain and Portugal's blocking votes can be overcome.

Nonetheless politics is a dynamic affair. It is certain that in the Fisheries Commission in Brussels, DG XIV, the ideal fishery would be a highly regulated, controlled, science-led, harvesting regime, utilising the latest in fish-friendly gear. It would be performed by a flagless EU fleet half the present size, without distinction as to nationality, policed by an EU inspectorate, also without national distinction. It is equally clear that political events can unexpectedly overtake plans drawn up in sealed chambers. It is the principal thrust of the analytical and detailed book *The Common Fisheries Policy* (1994) by Mike Holden, former chief of the Conservation Unit in DG XIV, that hard-working scientists' recommendations carried forward by the European Commission, were routinely overruled for reasons of political compromise by the Council of Fisheries Ministers. He gives countless examples. A simple one was the proposal for a 90-millimetre net mesh limit in the North Sea. It was presented in 1982, and after innumerable delays, with young stocks overfished then punitively overfished again, this rule was finally adopted in 1992. Holden explains that the Conservation unit was politically unpopular because it would curb fishermen and reduce their earnings, and how other policies which contradicted conservation, such as the fleet modernisation or structural policy, always took priority.

In 1996 Fishery Ministers in Britain were proclaiming that Britain will still have some sovereign fishing rights after 2002 though knowing full well that, saving a political volte-face and an adjustment to the original treaty, for which there is no prognosis, she will not. If politicians did not feel it to be such a sensitive issue in Britain they would not risk the recriminations which will follow when they are found to have been fibbing.

Curiously, this misrepresentation of the truth about the terms of access dates from Britain's entry. Edward Heath's government, desperate to get access for Great Britain settled, told Parliament that the 12-mile fishery limits were secure, having just read the text and signed the papers to surrender them. This deception of the House of Commons will sit uncomfortably in the history of British Parliamentary history until the day British people cease to catch seafish. In interviews since Edward Heath has tried to argue that nothing sovereign was lost because a sovereign gain was involved in the acquisition of a British vote in the Council of European Ministers – a partisan understanding of the word sovereign indeed! Edward Heath unsuccessfully tried to persuade Norway to join; the motive was to get a bigger fishing-ground into the common pool, thus masking the dimensions of Britain's fishing ground loss, and offering a sop to British trawlers.

The argument of Save Britain's Fish, the campaign dedicated to leading British fishermen out of Europe, is that Britain signed the January 1972 accession treaty under false pretences, and it is therefore invalid. One of its leaders, Tom Hay, a former trawler skipper who has turned his considerable energy to constitutional law, maintains that the only law the British Parliament cannot legislate is the national sovereignty. In his deep, resounding bass, he spells out the theme like an Old Testament prophet: 'We have temporarily surrendered administration of our sovereignty. No legal system can remove the right to repudiate any law. It is enshrined in the law of democracy.' He goes on to cite peoples who have reclaimed their sovereignty, such as Poland. Britain, he says, is different, 'because we were never a subjugated people. In a democracy people remain forever free.' One imagines Tom Hay's meetings with the British Fisheries Minister of the moment, Tony Baldry, also a lawyer, were uncomfortable for the politician.

The political strains of the CFP are a critical weakness. The animosities between member states rule out the co-operation between regulator and regulated which fisheries depend on. In no other fishery in the world is co-operative management, engineered by bureaucrats from outside the industry, unshackled from the discipline of costing its prescriptions, being attempted. The fact that member states, like Austria, with neither navy nor coast, can vote on fisheries, is patently absurd, and encourages the tendency for fisheries to be a political football.

Fisheries differ from other industries; no other entirely privately-owned industry is run by government. Furthermore, individual governments' attitude to fisheries differ. Spain acts for her industry, and maintaining fishing employment is a central plank of Spanish social policy. In Britain the government acts, not as in previous centuries to promote more fishing endeavour, but to curb it. Largely owing to the initial surrender of fishing rights, fisheries have become a matter of embarrassment to British politicians. The only fisheries cause that the Conservative government in the mid-1990s vowed to fight was to abolish 'quota-hopping'. This was the process by which fishing boats could be bought up by countries within the EU and not re-flagged. The new boat owner had bought not only the boat but the essential fishing quota from the vendor country. The highest bidders in the market for British boats whose owners had sickened of the whole charade of the CFP were Spanish, and to a lesser extent Dutch. So it happened that large chunks of the so-called British fleet actually passed into the ownership of rival fishing states. When the scale of the transfer became public people were shocked. In 1996 the government admitted that a fifth of the fleet was foreign-owned. In the case of the beam trawler fleet, raking the seabed for fish such as sole, forty-eight per cent of the fleet was foreign-owned. The only British aspect of the boat was the flag and the quota. Everything else belonged to the new foreign owner, and the catch was landed in the ports of the new home-state. For Spain and Holland, of course, an outlet had been found for their fishermen. The British government pledged to try and halt the abuse of quota-hopping, a pledge no one believed could be made good.

In such a vexed environment it is unsurprising that fishery science has been drawn into the imbroglio. Fishermen routinely accuse scientists of under-estimating stocks and applying TACs which are too tight. To some extent this occurs in all regulated fisheries. The tendency in the EU is for fishing research to move further and further into an enclave of its own, scientists increasingly resorting to test-fishing only in their own high-tech fishing boats. Previously scientists often travelled with fishermen and got to know them. The loss of contact has exacerbated things. Fishermen say, with some truth if one remembers Newfoundland and how the fishermen kept coming home with cod even as the stocks nosedived, that scientists do not know where to look for fish. Scientists argue the purity of test-fishing the identical area year after year to produce a reliable time-series. Fishermen reply, fish move. The argument goes on.

It is worth looking at the history of the North Sea. Concern about stocks started around a hundred years ago. Certainly when T.H. Huxley, a scientific lion of his day and then President of London's Royal Society, said in his speech to the International Fisheries Exhibition of 1883, that 'probably all the great sea fisheries are inexhaustible . . . and any attempt to regulate these fisheries seems consequently . . . to be useless', the practical fishermen themselves disagreed with him. Huxley's remark was made in the decade when the power of trawling doubled – and that was before the advent of steam! By the turn of the century British fishermen were urging the protection of young fish; in 1900 when a bill was brought before Parliament to introduce legal size limits on landed fish it was only narrowly rejected. Around this time most fishing nations in northern Europe had started up biological research stations, and to initiate co-operation at a national level the International Council for the Exploration of the Sea (ICES) was founded in Copenhagen in 1902. ICES still provides data and background science for Europe's fisheries management today.

During the early part of the twentieth century it became increasingly apparent that restrictions on catches were needed. Even during the first decade, landings of turbot, sole, rays, and plaice were falling. The big flatfish were caught less and less. In the 1870s they reckoned that there were ten to twelve times as many spawning plaice in the North Sea as forty years later. Now it was thought that with all the otter trawlers working the North Sea the sea-bottom would be fished twice over in a year. Ploughing the seabed had arrived. The fisheries biologists had studied the growth, reproduction, migration, distribution, and behaviour of the major North Sea species, and the infinite resource idea faded. Instead, general concern grew. In the 1930s several parliaments in European countries legislated to outlaw landings of small plaice and small haddock. Simultaneously America and Canada had agreed to halt the catches of halibut, another bottom-dwelling fish in the North Pacific, after a certain tonnage had been reached. In 1946 in the North Sea Convention most European states agreed minimum landing sizes and minimum mesh sizes. Control of fishing effort had entered the agenda for discussion.

Then came the fishing jamboree. The years around 1970 saw a huge rise in catches of haddock, cod and whiting in the North Sea, rapidly peaking then rapidly falling off. As catches pinnacled, spawning stocks plummeted. The most spectacular collapse was in herring. The fourteen member countries of an organisation called the North East Atlantic Fisheries Commission (NEAFC) failed to agree catching by quota until 1975. It was too late. By 1977, as mentioned, a total ban on herring fishing was applied throughout the North Sea. Mackerel suffered very nearly as badly.

People focused on this story. Herring and mackerel had constituted two thirds of the nine million tons of fish scientists said were in the North Sea.

Herring and mackerel had been levelled to an estimated two million tons, at extreme speed. Although the EEC started taking control of waters 200 miles from member states coastlines, the NEAFC continued to advise on the rest of the vast North Atlantic area.

Fishermen did what they always do, if it is feasible. They targeted new species. In the 1970s they turned their engines of entrapment to Norway pout, sprats and sandeels, low-grade fish used for converting into fish-meal. Scientists were faced with new questions. What were catches of this dimension – the 'industrial' fishery rapidly rose to an annual one million tons – doing to other stocks? This question is asked even more keenly in the late 1990s.

David Cushing was raking leaves on his lawn when I visited him. He believes that scientifically-speaking, regulating fish populations in the North Sea is a practical proposition, harder, but no less feasible than counting the leaves that were falling on his lawn. Dr Cushing's view is worth hearing. Now retired, he was once head of the Fisheries Laboratory in Lowestoft which, not only in his claim, is a world leader. 'In the old days' he said, 'Up to the 1960s, we recorded what was entering the fishery at a given age, or the recruitment, working out our calculations from fishermen's catches, studying the length of fish, where they were caught, and so on. That's not good enough now. We must understand the processes behind the recruitment.' He justifies the greater use today of specialised research vessels in three ways: there are fewer fishermen, they do not cover the North Sea as they did, and their gear has hugely improved and is still improving. Research scientists need to use the same equipment, to compare like with like.

The sample trawls done in the North Sea are described by Dr Cushing as 'comprehensive'. Working forwards from an estimation of the stock it is possible to estimate next year's stocks, in order to advise on a suitable TAC. The gap in knowledge today is not in counting adult or juvenile fish, but working out how recruitment operates, or how populations re-charge themselves. It is the variability in success at the larval stage which he thinks needs addressing.

Larval stages, highly vulnerable to predation, climatic events and weather patterns, temperature in the sea, strength of currents, and growth of the microscopic food supply on which they subsist, are very difficult to monitor. We simply know too little about what eats what. The North Sea has a greater variety of fish than Icelandic or Norwegian waters; each new species, living off its fellows, different fellows at different stages of development, makes the equations harder. If, as is known, cod are top-of-the-chain consumers of everything, should cod be fished hard to let others grow? Whilst it can easily be calculated what populations will develop to a harvestable stage in a closed environment – which salmon farmers are good at with fish in cages – it is a lot harder to see into an environment which is volatile, dynamic, and where human efforts are bringing home a catch only part of which is recorded. Illegal landings are a bane of the CFP, or any fishery which aspires to scientific management. They were reckoned in 1996 to be around 40 per cent of all North Sea catches. If the take is relative guesswork, how can the forecasts be accurate?

One of Dr Cushing's keenest complaints is that increasing catching efficiency of commercial fishing boats is not accounted for. He thinks this should be adjusted every one or two years. Another one, which he has pressed throughout his long career, is for the industrial fishery in European waters, but outside the embrace of quotas, to be brought within management. In favour ultimately of multispecies quotas, under which fishermen would go to sea to harvest several different types of fish, instead of today's system where many boats only have single-species quotas, where the 'wrong' fish is thrown back as waste, he sees no future or justification for the unbridled fishery conducted on a small, base-of-the-food-chain fish like the sandeel, whose inter-relationships with the other fish species is largely unknown. 'I rank the sandeel issue' he remarked tersely 'very high.' 'Sandeels are used by the Danes', his tone turned to disgust, 'to feed mink.'

During Dr Cushing's working life an event occurred, or in fact re-occurred but attached to a long time-cycle, the last occasion being within the era of industrialised fisheries, which severely discomfited scientists and tested to the limit fishermen's faith in them. It was the 'gadoid ouburst', so called because the fish affected, cod, haddock, and to a lesser degree whiting, are of the scientific family *Gadidae*.

The gadoid outburst was an unanticipated surge in populations of North Sea gadoids, the causes of which remain mysterious. All that scientists can say for sure is that conditions in the period 1963–1985 favoured the survival of young cod to a quite extraordinary degree. Recruitment was 163 per cent higher than in the previous nine years. For haddock the golden age was the decade 1965–1974, kickstarted by two huge recruitment years in 1961 and 1962. Fishermen's landings of these two species soared. They caught almost three times as many cod as in the 1920–1964 period, and over twice as many haddock.

Two unfortuitous events combined to make the gadoid outburst a painful embarrassment for scientists. The first was that the scientific model for these two gadoids used the start of the outburst, not the earlier decades, as the basis of estimates about recruitment. The base-lines set were very high. As catches started falling in the early 1970s, scientists recommended lower TACs. Politicians, being more expert on the political temperature than gadoid outbursts, over-ruled them; despite falling catches some TACs even went up. The CFP's conservation policy started just as catches in these two high-value species were tumbling off the plateau. Most fishermen had not been on boats in the earlier, 'normal' period, with more modest catches of cod and haddock. In Mike Holden's words the cod and haddock collapse in the 1980s, 'provided the industry with a stick with which to beat the donkey of the CFP which they detest.' The second unfortunate fact for British scientists and fishermen is that cod and haddock are the two species with which their destinies are especially entwined: Britain has 47 per cent of the cod TAC and 78 per cent of that for haddock. The previous gadoid outburst in the North Sea was thought to be in the nineteenth century. Scientists hoping another will save their reputations with British cod and haddock fishermen will have a long wait.

There is a reason, familiar to fisheries economists, why a failure to react quickly to falling recruitment is doubly calamitous. The scenario at the end of the gadoid outburst in the North Sea illustrates it. Fishermen had made money in the good years, and gained confidence. Capitalisation in the industry rose. When catches start to fall the fishermen risk defaulting on loan repayments and are put under pressure by their bankers. In turn politicians are lobbied. In the Council of Ministers fisheries representatives march in grimly to do their best for their fishermen. A TAC higher than scientific recommendations is hailed by politicians, beating their breasts, as a victory.

The effect on stocks is serious. Faced with heavier fishing efforts the age of fish caught reduces, and there is no turn-around time for fish to breed and provide recruitment. Most North Sea cod are mature by age four, all by age six. They were being caught after the gadoid outburst at age three, before having time to breed. Scientists have come up with a horrifying statistic. In these circumstances fish populations can be reduced to 10 per cent of the estimated number if no fishing had ever taken place at all. However, in the case of North Sea cod the stock was reduced to 1–2 per cent of what is called the 'unfished level'. Recovery will be slow; and the overcapitalised fleet is left high and dry. One of Britain's leading fishery scientists, Professor John Shepherd, of the Southampton Oceanography Centre, said: 'To some extent all fisheries are managed in the dark. My perception, though, is that most disasters happen when two or three very bad things go wrong at once. I would say that we are playing right at the edge of the precipice now with regard to North Sea cod.'

Outbursts, or fish appearing in new places, are erratic events which are impossible to write into fishery negotiations and international agreements. Yet they frequently occur, and often on a large scale. A hundred years ago there were no cod in the Barents Sea, presently a prime cod ground. Cod only showed up off Greenland in the 1920s and they have persisted there. Tunny or bluefin tuna were caught off Scarborough in the North Sea between 1929 and 1954 by British sport fishermen with rod and line, and immense equipment designed for the job. These monsters – some weighed over 800 pounds and pulled the tackle so taut it could not be touched – vanished and never came back. Theories for their disappearance include interception by Danish fishermen using electric lines, overfishing on the African spawning grounds, overfishing of the herring they ate, and the North Sea getting colder. The Atlanto-Scandian herring off northern Iceland took a thirty-year leave, terminating in 1996. Western Scottish herring departed some traditional sea-lochs where they once massed in stupendous numbers and have not come back. Pilchards disappeared off the British south coast in 1968, to be replaced by mackerel. Examples of major shifts made by stocks of fish the world over are numerous. For the fishermen they represent either a temporary bonanza, even a permanent bonanza, or a gift slipped from their grasp by the vagaries of nature. To scientists they are mysteries which remind anyone susceptible to the creeping cancer of certainties that he will someday, sometime, with regard to one stock or another, be mighty surprised.

A scientifically-led fishery, to which the CFP aspires, requires solid data. In the CFP the data is not in the least bit solid, indeed administrators calculating TACs write in generous figures for the fish which never become official. As with any illegal commodity there is a tendency for the scale of the problem to be maximised by the media, who love underhand dealings, and to be minimised by those protecting their reputations. The problems in the North Sea burst into the public limelight in 1996 when practising fishermen said on a British television documentary that 'black' or illegally landed fish were common, maybe up to 40 per cent of reputed catches. This 40 per cent figure was subsequently confirmed in a report by the Scottish Council for Trade and Industry.

Fish usually move onto the black market at the moment of unloading on the quay. As John Tower, a market development officer with Britain's Sea Fish Industry Authority (SFIA), put it: 'It's not hard to land 250 boxes into the auction and 50 onto the back of a lorry. The wagon drives directly to a fish processing unit.' Skippers in the modern fishery are in radio contact with traders as they head homeward from the fishing ground, and these sorts of deals can be set up with ease and efficiency.

Black fish skew the market; often it is economic for skippers to sell the best quality fish black, the remainder to auction. This leads to lower prices, grumbling amongst fishermen, and greater recourse to the illegal trade, in a spiral which ultimately devalues the product and turns fishermen to a sort of law-breaking which they profoundly dislike. The difficulty now is that as quotas bite the black fish market has become a vital economic lifeline for many. Those not partaking are disadvantaged. Black landings are a recent phenomenon, of only about seven years duration, but they have rapidly become part of the fishing way of life.

The other wild card which derails a science-led fishery is discarding, a practice which occurs more in the all-embracing maws of trawls and seine nets, less with more selective dredges, pots and long lines. It occurs mostly in places where quotas are tight, like the North Sea, and least where undersized fish can get swallowed up without a ripple at auctions in port, or where quotas are generous, such as in the CFP's central and South Atlantic waters off Spain. As noted, discarding has been made illegal in Norway and is not a major issue under Icelandic regulations. Professor Shepherd's estimate of CFP discards is 30 per cent of the fish stock, a shocking figure, yet some informed estimates are higher. In a 1994 conference organised by the SFIA one speaker cited a skipper fishing for top quality plaice who had rejected 85 boxes out of 100. They were 85 boxes of young, dead, high-value plaice, the seedcorn of future stocks. Unilever, the world's biggest corporate buyer of fish, reported to the House of Lords' report on the CFP that discards and the black fishery together constituted 'probably the greatest threat to the future supply of fish', a remark which must be interpreted in the light of their own heavyweight rôle in the market, which is naturally harder to control the more outlets there are.

Discards are, on the face of it, scandalous. Perfectly good healthy fresh fish, often from seas where such phenomena are in short supply, are tossed overboard dead, their air bladders burst from being hauled fast to the surface, or dead from suffocation under the weight of fish in the net. Few are returned alive, even fewer live on. In an era of acute awareness about the necessity to conserve natural resources such waste seems tragic and unnecessary, to some people almost criminal. It offends everyone alike, from fishery administrators to fishermen, and the language used to describe discards is often unprintable. The fact that wasted fish in the North Sea, reckoned to amount to as much as 600,000 tons a year, are estimated to support up to three and a half million extra seabirds, merely demonstrates how man's cavalier extraction policies have violently disrupted ecosystems to an alarming degree.

Discards, however, are a more complex matter than first appears. Fishermen have thrown back low-value fish in favour of high-value ones for time immemorial. Today it is called 'highgrading'. It makes economic sense to have the hold as full as possible of the higher-value catch. The CFP may institutionalise wasteful fishing practices, but it did not invent them. Records show that in the 1920s Scottish fishing boats sometimes discarded over half their haddock; the highest discard rates for haddock today touch this figure.

The worst discard rates anywhere occur in the American shrimp fishery, where a headline scandal in Florida in 1994 reported a discard mountain of half a ton of dead fish for just 90 pounds of shrimp. Since that time the scandal of discards acquired a sharper and more urgent profile, not only in the States but in the CFP too. One result is that discards in two fleets, those of Denmark and Britain, are now being measured.

Solutions depend on fishing practices in specific places. As described, Norway bans discards, moves boats on if they are catching small fish, and penalises them. It helps Norwegian policing rigour that there are only a small number of designated landing ports. Norway monitors catches at sea, which the EU does not. The EU monitors fishing operations, but not catches; the checkpoint in Europe is in port. Those close to CFP administration think it is highly improbable that catch monitoring will ever be introduced at sea. Many feel it is better to live in ignorance than prise open a can of worms. Also, the consistency of policing by different states is too variable. Observers permanently onboard are essential if the issue of discards is to be firmly tackled, and the CFP is in too fragile a state, politically and economically, to make this eventuality likely.

The EU has a different type of fishery from Norway, with many more species of fish. The known species range is 224, the stocks of around a dozen of which are scientifically examined. The North Sea has what is called a mixed species fishery for demersal fish, such as cod, haddock and whiting. The difficulty posed by this is that if the mesh size were big enough only to capture the oldest and largest cod it would be far too big to apprehend marketable whiting and haddock. Smaller nets ensnare younger cod. All these fish, along sometimes with prawns and flatfish, are often in the same place. If the skipper has space on his quota for more cod but not for more haddock and whiting, the latter may be jettisoned. To land them would be breaking the law.

Various suggestions have been made about solving the discard issue. Bans on discards in specific areas where monitoring is practical is one suggestion. On longer fishing trips by-catch fish which might have been discarded on the outgoing voyage, have the virtue of freshness if caught when sailing home; limiting time at sea could theoretically reduce this type of discarding. Seasonal closures could reduce discards where there is a high likelihood of catching the wrong fish. The main hope for reducing discards within the realm of practical politics and envisageable scenarios – and onboard observers fall outside this category – is improved fishing gear which focuses on fish size, fish behaviour when in the act of being caught, and fish species compositions, to attempt to ensure unwanted fish never get into the net in the first place. Closed areas and closed seasons can modify the composition of the catch to some extent, as fish adopt preferential stations at different times of year. These matters are part of what fishery scientists call the 'technical measures', and are considered further on. They are not efficient enough to offer an alternative to direct controls, nor do they have the direct benefits for the fish of reducing numbers of fishermen, fishing boats, and the amount of fishing. It is always well to keep in mind the timing of the two all-round historic bonanzas in the North Sea, immediately after each of the World Wars, during which the fish had been left to repair their stocks undisturbed. It may be unachievable, for most of the time, but at least it is perfectly clear: the best thing for overstretched fisheries is to be left alone. It is not fisheries that need managing, it is man.

Norway's Fisheries Minister was right when he said regulations were pointless unless they could be policed. Effective policing is critical in fisheries in a unique way. It must be remembered that in the days of £10 million fishing boats vast quantities of capital, walking hand in hand with vast expectations of profit and pay-back, push forward the man in the wheelhouse, the man deciding where to go next, the man watching his nearby colleagues to see if they have hit a shoal, the man who, when he hits a shoal, lives by his wits on the radio to other skippers, not exactly telling fibs, but making sure he gets his full share of the catch before heading for home and leaving the sea for others. The European fishermen fishing home-waters today is in a crowded and competitive environment. The financial backers want their man to perform against comparison with other skippers. The quarry is shrinking. Such is the friction between member states that a common attitude is 'Fish for all you are worth. If you don't catch them the next chap will. And he'll likely be a foreigner. Let's go for broke.'

The biggest reason for this attitude, which admittedly has a history long predating the CFP, is that policing is felt by fishermen to be ineffective, biased, and often invisible. It is well known to European fishermen that the resources deployed by member states differ widely. Spain, with its gargantuan fleet, has a patently inadequate fishery inspectorate. Mostly its staff are recruited from the main fishing ports like Vigo, in effect cousins and brothers being asked to scupper their relations. British fishery inspectors have visited Spanish fish markets and have never failed to find undersized fish, some no longer than a

Biro pen. Spaniards have always liked catching fish very young. Elver pâté is made from the young eels as they return from the Saragossa. There is little thought given in Spain to allowing fish to grow to maturity to optimise the resource; young fish consumption is part of Spanish culture. Culture, as many in fisheries management have observed, is part of the trouble. Behind the political differences between the key fishing states in the CFP – Spain, Portugal, Britain, France, Denmark – are chasms of cultural divide.

In 1996 the European Commission published a report on policing entitled 'Monitoring the Common Fisheries Policy'. Consisting of a compilation of member states' fishery protection profiles submitted by themselves, beefed up by a Commission overview, it is not as hard-hitting as an external report might have been. For example, it tactfully stresses direct comparisons between member countries are impossible, meaning of course, politically impossible.

Nonetheless it is as a table of comparisons that many have chosen to read it. Britain comes off best. Having long experience in fisheries protection and marine affairs in general, the longest coastline, the biggest EEZ, the most numerous non-British ships in its waters (up to a thousand at any one time), the report says Britain 'has a well developed national fishery control system, which is matched by the allocation of considerable resources.' Noting that British authorities said there was little point in making arrests if there was insufficient evidence to satisfy the rigorous standards of domestic courtrooms, the Commission nonetheless commented on the British shortcomings in fishery closures, the volume of black landings, and low number of prosecutions for infringements. In sum, though, the concluding remarks say Britain provides 'an example of how the CFP should be enforced.' Comments on other states vary widely.

France is criticised for connivance between inspectors and fishermen, Spain is said to have the most complex division of policing responsibilities between national and coastal autonomous regions, and neither Denmark nor Germany have any aerial surveillance of fishing boats, which the Commission considers essential. Most states were judged insufficiently resourced to carry out their duties. The report said the results of its enquiries were that policing was 'very poor generally'.

The report acknowledges the extreme difficulty of policing the CFP. It laments the absence of single-net rules, referring to the difficulty of monitoring catches against mesh sizes when the regulations are threaded through with waivers. It says undeclared catches are as large as they ever were and laments the practice of deliberately mis-identifying species, herring being confused with sprats, mackerel with horse mackerel, etc. The huge number of landing ports is also mentioned (over 450 in Britain) as a difficulty for inspectors measuring catch compositions. Catch compositions in Denmark, for example, are generalised for reporting purposes; an industrial catch of pout including some highly sensitive species is logged purely as pout. This must also have an impact on the purity of data used by scientists for the calculation of TACs. Most states have a complicated division of responsibility in fishery policing; every state has a different legal system; the standards of what constitutes evidence vary, as do the penalties. On top of these difficulties the Commisson comments on 'the permissive attitude towards the national fishing industries' with regard to breaking CFP regulations. This last problem would seem insuperable while national fleets persist, a fact the Commission is well aware of.

The hitch, or glaring defect, in CFP surveillance, which was beyond the Commission's remit to refer to, is that each nation state only watches its own fishermen. The exception is within the remaining, temporarily derogated national fishing limit areas, or EEZs, where the coastal state inspectorate covers everyone. Transgressors from other states are apprehended, theoretically, when they dock at home. If visiting fishing boats land in another country's ports, then indeed, they are subjected to inspection by that country's fishery officers. If they do not dock, and they stay in Community waters, they have only their own inspectorate to satisfy. This leads as one would expect to blatant abuses. Journalists visiting Spanish fishing ports have found an attitude of open connivance with rule-breakers, something European Commission inspectors either did not observe or failed to report. This orientation in turn contributes to a general attitude promoting selfish plunder – make hay while the sun shines. The spirit of Community co-operation is notable by its absence, and abuse of Community rules widespread.

The logical solution, a trans-national Community police, does exist, but with a staff of only twenty-two and a rôle restricted to monitoring and observing. If the Community Inspectorate intends to visit a particular sector it has to announce its intentions in advance. The spirit of this rule reveals the sort of toothless bulldog the Community fisheries police is meant to be. Operations can be sanitised before the inspectorate hoves into view. With these constraints the Community Inspectorate sees its role more in gingering up those responsible for policing to do their job. On a famous occasion off the Italian coast, when fifteen out of sixteen Italian drift-net boats were found to have illegal nets aboard, Italian officers were actually present. Despite this, the offending vessels had not even troubled themselves to hide the illegal gear. When Spanish and British boats clashed fishing for tuna in the international zone beyond the Bay of Biscay, the presence of the Community police vessel reduced tension. The Community fisheries police has, in the words of one of its chiefs, Tony Curran, ' a subtle agenda'. Sadly the subtlety is so refined that many people do not know of the unit's existence.

When EU vessels are outside the 200-mile limit, in international waters, the same rules about policing responsibilities apply, or fail to apply. Each state polices its own. Most EU fishery inspectorates have their hands full inside the common fishing area, let alone outside it, in zones where participating fishing states are apt unilaterally to award themselves slices of the cake anyway. The only ray of hope for the practicability of an international inspectorate is the precedent in the international waters loosely under the jurisdiction of the fourteen member states of the Northwest Atlantic Fisheries Organisation (NAFO). In NAFO waters enforcement is described by the European Community

Inspectorate as complex and sophisticated. A satellite monitoring programme is being tested, and there is an international inspectorate of mixed-nation officials. Contributing to it is an EU vessel which, as simply the representative of one of fourteen NAFO members, is seen as impartial. However, translating this co-operation in a very large area with pockets of stocks, to the tense cockpit of the North Sea and other EU waters, is hard to imagine.

Alternatives do exist to manpower-intensive boats and planes. In fact a quasi-perfect solution exists, to the dismay of the majority of active fishermen. Satellite monitoring systems and, in particular, global satellite positioning, can now relay information on any boat which is carrying the requisite transmitter. The signal which is bounced from the boat to the satellite to the control-room can reveal to the nearest metre precisely where the fishing boat is, where it is heading, and how fast. This has inestimable value to surveillance people, for example, in making sure closed areas are being observed. Streamlined in the Gulf War by America, and first used in the North Pacific drift-net fishery in 1990, it has enabled administrators in various walks of life to sit on almost godly knowledge about faraway persons who are being watched. On-site policing does not become redundant, but reliance on it is reduced. Since 1994, the EU has been testing satellite tracking in a pilot project. Satellite operators may know more about what is in store for European fishermen than they know themselves; at any rate, mobile communications operators say up to 7,000 vessels are due to be fitted with transmitters by mid-1998. Satellite monitoring is supposed to be operational on all EU vessels over 24 metres long from the year 2000. The appeal to enforcement officers is obvious. Officials see the principal benefit as being better able to estimate and understand fishing effort, the critical matter in matching fishing to resources. Satellite monitoring is presented to fishermen as having benefits for their safety. With the advent of low earth orbit satellite systems the technology is likely to get cheaper. The very latest tell-tale technology can picture the heat profile of a ship, showing whether it is towing a net or not.

Naturally, such eye-in-the-sky empowerment frightens the pants off fishermen getting on with the laudable business of making a living. Amongst the responses to European Commission efforts to have global satellite monitoring introduced was one which pleaded that the systems were not deployed in what were termed 'the most sensitive areas'. In other words, don't use it if it might produce useful results! Understandably fishermen are afraid of being disadvantaged if only some boats carry transmitters. Perhaps most of all, fishermen wince at the possible loss of commercial confidentiality. Down the ages skippers have been ranked on their 'nose' for finding fish. With satellites their secret hot-spots become public knowledge. During the December 1996 Fisheries Ministers meeting, deliberating the introduction of obligatory satellite monitoring, the negotiations became so grinding that the EU Fisheries Commissioner, Emma Bonino, actually passed out!

Effective utilisation of global satellite monitoring requires the monitoring agencies to act on what they find. Few northern fishermen believe the

Mediterranean states are capable of doing that. As so often in the CFP there are powerful vested interests for whom the existing mayhem has more advantages than a secure, long-term, sustainable fishery. Again we come back to the principal obstacle – politics. As with discards, black fish landings, unecological fishing methods and brazen carrying of illegal nets, false holds, and wily engineers' tricks on fuel injectors (to temporarily throttle down the true horsepower of vessels in horsepower-limited fisheries), the practical will to change the status quo seems lethargic. The politicians, apparently, have bigger fish to fry.

Fishery administrators and fishermen's institutions themselves seem to have focused on one avenue in particular to avoid confronting the insoluble fundamentals of the Treaty of Rome. This is the department known as 'technical conservation'; its aim is to avoid catching small fish and fish below market quality. Whilst technical measures embrace regulations on closed areas, fishing seasons, etc., by far the biggest hope for fish conservation has been in modified fishing gear, or what is broadly termed 'gear technology'. Gear technology consists of devices, or adaptations to existing fish-catching tools, which allow more of the fish, preferably more of the younger, pre-spawning stocks, to escape. Considerable research effort, led by Britain and Denmark, is centred on devising more fish-friendly, and fish-selective nets, pots and lines. Shadowing the industry, in the view of many of its scientists, is the quieter but no less determined industriousness of fishermen looking for chinks in the regulations with which to counteract selectivity. Cod-ends with escape panels are fitted with blinders or inner sleeves – the Spanish word for them means condoms – nets are filled with heavy weights, or fished faster, to make the meshes close up, and so on.

The first technical measures were started in the late 1970s, bulked up in 1980, and in 1986 when Spain and Portugal joined the EU amendments were introduced defining fishing grounds more closely. Gear regulations have always applied to towed gear, principally trawls and seines. Gill-nets are by their nature more size-selective; small fish pass through, and bigger ones cannot get in far enough to get their gills caught. The static gear sectors – lobster pots etc. – have always been controlled by coastal states.

From the start gear technology recommendations from the European Commission were watered down or neutralised by the Council of Ministers. One of the Commission's earliest recommendations was that if each boat could only legally have one size of net onboard the scope for abuse would be less, and the ease of policing would be greatly improved. It was designed to tackle the situation where inspectors were stymied by fishermen's claims that in a mixed catch each fish had been caught in the correct diameter net, where they suspected use of the smallest, most comprehensive net.

Fishermen leant on their political representatives pointing out that in a very mixed fishery, such as that over most of the common fishing area, it was essential to be able to switch target when at sea. To have to steam home to get another net, for another species, was a waste of time and money and quite unreasonable. No one disputed that a one-net rule would aid conservation; but it was attacked as typically unrealistic.

An illustration of the conflicting interests of member states was the late 1970s argument in the former British fishery area to the east of Scotland and up to Shetland, between British fishermen taking a traditional catch of haddock and whiting and Danish fishermen trawling fine-mesh nets to catch the much smaller fish used for industrial reduction, Norway pout. The conundrum was how to avoid catching large quantities of haddock and whiting, whose nursery grounds are in the same area as the pout. British efforts to get the Danish fishery closed resulted in Britain losing a case in the European Court of Justice, which deemed that Britain had behaved in a discriminatory way, to protect national interests.

The Norway pout box case concerned only a single fishery conservation problem. Most area-specific arguments involve a tangle of issues. To try and tackle these the CFP has accumulated a veritable jungle of rules. Each major commercial species has a minimum net-size ruling. For example, the panels in a mackerel or herring net should be no more than 32 millimetres. Demersal species being bigger, net-meshes are larger, but they vary – 100 or 80 or 65 millimetre mesh – from one geographical area to another. When Commission officials have tried to get larger meshes for demersals accepted, member states have always argued it would make it hard to catch whiting, the smallest common demersal; against this scientists have argued, usually unsuccessfully, that whiting are quite well caught by larger meshes. To accommodate the biological fact that in different fishing grounds fish grow and mature at different sizes, member states have different minimum landing sizes; this is used to vindicate different mesh sizes.

The Commission's recommendations are almost invariably subjected to endless derogations, with special interests prevailing on their politicians to plead for their exclusion. Even the seemingly unobjectionable regulation, making it illegal to carry gear designed for species which are not present in the fishing ground in question, has not become law. When the CFP is criticised for its Byzantine complexity – some officials say no single individual understands it all – it is the area of technical conservation that is the best example.

One reason is that the act of pulling a net through a moving body of water, with different currents, at different speeds, with different levels of turbidity, can produce a medley of various results. It is now routine for scientists using underwater cameras to film a net in action, watch the gradual narrowing of the mesh-panels as the boat's speed increases, and test out new types of twine which will react to different fishing methods with more or less flexibility. Fishermen being well-organised individuals had discovered from an early date that if they increased the width of the twine it would have the effect of compensating for the enlarged mesh.

Another problem for technical measures scientists is the wide disparity in types of fishing ground. Fishing off Gibraltar is different in 1,001 respects to fishing in the Skagerrak. As a result the Baltic and the Mediterranean have rules of their own, and are managed by coastal states. These rules too are riddled with derogations.

A great volume of scientific work has been done by EU scientists to bring discipline to Europe's fisheries and to protect young fish both in their spawning areas – which can be anywhere – and in their nursery areas, which tend to be close to shore. Not only has mesh size been closely studied but so has the colour of netting. Fishermen long ago understood that different colour nets become more or less visible in different waters at different depths. Red and orange radiation is absorbed by the water in the shallows and therefore in the Mediterranean's clear waters red and yellow nets are prevalent. Black tunnels in the cod-end of nets discourage the passage of fish and have been used in tests to achieve better selectivity.

Pioneered by the Americans, with devices to exclude a by-catch of turtles in the shrimp fishery, one of the concepts which has gone furthest is the separator panel, now a compulsory adaptation to trawls in Norway. For fifteen years scientists have been investigating how different fish react to being caught up in a trawl, and how quickly fish run out of oxygen when enmeshed in a trawl towed at different speeds. They discovered that most fish, cod excluded, swim uppermost in the trawl once inside the net, and escape panels in the top of the cod-end could theoretically let smaller fish out. By experiments in flume tanks, simulating real fishing, scientists discovered that successful escapement was affected by a whole range of factors, ranging from towing speed and water clarity, to cod-end design and the type of twine of which it was made. Particular efforts were directed to a selective trawl for whiting because, although small, whiting is thought to be the biggest predator of other important commercial species when young – cod and haddock – as well as of herring, sandeel, sprat and Norway pout; on top of which whiting is cannibal. Whiting is a difficult target fish in a trawl with separators because of its peculiar energy in escaping capture.

It is not only different species of fish that react differently to trawls. Seasonal changes in a fish's bodily condition, and temperature changes in seawater, can alter selectivity too. An EU research programme using haddock has found that logical expectations can be up-ended. Fat and fit, pre-spawning fish nip out of the way of the path of nets quicker than thin fish. In warm water they evade capture better than in cold water. It was calculated that in April, after fish have spawned and they are thinner, 15 per cent more whitefish will be needed to give the same weight of catch as in September, when seawater is warmer. If CFP areas are to be shut off seasonally, a perennially-mooted idea, this sort of knowledge would help to catch the right fish at the right time.

The senior European laboratory researching gear technology is the Marine Laboratory in Aberdeen. Within the realm of fishery science it is quoted as employing several of the current world experts on fishing gear adaptation. One of them is Dr Richard Ferro, and he is a lucid explainer of the role of gear technologists.

His purpose he describes as 'about simple regulations to improve conservation in fishing gear.' He believes regulations should aim at simplicity and comprehensibility, for a court judging transgressors. He is highly conscious that tightening up too much on fishermen, and alienating those meant to put researched gear improvements into practice, is self-defeating. Most fishery managers energetically concur with this: acceptance of new methods by fishermen is crucial. Dick Ferro warns that verbal agreement does not always equate with compliance when at sea: 'When you speak to them they all agree it is sensible. Then they go out and break the regulations.' Another fishing gear specialist said that if gear modifications were happily accepted by fishermen it usually meant that they had fathomed a way to work round them. Most fish-selective gear needs to be used sympathetically. A further complicaton is that the CFP is a blend of national rules and community rules; gear modifications need to apply uniformly. Under present derogations there may even be two different cod-ends on one trawl. Who is to say in which the fish was caught?

Gear regulations are notoriously hard to police. Dr David Wileman of the Danish Institute of Fisheries Technology and Aquaculture (DIFTA) in Hirtshals, the other main gear technology research station, went so far as to say, 'Acceptability within the industry is the key thing.' 'You see' he continued with a note of resignation, 'It is very very hard to get fishermen to swallow something they don't want to.' Gear scientists are not only looking for fish-catching methods which achieve good selectivity, they are looking for designs which fishermen will permit to be worked. Their job is far from straightforward, or abstract.

I asked Dick Ferro if the research efforts in Aberdeen and those done at DIFTA kept pace with new fish-catching improvements. Developments like the rockhopper trawl, for example, enabled trawls on sprung legs to walk over rough seabeds which had become sanctuaries; rockhoppers jumped over centuries of involuntary restraint. 'The equipment they use can change so fast' replied Dr Ferro, 'And it is hard to measure the rapidly accelerating efficiency of fishing gear. It is impossible for scientists to write this into their calculations.' The separator panel, he believes, has specific uses in specific places; in particular it could be useful for conserving cod in the North Sea. One of his laments is that effective gear improvements are frequently stymied because they conflict with other regulations, for example on by-catch, already in place and therefore very difficult to amend.

On the general purpose of technical measures in general he is expansive. He sees an increase in the breeding stock as paramount in the pursuit of long-term sustainability and the maintenance of fishermen's incomes. He sees gear improvements as sparing young fish from becoming dead discards, quoting an ICES figure for the horrific annual discard total of Atlantic hake – up to 130 million individuals. In the North Sea more small haddock in a haul are often thrown back than are kept for sale. He is acutely conscious of the irony that sometimes the species being protected goes on to act as a major predator on other valuable stocks.

In the long term Dr Ferro argues that it is obvious that bigger fish are financially beneficial to fishermen; in other words larger mesh sizes, separator panels, protected nurseries etc., are a good investment. He quotes the example of hake, a large fish which, capable of multiplying its weight by more than six times between the age of two and six, is then worth twice as much per pound. The same fish, he argues, is worth twenty-five times as much captured by a selective long line at full maturity as it would be prematurely scooped into a fine-mesh trawl when young. Warning that small mesh sizes lead fishermen logically to concentrate on fishing-grounds where fish are massed and small, and sometimes to cause a complete stock collapse as happened to hake in the Bay of Biscay, in a paper written for the European Commission Dr Ferro concludes: 'An increase in mesh size is an investment whose dividends can only be reaped after the fish has grown sufficiently.'

No gear scientist is foolish enough to worship only at his own church. Gear improvements and technical measures to protect nursery grounds are partial management motions, not a whole policy. After all, wider meshes and separator panels may do something to alter the composition of the catch; they may not affect its magnitude. They have no bearing on fishing effort. In the CFP area in order to ensure cod reached the age at which they breed, new meshes would need to be something like 180 millimetres. Many haddock, whiting and sole would swim right through them. In a mixed fishery with fish of very different sizes and types gear technology alone is pointless. The recent discoveries that what used to be compartmentalised as pelagic, mid-water and demersal categories of fish are loose generalisations, and many fish move up and down in the water column, is an added difficulty. The classic modern fishery collapse using minimum mesh sizes alone occurred on America's Georges Bank off New England. Haddock, cod and yellowtail flounder ended up at pitiful levels and fishing in a once-great fishery ceased. By the same token, in the North Sea, where it is reckoned the waters are trawled at least twice a year, the giant fish of the past, halibut, ling and turbot, which were also the most valuable, are now extremely rare. 'Ultimately' wrote John Shepherd, 'the only way to conserve fish is to kill fewer of them.'

It is one thing to know what war is being fought, quite another to get right the ground on which to fight it. New findings at DIFTA threaten the rôle of gear technology as one of the principal means to fish more selectively. DIFTA scientists have recently discovered that mortality amongst fish which get through net-meshes and escape panels is far worse that hitherto suspected. Biologists who comforted themselves with the thought that any larger fish which get through the meshes would escape and breed have been reduced to thoughtful silence by the discovery that 10–30 per cent of escapees die within the first two weeks.

Apparently fish which get through mesh panels suffer an assortment of abrasions. These can be from the mesh-twine, or from other fish. Abrasions from scale-loss have always been recognised as dangerous for fish – anglers returning fish to the water try to avoid scale-rubbing, as do fish farmers with fish pressed up against fish cages in storms – because it is normally followed by fungal growths. Cock salmon in spawning rivers are often covered in fungus post-spawning, and it spreads from the head which is used for burrowing into the gravel in the preparation of redds. Scientists tracking fish which have passed through escape hatches of all sorts have found that herrings and sprats are particularly prone to disease, flatfish are less susceptible, and of demersals that cod survive better than haddock. Fish which are alright after two weeks are generally safe. The novel tracking and monitoring which unearthed these depressing findings went on for up to three months.

The significance of this discovery, which at the time of writing is recent, looks large. If a notional figure for fish dying from negotiating escape panels is added on to a notional discard figure, and also, to the small figure for fish landed and thrown back into the sea after failing to sell at auction, a little-mentioned wrinkle in the marketing of fish, then the losses to fisheries in CFP waters become immense. The sums for TACs would need to be re-calculated to approximate better to reality.

If as many as a third of fish surviving the encounter with some fishing gears go on to die, the whole role of gear technology may need re-assessing. It may mean that enlightened management should be looking at complete-catch policies on carefully-targeted stocks. Small meshes would be used, with no pretence to allow escapees, and the fishing boat would retain everything that came within its embrace from sea urchins and mussels to sunfish. If fishing gear mortalities are as high as these findings suggest, we may need to move to a no-discard, no-escapement policy, with rigidly enforced marketing of every fishy article that is hauled aboard. This would silence those crying out against discards, simplify nicely the work of fishery inspectors, and reduce in significance the often only partially viable efforts of gear technologists in favour of much greater and more precise targeting of exactly-defined stocks. Debate about these matters has been taken much further in other places, notably America. Implementation of modified gear programmes is further advanced and thinking on the wider effects of selective fishing is more evolved. The larger arena of gear modification is returned to later.

There are those who think that the common fishing area in Europe will suffer major stock collapses before sufficient disciplines are accepted by the Council of Fisheries Ministers. Others, accepting that almost all stocks in European waters are under threat, believe that time can be bought for Europe's fleet and in that time a rescue operation can be effected. If decommissioning vessels and slimming down the European fishing fleet is fast enough, and TACs kept low enough, stocks have a chance of holding their own before some sort of equilibrium between the resource and its exploitation is achieved. This latter perspective recognises the need for a safety valve, maybe temporary, but nonetheless somewhere EU vessels can fish while the painful process of downward adjustment creeps along.

The safety valve is fishing rights in the seas of third world countries. For sure, European fishing interests have negotiated access in Greenland, Icelandic, Faeroese and Norwegian waters (in return Norway is a 50/50 partner in the principal EU fishing-grounds), and bilateral arrangements worth rather less with Poland, Lithuania, Estonia and Latvia. The fact that in 1996, 6,500 tons of wild Atlantic salmon in the Baltic (Baltic salmon are a different, more localised stock to mainstream Atlantic salmon) were allocated in bilateral arrangements between Denmark, Finland, Germany and Sweden in exchange for, among other things 1,000 individual salmon for Latvia in EU seas, would amaze many outsiders to the strange realities of bilateral fishing deals. However, the fact is that all of the above mentioned countries are doing deals about fishing for the same fish as European fishing boats catch. The arrangements which really make many freethinkers hot under the collar are those negotiated by the EU, principally for the Spanish fleet, with poorer countries for fishing in their EEZs. These usually involve targeting more exotic species, often fish used for rendering down into fish-meal, where the host state has the appearance of being a victim, with weak governments wracked by internal problems, debt-ridden, desperate for foreign currency, and completely without the means to check on whether or not the visiting fishing fleets comply with the conservation aspects of the agreements they have signed up to. Negotiations are generally carried through with the 'third countries' finance ministry, not the fishery ministry.

In the opinion of many the EU has exported its embarrassing fish-catching over-capacity and, in particular off the west coast of Africa, one of the best multispecies fisheries anywhere, is practising some of the worst colonial-style abuses imaginable. The EU's fishery is in the spotlight about many matters, signally fish conservation in EU waters; the deals with African states, although accounting for a huge percentage of the EU catch and CFP budget, have been kept under wraps.

In 1996 the EU had agreed fishing deals with Mauritania (for five years, under which an astonishing 240 EU fishing boats will be accommodated) for ECU 267 million; with Morocco worth ECU 500 million (for five years, twinned with access into EU waters); with Sénégal (for two years) worth ECU 18 million; with Angola (for two years) worth ECU 18 million; with diminutive Guinea Bissau a two-year deal worth ECU 12 million, and with the Seychelles Islands a three-year deal for tuna worth ECU 10 million. With Argentina the fishing agreement is in joint ventures and is a new form of deal called a 'second generation' agreement; it is for five years and worth ECU 162 million. At present the EU is trying to win access to the fishery zones of Chile, New Zealand and Peru.

In addition to simply paying for the principle of access, with vessel owners paying supplementary licence fees, the reciprocal-type fishing arrangements with other north Atlantic states, and the new-breed second generation agreements, there are access arrangements with other states in international waters, as in the NAFO area with America and Canada. However, the contentious subject, without doubt, is the first of these.

Often the agreements with 'third countries' contain other payments, for example towards fisheries research and training. However, European

Commission officials have admitted they cannot track the funds and ascertain that they have been used for the correct purposes. It was discovered, for example, that on one training programme there was a 90 per cent failure rate owing to trainees being unprepared and training being unsuitable. Bribes accompanied the transfers of funds, the relations of government officials were frequently beneficiaries, and in the words of one official in DG XIV specialising in fishery access agreements in Africa: 'Pious Europeans talked of good governance, whilst in effect bribing African officials to pursue irresponsible policies.' Euro MPs have claimed that in the European Parliament there is little or no opportunity to scrutinise these deals, to see if the management plans are sustainable, or to understand how the agreements really operate. In a 1994 access agreement with Sénégal, European Parliament members were told of the deal four months after it had been signed. Those who have made special studies of fishery access agreements have frequently opened a can of worms.

Large-scale fishery access agreements started as the escape route for distantwater fleets pushed out of fishing-grounds as states declared their EEZs in the late 1970s. Prior to that, European vessels had fished off Africa – France had maintained fishery access agreements with several of her former French West African colonies, such as Benin, Gabon, the Ivory Coast and Mauritania – for some time, but after the United Nations Convention on the Law of the Sea (UNCLOS) in 1982 (which took until 1994 to bring fully into force), the need to find new fishing-grounds became urgent. Following the entry of Spain and Portugal into the EU in 1986 the process became accelerated and more aggressive. Cheaply-caught fish helped keep fish prices down for European consumers and the unwieldy Spanish and Portuguese fleets had somewhere to go.

In some countries the EU fleet rapidly became a major player in coastal life and compensation fees for access paid by the EU, added to the licence fees paid by vessel owners, came to seem indispensable to national income. Most West African coastal states at the start of this period had minimal capacity in industrial-scale fishing, the art of transferring fish to dry land being performed from dug-out canoes. While the EU secured more and more fishery agreements for its fleet, the bulk going to boats from Spain, Portugal, and France, fish consumption in Europe was rising. By 1994 the EU was importing more than half the fish its population consumed. Not only had home catches halted or dropped but its citizens were eating more fish, the average annual consumption per person rising from 19 kg in 1984 to 23 kg in 1994. Although much of the increased import consisted of white fish from traditional fish suppliers such as Norway, the fastest increase in contribution was from Namibia. In 1992 Namibia supplied Europe with 40,000 tons of white fish; by 1994 the figure was 165,000 tons. African fish like hake, hitherto unknown in much of Europe, had become a familiar foodstuff. By this time fishery access agreements were not only accounting for half the fish eaten in the EU, the cost of the agreements by 1987 had climbed to over half the budget; in 1993 this had fallen to 40 per cent, but the budget had steeply risen. In 1992 the deal with Morocco gobbled up 49 per cent of all fishery access expenditures. To cap it all, in a thoroughly confusing and insufficiently accountable programme, the EU's Court of Auditors produced a critical report on fishery access agreements showing that whilst the budget had soared the benefits to Europe had been incommensurate, with vessel owners paying very small proportions of the cost (in the Sénégalese agreement only 11 per cent), and with too much of the compensation money disappearing without trace. The Court of Auditors did not focus on this, but it is generally true that the compensation or access payments made by the EU were way above true economic values, whilst licence fees paid by vessel owners were too low, a combination which had an especially unfortunate effect – it exacerbated overfishing.

It is important to understand the impact of big modern European vessels on West African coastal waters. Traditionally the West African fishing grounds produced a wealth of valuable food fish of almost every type, from grunts and fatlips, octopus and shrimp, to lobsters, tuna, barracuda and billfish. Big rivers debouched from the interior bringing surges of fish-boosting nutrients to inshore nursery areas. The way of fishing was by canoe. With strong inshore winds and large waves the canoes had to be long; for example, a 15-metre canoe was typical on the coast of Ghana. They fished all manner of gear, often preferring purse seines. Weighing up to 8 tons when laden these canoes took some skill and considerable manpower to beach. West African fisheries were labour-intensive, and the scene onshore as canoes unloaded was a hive of industry and activity. In Sénégal, for example, the 'artisanal' or traditional fishery was estimated to employ 50,000 fishermen and 200,000 shore-based workers. This is a substantial part of the population. The seashore in the busiest Sénégalese fishing villages is an open-air processing centre. The catch is salted and dried and smoked there and then on the beach. Smoke fumes waft along the coast. Exhausted porters, carrying fish from canoe to land, rest supine on the sand. Fish provides the Sénégalese people with most of their protein and the fish-strewn beach testifies to this.

When the modern, resource-hungry age hit West Africa the first disruption to the artisanal fishery was caused by deforestation on the coast. One of the big coastal hardwoods had been used for canoe-building. The cost of traditional canoes rocketed as the great tree-trunks had to be brought from further and further afield. Eventually the Sénégalese government was importing big lumber from as far away as the Congo. The artisanal fishery was being weakened from within at the same time, in the 1960s, when the first industrial fishing fleets, from the Soviet Union and eastern Europe, began to buy access to the fishing-grounds.

The West Africa fishery became subject to what Serge Garcia, a senior statesman of fisheries, of the UN's Food and Agriculture Organisation (FAO), calls 'the sheer force of development.' Assisted by the World Bank, fisheries which had carried on unperturbed for centuries were suddenly given new muscle. Canoes were equipped with ice-boxes and outboard motors. At a stroke the range, speed and power of the fish-extractive machinery hugely increased. The replacement of polypropylene nets with monofilament multiplied their catching efficiency by up to seven times. The distinction between 'artisanal' and 'industrial' fisheries became blurred. Artisanal fisheries could now be overfished, and were. The canoes and beach-operated seines exploited young fish shoaling in the spring when close to shore, then, when young fish joined the main fish stocks later on, trawls exacted their toll further out. In addition to recruitment over-fishing, and mature stocks overfishing, there was usually no scientific monitoring and minimal or non-existent policing control. The statutory reporting aspects of the access agreements – of catch records, time of entry into the fishing zone, etc. – were frequently not observed. In addition to pressure racking up on stocks, artisanal fishermen were physically at risk. Big European fishing boats sliced by night through artisanal nets and capsized canoes. In 1991 alone twenty-four local Sénégalese fishermen died in collisions of this sort.

The most naked and unjust fishing exploitation off West Africa took place off fish-rich Namibia. Between 1990 and 1991 freezer trawlers of exclusively Spanish origin, sometimes thirty-five in one day, continued to do what they had been doing since 1970 – dig deep into Namibia's key fisheries capital, the hake stock. In the twenty-year period up to 1990 it is reckoned over eight and a half million tons of hake was netted by these boats, with a value of some £5 billion. Then in 1990 Namibia became independent. By this time the plunder was having its effect: 83 per cent of the hake being landed were classed as juvenile.

The Spanish boats, having been asked to leave, continued to fish, erasing their identification numbers. Namibian fishery managers went through a baptism of fire. They had no vessel capable of apprehending the swift trawlers. They resorted to commissioning a helicopter. In surprise landings fishery inspectors suddenly materialised on the decks of the Spanish poachers, brandishing tickets of arrest. The Spanish were dumbfounded. A full-scale real-life adventure story unfolded, ships swirling in circles to avoid being landed on, and then skippers rigging their decks with wires to bring down a descending helicopter. The Spanish reaction to being policed was classic moral outrage; 'monstrous', one skipper expostulated. Gradually the Namibians took control of the situation and their fishery has now become the main engine of economic growth. They discovered fisheries were worth fighting for, but extreme toughness was needed.

Several points stand out. Why did Spain do nothing to control its fishermen? Why were fishing licences not subsequently withheld from the offending skippers? Why did the EU, which had a fishery access agreement with Namibia, not intervene? Interesting questions, and the absence of answers reveals the way 'third country' fishery access agreements work and are regarded in Brussels.

It might be said, with truth, that back in the 1960s Russian and east European factory ships had fished West Africa without regard to stock punishment, or the impacts on the coastal state, where very large numbers of people lived principally from the sea. The difference in the EU's relationship with West Africa (the first agreement, with Sénégal, was signed in 1981), is that it purports to be offering development assistance to the host countries. It presents itself as a generous benefactor. Yet sometimes – one case was the 1992–94 fisheries agreement with Sénégal – quotas have been agreed which substantially exceed the recommendations of the research institutions which are themselves co-funded from the CFP budget. The EU on its side of the deal guarantees a part of its fleets' catch will be landed locally, and to employ on its boats some local crewmen. The latter stipulation has often been conspicuously unsuccessful, and the former fell into disrepute in those agreements which failed to specify what part of the catch was landed locally. Host states often got the fag-end of the catch. Some local landings merely fuelled existing European-owned fish processing plants.

The International Institute for Environmental Development in London studied the case of Sénégal with particular reference to the synchronisation between access agreements and EU overseas fisheries development policy. Absurdly contradictory policies were numerous. The rights bought for EU freezer trawlers were found to be in conflict with development assistance to local fish processing. The market for artisanal catches was undermined by large catches of fish which commanded a value in Europe which could be twenty times as high. Fish prices, often to people who had hitherto lived without a money economy, rocketed. This left coastal populations mystified, and beach-based communities injured. The contradiction is that EU development projects of a more general nature were pledged to alleviate Sénégalese poverty.

High discard rates by foreign trawlers had an adverse effect on artisanal fishing, wastefully eliminating the smaller, younger stocks. This is particularly relevant in light of the FAO figure for discards in trawling for finfish – as high as one and half kilos for every kilo retained. In the case of Sénégalese trawled shrimp nearly three times as much is discarded as is kept. Compared to other shrimp fisheries in the world this by-catch figure is low, but volumetrically the incidental catch is enormous.

If it is at first awkward fully to appreciate the significance of these conflicting policies it is only necessary to consider what an outcry would arise if there was such a degree of manifest inconsistency in fisheries within the European fishing area itself. At least within the EU scientists are called on to justify what is happening. The treatment of Sénégal is indefensible. As a British research scientist visiting Sénégal put it: 'The question arises: should the EU sign types of agreement considered unsuitable for its own waters?' The Sénégalese coastal shelf fishery now bears every sign, in several species, of being chronically over-fished. As the Sénégalese have learned that almost every fish commands a market somewhere, there are now, according to the Institute, no species which are under-exploited. Canoes have had to travel further and further to find fish, some even going as far south as the coast of Guinea Bissau. Sénégalese fishermen claim shrimp trawlers have raked the seabed again and again, not only killing a huge by-catch, but damaging the seabed. Still, however, the fisherman in Sénégal earns more than his compatriot working the land. Fishing communities are better off; which is one reason the access agreements have survived.

The unacceptable interface of EU trawlers and traditional fishery canoes was recognised by a delegation of EU parliamentarians in 1995. Sénégalese fishery workers, along with a group of non-governmental organisations in Brussels called the Coalition for Fair Fisheries Agreements (CFFA), had already alerted the EU Fisheries Ministry to the raw deal being forced on Sénégal. Although access to the Sénégalese fishery did start to shrink in 1994 the effect on the fish may remain the same. One of the findings of the European Court of Auditors was that earlier catch ceilings on trawlers allowed into the fishery were under-utilised, in the 1990–92 period by 45 per cent. The 1994 agreement, although nominally for less access, allowed for approximately the same amount of fishing pressure. It would be an exaggeration to say the EU agreements with Sénégal are entirely exploitative, but it is also true that EU negotiators are aware that should agreements stumble Sénégal has a resource which would be of interest to others, for example, Japan. The non-exclusivity of the agreements is a structural weakness, and puts a low premium on conservation.

The research on Sénégal, done by Nick Johnstone of the International Institute for Environment and Development, concludes that the fishery access agreements benefit nobody. From a fisheries view they may not. The politics are easier to read. Sénégal has a large component of the workforce engaged at above average wages, and has gained a foothold in EU markets; the EU has found a resource for Spanish fishermen. It could be said that the EU's early fishery access agreements purchased the compliance of the Spanish in a wider political theatre, the EU taxpayer unwittingly picking up the bill.

Since Nick Johnstone's report, crunch-time arrived for the EU fishery off Sénégal. In late 1996 negotiations finally collapsed on the details of a future fishery agreement. As one commmentator said: 'Sénégal is close to France and Spain; the Sénégalese twigged to the benefits these countries were squeezing out of them. Finally, they worked it out for themselves.'

As the smell of something bad has permeated EU access agreements with West African states, and as some countries – Namibia was the first – have got wise and become much tougher to deal with, EU policy has evolved in favour of the second generation agreements. These continue to secure access for EU fishing fleets, but the pill of stock over-exploitation is now coated with the succulencies of joint ventures, technology transfers, scientific help, and the establishment in host countries of local subsidiaries of EU fishing companies. The Commission has openly declared that it has to do more 'to safeguard its position as a privileged partner' with third countries in order to compete with their own more modern fishing capacities. The aims are patently contra-dictory.

The initial second generation agreement was signed with Argentina in 1992. It was a peculiar agreement because the joint ventures are exclusively with the fishing companies of one EU member state, Spain. The generously-funded fishery agreement department in the EU Fisheries Ministry seems to have become a department of the Spanish fishing ministry. The EU signed up to ECU 162.5 million for a five-year term in what have been rudely but not inac-curately called 'exit grants' to get Spanish boats out of EU waters. The exit is in the form of the condition that as part of the joint enterprises the Spanish boats are transferred onto the Argentinian fishing register. Hey presto! The EU reduces its fleet; and Spain gains a rich fishery, and for free. Seventy Spanish boats have been allocated 250,000 tons of fish a year, a substantial prize. Almost half the catch will be hake, a special favourite on the Spanish fish market. It is tacitly assumed that the Spanish quota will be exceeded, CFFA suggests by at least a third. A further gift of ECU 28 million was provided to Argentina for scientific work and training. It has been pointed out that the most immediate beneficiaries of this seemingly cosy deal are the Spanish banks behind the vigorous Spanish fleet reconstruction programme, itself at EU tax-payers' expense, now getting reimbursed once more by EU taxpayers. Again, the EU agreement with Argentina had painful impacts for local fishermen. In order not to overpressure stocks the Argentinian government is demanding catch reductions which affect 700 vessels from the home fleet.

The EU's deficit in fisheries has been steadily increasing, sometimes by 15–20 per cent a year, in both quantity of fish and value. The reliance of the EU on its 'third country' agreements has become steadily greater. A CFFA document has estimated that 30,000 EU fishing jobs could hang on its continuation. The European Commission has itself written that fishing agreements have become 'a basic element of the CFP'. In response to the Court of Auditors' critical report the Commission admits that the agreements are to compensate for over-capitalisation in the EU fleet and warns that any cessation of them would be socially, as well as financially, damaging.

The European Commission realises that the world is shrinking for its roving distantwater vessels, for the gatherers who are hunting far from home. The fact that the overriding beneficiary of fishery agreements is Spain is an embarrass-ment, particularly when conservation ideologies, in which Spain barely bothers to articulate even token interest, have started to dominate the sister EU busi-ness of agriculture. The late 1996 figures for vessel reduction in the EU fleet contains the acknowledgement that browbeating natives on the shores of Africa or South America is a time-limited expediency. By the day of judgement probably quite a few vessels will have crept onto other non-EU countries' fishing registers, whilst still repatriating profits to the mother state. However the bulk of Europe's top-heavy fleet will remain, yet to be pruned. If the propor-tion of European-consumed fish caught off third countries begins to diminish it will have interesting effects. To date these catches have helped suppress fish prices. If the market ever has to pay the true environmentally-audited cost of catching fish from Europe-only waters, the price of fish will reach unheard-of levels.

The big fishing companies in Europe, of which Vigo-based Pescanova, which has the largest independently-owned fleet (over 100 vessels) in the world is the kingpin, think hard about the future. Pescanova, which in 1961 had only one boat, has demonstrated that forward thinking can work. The Spanish company has set up joint ventures in Australia, South Africa, Chile, Canada and Namibia. Perspectives for the company which pioneered freezing the catch at sea are global. One of its British managers told me the efforts to manage stocks in Europe were regarded as 'a sick joke'. 'It may be too late for Europe to get its act together' he lamented. 'Something' he said, with the unflustered air of someone with irons in several fires 'needs doing. Or the European fishing industry will disappear.'

There are many doom-mongers in the EU fishery. But doom-mongering from the market-place is different. Pescanova's diversification out of Europe is putting actions behind words. There is one area in the EU fishery, however, which Pescanova is eyeing closely. This is the fishery for deepwater species. Pioneered off New Zealand, which built up a major industry around species like the orange roughy, which live deep and have only recently become a food species, the search for new fish to catch, and new EU waters to explore, has intensified. Pescanova, among others, is highly interested by reports from skippers of good catches of sizeable fish. The fact that most of the deepwater species have an almost indiscernible taste is not seen as a marketing problem. In fact it may be turned to advantage. Aquaculture is busily proving that blander tastes in fish are readily accepted by a public that is shifting from red meat to white.

The fishing-grounds for deepwater fish are westwards, beyond the Outer Hebridean Isle of Lewis and onto the north-eastern Atlantic slope past territorial limits, at depths of 700–2,000 metres. The mainstream deepwater species are orange roughy, blue ling, roundnose grenadier, black scabbardfish, and black shark. These fish are neither subject to quota in the CFP, nor in any international zones beyond. Their capture is still a free-for-all, uninhibited by scientific research programmes.

The majority of species hauled from the abyss, where pressures are greater and bodies therefore less rigid, are too watery-fleshed to command a market. They are then thrown back dead, so-called 'trash' fish. This contrasts with fish protection and care in Australia where, for example in Queensland, sport anglers are issued with complicated instructions on how to release the pressure in air bladders by careful puncturing, in order to maximise the survival of deepwater reef fish.

The fleet which first started to go for Atlantic deepwater fish west of the Scottish isles hailed from France, many of the trawlers steaming out of Boulogne. The catch is almost entirely taken back to France and eaten there. At time of writing many fish marketeers manage to smile, although a little half-heartedly, when warming to the deepwater subject. Their hesitation is explained by the facts, widely understood, that these deepwater species are found to be in much smaller congregations than pelagic and white fish (it is thought that the abundance of fish is a hundred times greater on shelves and banks than it is at a thousand metres), to breed much less successfully (black sharks produce especially slowly), and to live long, slow, mysterious lives. In other words, they have every characteristic of fish which could be extremely adversely affected by too much fishing pressure.

There are no scientists to help much with the deepwater species. French scientists from their marine research institute, IFREMER, have warned of the limitations of this fishery and pointed to the unsuitability of the target stocks for the fisherman's ideal of steady yields. The European Commission has acknowledged the vulnerability of deepwater fish, but seems notwithstanding to regard the 'Atlantic Frontier' as a handy safety valve for disgruntled fishermen, and has encouraged deepwater exploitation by granting development funds for suitably-placed ports such as Lochinver in north-west Scotland with which to build accomodations for the new industry.

One of the alarming facts recently discovered is that some of these individual fish are over one hundred years old. It is a strange thought, and perhaps one capable of arresting your fork halfway into the white soft flesh on your plate, that the article you are about to consume may be three times, even five times, your age. We eat nothing else so long in the tooth. Elementary biology dictates that removing creatures of this seniority from the population is going, unless very carefully conducted indeed, to turn into a one-off stock removal, painfully to be followed by a protracted period of inaction whilst it recovers, with unknowable consequences for the general ecology of the deeps. The concept of recruitment applied to such slow growers and slow maturers is something with which European fishery managers are quite unfamiliar.

To some extent, whilst the rejoicings of new-found stocks drown out more ordinary developments, the downturn is already in motion. French boats have been fishing for blue ling on the continental slopes between March and June since 1973, for grenadier, originally a by-catch in the blue ling fishery, since 1989, and for orange roughy since 1991. At a conference held in Hull, England, in 1994 the representative from France said landings of blue ling were decreasing annually. He said the only stock which looked as if it could provide steady catches at present levels of exploitation was grenadier, from which French boats were taking a modest harvest of around 12,000 tons a year. As with many deepwater species the edible proportion of the grenadier's bodyweight is small, around a quarter only. This points to utilisation as raw material for processing into meal.

In an ideal world science would precede the fishing trawls when the target species is a complete biological unknown. It is desirable to know, for one, and how these new fish react with other, more familiar ones; how the ecology of the abyss relates to the waters we know better. The precautionary principle, unassailable in logic, is, equally, unrealistic in practice. No one is going to pledge financial resources to a species returning nothing. As has already happened fishermen and fishing catches will give us the earliest knowledge we will get. A cautionary tale in the frantic over-exploitation of deepwater fish species has already been provided. The New Zealand scientist present in Hull said he hoped his own country's experience with the orange roughy might be of use in Europe. New Zealand was the first country to build a major edifice around a deepwater species and their experience is an educational one.

The background to the story of orange roughy is that sea-girt New Zealand has one of the world's largest EEZs. However, fringed by submarine ridges and plateaux which shelve off quickly, three quarters of New Zealand's water is deep. The country is exceptionally well-blessed, nonetheless, with stocks of fish. It is fair to say that this small nation, in the throes of one of the most capitalistic experiments with a liberal unsubsidised economy anywhere in the world, has made a creditable effort to use its fish stocks to best advantage and in a short time has built a very large export industry out of them. Orange roughy has been both the initiator of New Zealand's accelerating success with seafoods, and also its principal victim.

The temperate ocean orange roughy is a strange beast indeed, small (up to two-feet-long), white-fleshed (it is marketed as the fish for non fish-eaters), found as deep as 1,800 metres and usually densely shoaled, its ugly pug-nosed appearance gives no hint as to its exceptionally finely-tuned sensory system. The rare feature for which it is famous is at once its Achilles heel as the target of a commercial fishery: it is one of the world's longest lived animals, surviving up to 150 years. It matures when over 20, and spawns at around 30, then only once a year, releasing tens of thousands of eggs rather than hundreds of thousands. It is reckoned the orange roughy's natural rate of increase is only 1–2 per cent a year, fractional compared to most fish. It was a lack of understanding of the fish's age and reproductive rate that so wrong-footed New Zealand's fishery managers. The orange roughy's low fecundity may mean it is unsuitable for any industrial-scale fishing at all.

Alastair MacFarlane, the deputy chief executive of the New Zealand Fishing Industry Board, has admitted that 'The knowledge of that species lagged behind the fishery.' The orange roughy was being fished hard, in a bonanza atmosphere with trawlers queuing at the hot-spots, frantically shovelling fish off the decks for the next tow, long before anyone knew a thing about its biology. Often inadequate nets resulted in more fish being lost than landed.

Discovered by Russian trawlers in 1977, a year before the 200-mile limit was declared, the orange roughy was fished without restriction until 1980. Often it was captured when aggregating to spawn. The first TAC was set in 1981, for 23,000 tons (it was never to exceed 38,000 tons), and although New Zealand is reckoned to have one of the best fishery management policies today, policed with satellite monitoring, in the 1980s reported catches were routinely exceeded by a third. A combination of poor scientific assessment (in 1982 it is now thought scientists trawling samples were moving with the migrating fish and logging them more than once), and heavy pressure from the fishing industry during a time of economic hardship in the country as a whole, meant that quotas were 3–5 times as high as government scientists were recommending. By 1989, on the admission of a document from the Environment Ministry, TACs needed reducing by about three quarters. All this had taken place against a courageous attempt by New Zealand to run its fishery on novel lines, initially with a high-risk fixed ITQ system, in which the government could be required to compensate fishermen for stock collapses. It was a time of structural trial and error in an industry in which the overpressurised supply stock had retreated from the main shelf, the Chatham Rise slopes, to deepwater refuges around submarine pinnacles.

Meanwhile, the orange roughy became a rarer and rarer fish. Over much of the original spawning grounds it was seldom found in the heyday numbers. The estimated total stocks of orange roughy on the Chatham Rise was 400,000 tons in 1978 and only 30,000 tons by 1990. Despite the knowledge that recruitment at best was a mere two and a half per cent the 1993/94 season quota recommendations of around 6,000 tons were translated, after the politicking, into a TAC of 14,000 tons. The government was presumably aware that much of this catch would come from outside New Zealand's EEZ, aware too that even there orange roughy were harder to find.

The orange roughy had, maybe, performed its role: to stimulate an industry in New Zealand, principally for export, worth about $100 million in 1980, into becoming a major economic engine of growth with a value of over $1 billion by 1992. There had been a learning curve to mount. Because of the stock depletions in orange roughy, amongst other species, the government was forced after 1990 to abandon fixed ITQs; paying fishermen for mismanaged stocks was obviously unaffordable. ITQs from then on were defined as proportions of TACs not as fixed tonnages. So orange roughy has been a seminal player too in the evolution of fishery management in New Zealand.

The legacy for the orange roughy is an extreme population reduction accompanied by, in one of the perversities of consumer demand, an extremely high market price. On the west coast of America orange roughy is a prized delicacy ensuring that, surprisingly enough, it is still New Zealand's most valuable fishing export. Philip Atkinson of New Zealand's Fishing Industry Board in London called it 'a Cinderella story'. Certainly it has a similar poignancy.

The fishery access agreements with 'third countries', and the deepwater fishery off western Scotland, have remained little known corners of the CFP. The deepwater fishery only reaches the public domain when curiosities from the deeps are hauled on deck, reforming our view of the abyss; access agreements worth huge sums of EU taxpayers' money are negotiated behind closed doors, by officials not elected politicians, and never reach the mainline press. Since 1996 the arena that had drawn all the fire from conservation groups, and seen fishermen at odds with fishermen, and European nations at odds with each other, has been the 'industrial fishery', so-called because the catch is processed and sold as oil, fertiliser, animal feed, or feed for fish farms.

1996 was an extraordinarily high-profile year for a tiny fish with a population numbered in billions, which burrows beneath the sea-floor on the sandbanks of the North Sea, and with which almost no-one was conversant before. Greenpeace activists dressed as puffins – to highlight puffins' sandeel diet – festooned lorries transporting biscuits which use 'industrial' fish oil, and in a remarkable publicity stunt emphasising its green credentials the food giant Unilever, which provides a quarter of Europe's frozen fish, declared it would cease using the oils from industrial fish or anything else sourced from non-sustainable fisheries, a move rapidly copied by the British food companies Sainsburys and United Biscuits. The sandeel is not the only fish macerated into industrial oil – in the north Atlantic there is an equally large fishery in another small fish, the capelin – but sandeels are the main industrial fish in the North Sea.

In global terms industrial fisheries have overtaken white fish harvesting as the principal fishing activity measured by weight. Anchovies, mostly for rendering down, are caught in greater volumes than any other fish in the world. The 1994 catch was a mind-boggling twelve million tons, landed mostly by Peru and Chile. As aquaculture has continued to boom so industrial fishing has risen to supply it. Processed small fish is what is fed to farmed big fish. The global catch of industrial fish is around thirty million tons, of which one million tons, or three per cent, is caught in the North Sea.

Sandeels have been the focus of protest. Greenpeace says the sandeel offensive, involving trying to deter fishing boats from entering sandeel grounds, was one of its largest-ever direct actions because the size of the catch is immense (1,134,000 tons in 1989), and because the sandeel has been championed by the widest array of interests. Salmon fishermen point to smolts' need for nutritious sandeels, or the larvae of sandeels, when they swim from freshwaters to salt. Mackerel, cod and haddock fishermen see the removal of the bottom link in the food-chain cutting into the diet of their quarry. Nature conservation movements, like the Royal Society for the Protection of Birds, object because sandeels are a vital diet for European seabirds; and extreme greens, who have noted the lack of scientific understanding of sandeels, see industrial fishers as a soft target. It is a curious fish to have excited so much controversy. Sandeels are, in addition to being one of the largest European catches, the only fish harvested on a large scale which is not subject to quota. With the exception of periodic restrictions applied by Britain on certain parts of the coast, for example, the Shetland Islands, the sandeel fishery is entirely unregulated, an obvious anomaly in a CFP which aspires to scientific management. The reason given is nothing to do with the sandeel stock *per se*, but that because of their protective shoaling habit sandeels are netted in a 'clean' fishery, or one without by-catch. The sandeel fishery has a last unusual characteristic: it is almost entirely prosecuted by one country, Denmark. In the North Sea Denmark takes 84 per cent of the sandeel catch, from early April through May to early June. Norway as part of its bilateral agreement with the EU lands most of the rest. Denmark is a major player in industrial fisheries with an industrial catch that has risen (atypically in general European fishery terms) between 1990 and 1995 by fifty per cent. Also Denmark is a participant in six of the total of seven species which are put to industrial uses. These include Atlantic herring and blue whiting. Under EU rules the former must be offered for human consumption first; only poor quality and over-supplied herring can go for processing.

The object of so much current interest is the least dramatic looking creature, known only to most Europeans as the little, slim, silver fish held in the beak of much-photographed puffins. The sandeel's range is localised and it lives, as its name implies, only on the sandbanks, from which it emerges for spawning from January. The principal fishing grounds are off Norway, the south-eastern corner of the Dogger Bank, the Wee Bankie and the Mar Bank off Scotland, and waters to the west of Jutland. When it is fished the fine-mesh nets are trawled just above the sand. On landing it is subject to exhaustive scrutiny, the critical measurement being the oil content. Sandeels, of which there are five species in the North Sea, despite the smallness of those normally seen, can grow to a foot long; but to do this they need to achieve their full life-span of eight years. Most are caught young, and at 2–9 inches. Their diet is worms, euphausiids and the larvae of crustaceans, and the largest species eats fish and other sandeels too.

The processors and catchers of industrial fish have a case, of sorts, to justify their fishery. This is usually presented by their representative organisation and research arm, the International Fish Meal and Oil Manufacturers Association, more memorably known as IFOMA, based in St Albans, England, Britain with its strong stock rearing and fish farming sector being the largest European consumer of fish-meal products. Dr Ian Pike from IFOMA gave a robust defence of the industrial fishery. On the stock situation he is emphatic. ICES, he says, reckons there are twice as many sandeels as are needed for the minimum biomass. The annual catch is about a third of the stock (he is not slow to point out that some catches in demersal fisheries, for fish for direct consumption, are double that). Although there is no TAC, he says the fishings are subject to a form of agreement on suitable limits, guided by ICES stock assessments. All depends on the breeding success of the thin, silvery sandeel. As Dr Pike puts it: 'If we're not fishing sustainably, it's suicide. Remember: we have been in the North Sea for 40 years.' This is a plea which fishermen being curtailed in their customary freedoms are fond of trotting out.

He ratchets the defence into attack. 'We are more sustainable than most human-grade fisheries. Look, there are no industrial-grade fisheries under threat.' It is a fact: the three main industrial fish species in the North Sea, sandeels, Norway pout and sprats, are all categorised as adequately stocked. He points out that the menhaden fishery (the menhaden is a small, bony, pelagic fish) in the USA is not only also unregulated, but fifty per cent bigger than the European sandeel fishery. On the impact on other valuable stocks he quotes stomach-content findings showing that sandeels are under ten per cent of diet in cod, haddock and Atlantic salmon, twenty-five per cent for whiting. Sandeels living buried in sand are, after all, available to predators for only a short time. IFOMA has calculated that if sandeel fishing was cut by a fifth it would only increase the spawning numbers of haddock and cod by two per cent and one per cent respectively.

Three times as much industrial fish is processed into meal as into oil. The oil is mostly for humans, fish oil being good for the brain. Omega 3 in fish oil is not present in vegetable oil. Its absence can lead, Dr Pike told me with heavy meaningfulness, to brain abnormalities. The meal, to which white fish offal is added, is astonishingly rich in protein – around sixty-five per cent. About half the world's meal from industrial fish goes to feed poultry, an end-use, he expects, which will be supplanted by aquaculture. As he remarked, it is now possible to convert a unit of fish-meal into a unit of salmon, a veritably marvellous conversion rate. If, as the FAO predicts, the scale of aquaculture is to double by the year 2010, fish-meal has an assured market.

Processing of industrial fish, Dr Pike said, was technically complicated but involved no additives. 'In their natural form industrial fish are useless to humans. We extract them without damage to ecosystems and have a good co-existence with wildlife', he said. Like many in fisheries, on a wider front Dr Pike thinks Greenpeace, which was principally responsible for the heightened profile of industrial fisheries, has a bigger and more dangerous agenda: 'Greenpeace are against harvesting any wildlife. They want to teach a lesson to fishermen by targeting industrial fishers.' This view is borne out in Iceland, the main catcher of capelin, another industrial species, where Jon Reynir Magnusson, a former president of IFOMA, told me: 'I am sure one day Icelandic capelin will be on the hit list. The industrial fishery is only the first target. It's the most vulner-able. Industrial fishing is what they want to eliminate first.'

Jon Reynir Magnusson, a chemist by training, had expansive views on indus-trial fishing. On the possibility of getting a TAC established for North Sea sandeels, he said 'I don't think it is a necessity for the fish, but it's good for the pol-itics.' Rejecting charges of overfishing and wastefulness by industrial fishermen, he says 'We are trying to utilise the resource to its fullest.' He believes there are numerous, so far unexploited, uses for fish-meal – in tablet form, to enhance ath-letic performance or to help revitalise the elderly, or as oil, in further processing and perfecting of butters and fats. The Greenpeace star, he feels, is waning: 'When trying to save whales they looked like heroes and attracted lots of donations. That era is past.' He laments the gaps in knowledge about marine interactions and about biological cycles and would like to see knowledge improved on, say, how many capelin whales and seals eat; and he is equally capable of dilating with lyr-icism on the capelin at breeding time when males cosy up to females performing what he calls a 'dance', before sticking to their mates by means of a hairy velcro-type strip down their sides which is developed for the purpose prior to pairing. The industrial fishery has some eloquent and persuasive spokesmen.

Some of the opponents of industrial fishing think the immature herring catch by Danish fishermen, described in chapter 2, is so indefensible that it will be the first to get restricted. However, the main target will still be sandeels. They are in the public eye, their fishery can be associated with deprivation suffered by puffins, and more interest groups are involved. The case against sandeels has been drawn in considerable detail.

The argument centres on what effect the sandeel catch has on other fisheries. If a million tons is being removed, of a small fish which is part of the diet for many larger species, is it not reasonable to suppose that the fishery is having a downstream impact? Defenders of an unregulated sandeel fishery say there is no proof their actions have affected numbers or condition of salmon, cod, haddock or anything else. They hark back to this theme – the absence of proof against them. They know that proof in such an elusive species, part of such a complex food-chain, may be a long time coming.

The plea to await scientific proof is too often used as a means by which man-agers can avoid taking actions which they know will be unpopular. The case of the Atlantic salmon illustrates this best. The worldwide catch was four million individual salmon and grilse (salmon which spend only one winter at sea before returning to the natal river) in 1975. It steadily declined till in 1996 it dipped below a million. The plea for more science has prevailed and during this twenty-year period of the Atlantic salmon's decline scientific papers about its biology have poured from an ever-increasing number of salmon analysts. The fate of the last Atlantic salmon alive is to be scrutinised by a hundred scientists, each arguing his own theory for its disappearance.

If science is to clear the misty situation regarding sandeels the prognosis is poor. Scientific statements about this fish and its rôle in North Sea ecology vary wildly. Where defenders of the sandeel fishery argue that its substratum life-style means it cannot support the main white fish for much of the year, others have found sandeels in cod stomachs when they are supposed to be still buried, and conclude that cod extract them from their sand-holes by digging. The importance of sandeels in other fish's diet has been variously estimated. Old fishermen recalled the days when mackerel and codling, much larger specimens than found today, had their bellies crammed with sandeels; these old timers say the industrial fishery is responsible for the present 'marine desert'. A report commissioned by Unilever, backing its withdrawal from the industrial fishery, claimed that the proportion of industrial fish in cod diet is 21 per cent, in haddock diet 25 per cent; and in whiting the figure is 60 per cent, of which most is sandeels. Mackerel diet, the report continues, is 37 per cent industrial fish, also mostly sandeels. Sandeels and pout, on this analysis, are the two single most important foods for the high-value fish of the North Sea. Other reports, as you will have noted, produce different figures. Unilever's reporters even cite studies arrived at by scientists bent over their logarithms and computer screens which show that what with industrial fishermen catching sandeels, and other fish apparently consuming well over the one million tons human take, those eaten by seabirds, in addition to those consumed by porpoises, more sandeels are eaten each year than are supposed to be there! Something is wrong. Certainly ICES scientists think so; they savaged the report. Not for the first time the extrapolations done by fishery scientists, based on some questionable assumptions, have proved a nonsense. If so many sandeels are removed, for example, why is the stock judged healthy? What exactly is this fish's repro-ductive power? One's sympathy must lean a little to the scientist; after all, this fish is not only under the sea but for most of the time under the seabed too.

Fishermen's catches have been rising, for sure. Is it the case that this has played a part in the shrinkage of cod and haddock stocks, or have they simply been overfished, or not bred well? Or has the heavy fishing of sandeel preda-tors, like cod, mackerel and haddock, as some studies suggest might be possi-ble, actually had the effect of counteracting the impacts of industrial fishing? The sandeel fishery, perhaps as well as any, shows what a suppositional world fishery scientists inhabit. The case of capelin, at any rate at the theoretical level, is simpler, because the capelin is short-lived and dies after spawning. Catching post-spawned capelin should do no harm to the stock itself; its effect on adja-cent stocks, again, is merely a question surrounded by difficult answers.

Salmon interests have argued that if sandeels are the choice target of smolts just run to sea then the industrial fishery not only scoops from under smolt noses their survival ration at a critical time, when they have recently arrived in the dangerous saltwater arena, but the fishery catches these vital smolts along with sandeels. Jim Slater, of Britain's pelagic fishermen's organisation, has emphatically stated that adult salmon turn up in his members' nets only once in a blue moon, and smolts the same.

The Danish Institute for Fisheries and Marine Research says that in forty years of sampling they have yet to find a salmon smolt; but the value of this evidence was not enhanced by the fact that Denmark continued to deny some of its oil-rich sandeel catch was being used as fuel for power-stations until photographs of sandeels being delivered to power-stations were published in a British newspaper! It was a dream scoop by the environment journalist Charles Clover. Greenpeace took up the cudgels, and industrial fisheries have been under the spotlight since.

Salmon bodies responded to the denial of any by-catch of smolts by saying, well, who would notice the odd smolt in a catch of tons of small sandeels anyway? No one is sorting them individually. The Danes seemingly did not notice their coastal 'sprats' were young herring. And only a few smolts, of one of the most endangered fish in the sea, could be the critical issue of the stock in one particular imperilled catchment or natal river. The Scottish Council of the Salmon and Trout Association pointed out that what little research British scientists conducted on the by-catch in the sandeel fishery was done too late in the summer, rather than in April and May when the smolts were feeding on sandeels. The Association contended that scientists only looked for incriminating evidence when they knew there was nothing to be found.

Opponents of industrial fishing suggest the sandeel fishery is subjected to major and definitive research to see how it affects other stocks; that meantime precautionary quotas are applied to Denmark and Norway; that sandeel fishing, particularly in the western half of the North Sea because of the value of sandeels to the more densely congregated food fish there, should be curbed; that by-catches of the industrial species are reduced by gear modifications; and that the loopholes in the CFP of misreporting and highgrading and discarding are borne down upon. The argument goes on that more industrial fish left swimming will mean more high-value fish. Don't knock the base of the food-chain, is the message. There would appear to be a lot in that contention.

Until, that is, you think about it through the perspective of Professor Tony Hawkins, Director of Fishery Research for Scotland. He has written on the industrial fishery in a refreshingly non-partisan way, challenging a number of easy assumptions. Sandeels, he gently corrects us, are not the base of the food-chain. Tiny plants dividing up to form phytoplankton are the base. The separate consecutive strata of shrimps and fish larvae; of molluscs, pelagic worms, krill and large jelly-fish; of crabs, snails, bivalves and burrowing worms, all precede sandeels at the fecund base of the food-chain.

He has a fascinating conversion rate table which graphically illustrates life processes beneath the sea. He says that in broad terms ten kilos of phytoplankton (plants) are needed to produce one kilo of zooplankton (the smallest fish), which can produce one hundred grams of sandeels. This sandeel *bonne-bouche* will add ten grams to the weight of a hungry cod, and a seal (top of the predator pyramid in the North Sea) which eats ten grams of cod will in turn add to its own weight by only one gram.

'Man as a fisherman', says Professor Hawkins 'can derive his food from any level of the production pyramid.' Then, why not focus on fishing precisely at the industrial species level? It supplies great bulk, and it saves killing slower-growing fish like cod. He writes, 'There is a case for pursuing a variety of fisheries, based on several tiers of the food pyramid, to provide for diverse tastes in food, and to avoid over-exploitation of any particular part of the marine ecosystem.' Tackling the shrill voices of the conservationists he points to the excellent and growing numbers of seabirds in the North Sea, of rising grey seal populations, of more killer whales. He suggests there is no shortage of food for these species, none of which are harvested by man. He maintains removing some parts of the food-chain in large numbers is most likely to result in fish finding a substitute. Salmon, he reminds readers, are catholic feeders. They are known to partake of thirty different foods, mostly mid-water schooling species, or bottom-fish when in the pelagic phase. The life of the smolt when it reaches the sea, and until it arrives off Greenland, Dr Hawkins acknowledges, is a major mystery, and keenly in need of better knowledge. He ends his provocative treatise by suggesting the sandeel fishery is regulated where there are proven localised and urgent wildlife demands on the resource, be they seabirds or smolts. When the British government did ban a sandeel fishery it was off Shetland, in order to protect nesting colonies of seabirds which ate sandeels. The government, Professor Hawkins' employer, has recently called for sandeels to be brought within regulated fisheries and fitted out with a TAC.

The last twist in the sandeel debate was in summer 1997. Prompted by the alarming fact that now only 10–20 per cent of Atlantic salmon smolts going to sea return as adult fish, compared to 50–60 per cent in the 1970s, scientists focused on the great unknown of smolt survival. The findings of 1997 corroborated those of 1996. Trawl surveys done very close to the surface finally located the little smolts at sea as they headed north for the winter feeding. They were located in dense shoals on two shelf-edge currents, one off the Norwegian Trench and the other off the Wyville-Thomson Ridge, a shelf-edge between the Shetland and Faeroes Islands. Gut sampling showed not sandeels themselves, but sandeel larvae, along with many other larvae, including those from capelin. Tentative assumptions are that salmon post-smolts travel in shoals, use the narrow band of shelf-edge currents for propulsion, occupy a niche in daylight hours in the upper sea layers at no more than ten metres depth, and fatten on larvae. Tony Hawkins, pondering these fresh findings, doubted that post-smolts ever eat sandeels. This does not mean sandeels are irrelevant to salmon smolts; their larvae may be a key dietary component; but it explains the absence of smolts in sandeel trawls. Whether salmon use the huge sandeel resource again, later in the lifecycle on return from the feeding grounds, remains unknown.

If there is a formidable slipperiness in information data on sandeels the politics of the industrial fishery are much clearer. Despite the fact that catches reached a plateau in the 1970s Denmark desperately needs the fishery to continue. Of the four main fishing nations in the North Sea (Denmark, Britain, Germany, and Holland) Denmark has the largest number of fishermen as a proportion of national employment. The smallest of the Scandinavian states, Denmark punches above her weight in EU politics and is very much a major fishing nation. Her fishery science institute is world-famous, and along with Britain she is the only country to boast a fishery research institute over a hundred years old. When national allocations of the seven main fish species were decided in the formulation of the fisheries conservation policy back in 1983, which gave birth to the pivotal and now contentious concept of 'relative stability', Denmark received 23 per cent of the value of the whole EU catch. Denmark is a particularly big catcher of herring and mackerel; the self-same fishing boats which comprise the Danish pelagic fleet are suited to fine-mesh trawling for sandeels.

Seventy per cent of Denmark's catch today is of industrial fish; the 1996 catch of 1.25 million tons was worth £75 million. Furthermore, Denmark has developed a very big fish processing industry, and fish products represent over five per cent of all Danish exports. Denmark was originally allowed a free hand with industrial fish within the CFP because it was the traditional user of fish for reduction to oil and meal. Interestingly, from the outset the other states disapproved of industrial fishing.

If the industrial fishery is to be restricted in any way – and one possibility, maybe remote, is that consumer pressure against any animal or fish protein in livestock or farmed fish diets will disrupt the market for fish meal and oil – Denmark will assuredly demand some mighty slices of TAC in other stocks. Relative stability will require her proportion of the total catch to be maintained. With Spain and Portugal impatient to get into the North Sea as well, this is the very last scenario fishery managers would wish for.

It is not only the effects of humans on fishing boats that concern conservationists and ecologists. It has been loosely said that there is spreading marine pollution. In the context of Europe, and leaving the self-governing Mediterranean and Baltic aside, this means the North Sea. Some scientists consider the issue of the environmental health of the sea a sideline; but they tend to be engaged in the field of stock assessment. No profile of the European fishery would be complete without some consideration of the marine environment, particularly because, in linkage with climate, it has a direct effect on recruitment, or how many young fish reach the fishery as reproducing adults.

The reason the North Sea is one of the most fertile wildlife areas anywhere is because many large rivers debouch into it bringing with them mineral enrichment from the land, and because it is constantly refreshed and invigorated by nutrient-rich water from the Atlantic. Most of this relatively warm, relatively saline water streams into the North Sea from the north, sweeping in from between Orkney and Shetland and from the north-east of Shetland, then bearing southwards from the Norwegian coast into the Skagerrak where it mixes with Baltic water, and into the centre of the northern North Sea. In addition Atlantic seawater is borne on currents and with prevailing south westerlies through the English Channel refreshing the heavily populated coastlines of France, Belgium, Holland, Denmark and Germany. Strong tides help to mix these waters together. The North Sea's solitary outflow is in the north-east corner, commencing in the Skagerrak and surging northwards along the coast of Norway. The high degree of biological activity in the North Sea is profoundly affected by its happy geographical position, fed by currents and rivers all-round.

Countering some of these benefits from the point-of-view of the fish is the extremely dense human activity in the North Sea. The human impacts are multiple. Freshwater run-off is affected by agricultural irrigation, reservoirs and hydro-electric schemes. This affects the structure of surface seawater and water circulation, which can in turn affect the dominance of fish species.

Sea defenceworks, land reclamation and dredging are common round the North Sea, and widespread effects of these are only just starting to be understood. Sea defences in one place can lead to deficiencies of natural deposition from the land and cause erosion elsewhere. Dredging removes gravel and sand banks which are the nursery areas for many species in the sea. Between the pebbles small crabs and lobsters, and the larvae of many fish when they have drifted down the water column to find shelter, bury themselves. Here bigger fish come to find them and a special habitat of intense fish activity is developed. Dredging not only physically removes the gravel and sand, a vital habitat for fisheries in general, but the so-called 'plume' from dredging operations, normally conducted by vast suction hoovers, smothers productive areas in the neighbourhood. Dredged seabeds are reported by divers to look ghastly. As Phil Lockley, an experienced North Sea diver and former fisherman, remarked, 'If you've been under water for twenty years, you know how it should look. In dredged areas it is horrendous. When you see it under there, you know it's wrong.' Not only are there none of the normal fish but the remnant deposits of aggregates become infested with unnaturally high populations of scavengers, inhibiting the establishment of new benthos. Some have likened dredging to taking the topsoil off farmlands. The trend is for the growing need for aggregate to be met by marine extraction in preference to land-based mining. The dredgers, which only started creaming off the aggregate banks in earnest in the 1970s, work offshore, far from prying eyes, away from associations of irate local residents loquaciously protecting their environs. As one fisherman, said, after the dredgers have gone, the sea is flat again.

Oil and gas platforms, rigs and pipelines also have localised effects, some complicated. From the fishermen's angle mid-sea structures have reduced the fishable area, as well as acting as fish sanctuaries; specialised ecosystems have grown up around rigs. Round-the-clock artificial light from oil platforms affects local photosynthesis, and attracts fish.

Something that is only recently being looked at more closely is the effect of nuclear power stations on fish stocks. Nuclear power stations dot the coasts round the North Sea and the English Channel. The mechanism which harms fish is bound up in the need for power stations to cool down by using seawater; the water is used to cool off the steam. In order to avoid debris hitting the power stations' condensers (small-diameter tubes), the water is screened. Anything larger than three centimetres is sucked towards the screen and dies. Smaller fish material, like eggs and larvae, which gets through is either killed by the chlorination, or by the higher temperature inside the plant, or by the extra pressurisation. According to Dr Peter Henderson, from the Zoology Department at Oxford University, himself formerly employed in the power industry, 'a diluted soup of dead animal and plant matter is then dumped into the sea.' This lot kills anything it contacts because it is hotter and it is chlorinated.

Nonetheless the principal problem is the quantity of fish life killed in the inflow pipes. Dr Henderson has done some calculations. An average small nuclear power station sucks in about thirty tons of water per second. A large one, such as Graveline in northern France, with its six pressurised reactors, takes in about 240 tons per second, the equivalent of a large river. This water is inshore, usually estuary, water, from precisely where the nurseries of most marine fish life are located. The figures Dr Henderson has done should be well-known amongst fishery managers, yet the subject is virtually unrehearsed. Graveline alone, he reckons, kills around a hundred million young herrings a year. Of fish over three centimetres squashed on screens, Dr Henderson believes there are about 300–500 million annually destroyed by British and French power stations. The larvae, very young fish, and eggs deficit is in the region of 100 billion. Faced with this scenario, energy industry representatives retort that big trawlers kill a lot of fish too; but they never compute the accumulated effect of all the estuary power stations, nor the fact that power station screens are hitting stocks at the base of the pyramid prior to all those fish life forms realising their full size and potential. At the least it would be prudent to factor such extravagant subtractions from the fish biomass in the North Sea and the Channel into fisheries models. It does not happen.

Better still, there should have been obligations for British privatised power utilities to continue the work started by the state in keeping fish away from intake screens in the first place. Scientists have been working for twenty years on this, not without results. There are a number of ways of making adult fish keep a safe distance. Both electric force fields and underwater lights (less good when the water is murky) have some effect. The most promising leads have been in the deployment of underwater sound. Sending out sound waves, on low frequencies, is effective to between 50 and 100 metres, providing there are not extremely strong currents. Some fish, especially those with swim bladders, are more sensitive to sound than others. In delicate species like sprat and herring the former state scientists, now in the private sector, claim success rates of 80 per cent. Flatfish such as plaice, sole and flounder are less sensitive. The boffins have developed different sound signals for differing species mixes. The norm is a siren-like repetition. In order to monitor fish reactions to the sound waves they have been fitted up with acoustic tags. It is ingenious stuff, and has the capacity to greatly improve things. The problem is that the only one fitted and in operation is in Belgium. In France, where the stations are state-owned, there is little operational transparency and the stations are often situated, ridiculously, in known nursery areas. In Britain, similarly, there is no obligation to use deterrence techniques. This is simple sloppiness, spiced by the desire to make the privatisation issues uncluttered. For a large power station sound systems would cost a mere £100,000.

Legal constraints have altered the amount of foreign bodies in the North Sea. For example, incineration of waste at sea stopped in 1991; dumping of industrial waste became illegal in 1995; and dumping of sewage sludge becomes illegal in 1998. The 'polluter pays' principle is now established in law and most European countries have signed conventions pledging them to 'Best Available Techniques', 'Best Environmental Practice', and clean technology. Various indices of base metals in the livers of fish, and other similar measurements show North Sea wildlife populations in an improving state of health, not the reverse. As chemicals such as DDT have been banned, so their presence in, for example, mussels, which particularly concentrated DDT, has declined. European rivers are generally becoming more fish-friendly. Atlantic salmon now swim the Thames once more and the sensitive allis shad has returned to the once filthy River Medway in Kent. Rivers throughout Europe are cleaner and the North Sea is not the putrid sink it is often portrayed as.

The act of fishing itself is one of the main agents in altering the seas. Beam trawling digs up to six centimetres into the sea-bed. Heavy tickler chains in front of trawls, which chase flatfish from their sandy refuges, also harrow the seabed. As much as one hundred per cent of bottom-dwelling or benthic creatures can be caught up in a trawl, and beam trawlers in one survey on a limited number of species killed up to fifty-five per cent of benthic fauna as they passed. Fishermen themselves have joined the hue and cry over beam trawlers, one fishing association in the west of Scotland demanding that trawling be banned from certain areas which could then be reserved for fishermen using less harmful lines and pots. In really developed fishery management systems, like Japan's, fishing areas at sea are carefully divided to optimise their productive return. This is explored in Chapter 6.

However, the international group of scientists who produced the 1993 North Sea Quality Status Report warned against logical extrapolations with regard to unproven side-effects of fishing. Calculating that damage by bottom trawls to the benthos should theoretically be most apparent in long-lived species they looked at southern North Sea populations of burrowing sea urchins which live an estimated 5–10 years, and at those of a certain bivalve which can survive a century. Both creatures were far more common than expected, despite being in areas repeatedly ploughed by trawls. As has been said by many, undisturbed sanctuaries serving as reference sites are needed to illumine this remote ecology about which, ultimately, very little is yet known. The authors of the North Sea report are categorical: 'The changes wrought by fisheries should be viewed as one of several anthropogenic effects in a system that has no long-term equilibrium.'

Then what constitutes pollution is a subjective matter. The best illustration of this in European fisheries is in eutrophication, the term given to the results of nutrient enrichment. Spoken of with disapproval by fly fishermen on English chalkstreams, because it leads to the wrong sort of weeds proliferating in trout streams and takes away the fish's oxygen, eutrophication can have an altogether different meaning for cod in the southern North Sea.

The totality of the effects of eutrophication are unknown, which means many scientists will not stick their necks out. But in the North Sea firm things can be said, although in a brief time-series, of changes that have happened in the water. Because of more intense human activity there is more nitrogen and phosphorous in the North Sea. The largest rivers, like the Rhine and the Elbe, carry nitrate loads which are about fifty times that which they meet in waters from the Atlantic. The large rivers flow into the southern part of the North Sea and it is here that nutrient changes have been logged most conspicuously. Because of new laws restricting chemical inputs to rivers, total nitrogen levels are often held down by the falling quantity of ammonium. Phosphorous levels in the Southern Bight off Belgium and Holland in the late 1970s exceeded by up to seven times the levels recorded in the 1930s; for nitrogen the increase was four times. From 1985 levels of phosphorous fell sharply. Airborne nitrogen, mostly from motor vehicles, is in some places equal to, or even greater, than river-borne loads. Again the northern North Sea is much less affected.

The effects of these nutrients are to promote primary growth. Proceeding through the food-chain to bigger shrimps the accelerated plant growth leads, and has led in the opinion of many scientists, to the huge increase in fish numbers in the southern North Sea whilst eutrophication was at its peak. In the words of Dr R. Boddeke, of the Netherlands Institute for Fisheries Research at Ijmuiden, the entire modern Dutch fishing fleet has been built on this increase. For a cod in the North Sea eutrophication had much to commend it.

Dr Boddeke has expounded his somewhat unfashionable thesis in a long paper written for the Proceedings of the World Fisheries Congress. Firstly, he establishes the sharp rise in fish landings from the southern and central North Sea starting around 1955. Plaice was the most striking example: excepting a post Second World War peak catches of plaice had been steady for a long time. After 1955 catches multiplied by eight times in Dutch waters. Cod, whiting, sole and turbot hugely increased too. Holland itself, which had historically specialised in this fishing area, benefited most. Dr Boddeke holds to the Dutch experience, but the wider frame was as breathtaking. By the mid 1970s the total North Sea catch had reached three million tons, double its level fifteen years earlier; and this was despite a dramatic decline in the two species of which the largest finfish catches were made, herring and mackerel. They had peaked by the mid-1960s.

The great increase in populations of the main demersal and flatfish species has been labelled by other scientists the aforementioned 'gadoid outburst'. Dr Boddeke remarks on the fact that other scientists take the outburst as their starting-point, attributing it to a very good breeding period for haddock, rather than explaining why. Because he does have an explanation for why, he also has one for why it ended. It was simple: Germany, where the principal rivers in the area are sourced, instituted the purification of sewage from 1984. Phosphates were outlawed in detergents. Phosphorous levels dropped, eutrophication lessened, fish ceased to outperform themselves. The answer to that old schoolboys' conundrum – how to enrich the North Sea – was, apparently, to let the phosphorous-rich sewage from Continental Europe freely flow in. Dr Boddeke's argument is a degree more particular. Nutrients which fall inert proliferate choking algae; it happened with disastrous results in the Black Sea. The nutrients must move. Pointing to the widdershins circulation of waters in the North Sea, and the fact that this nutrient-rich water was shifted onwards quickly by Atlantic water pushing in from the Straits of Dover, Dr Boddeke explains away the absence of any of the normal effects of eutrophication – loss of oxygen and consequent algal stagnation. The specific hydrological character of the Southern Bight helped disperse the nutrients before they became noxious.

Citing the findings of numerous colleagues, he builds his theory. Many creatures thrived. The Dutch catch of coastal cockles rose from an average 195 tons in 1955 to an astonishing 65,000 tons between 1987 and 1989. By 1990 recruitment had collapsed. In the case of mussels, not only was the spatfall (egg production) increased but the mussels were quite different in size. In the German Wadden Sea three year old mussels had doubled their length from three to six centimetres by 1986. Brown shrimps, the main shrimp species on Europe's soft-bottomed coasts, also grew in range and number between 1969 and 1982.

For the reason that eutrophication is linked substantially to freshwater discharges, Dr Boddeke wishes to demonstrate that those species not reliant on nursery areas in estuaries, or in parts greatly influenced by freshwater surges, showed no population increases to compare. Catches of skates and rays declined in the period. Haddock, also not a user of shorelines as nurseries, was excluded from the gadoid outburst. The pelagic populations of herring and mackerel did not behave in this period in a way that could be correlated, as expected. He then chronicles the subsequent decline of the main beneficiary stocks, showing that not only landings collapsed but the size and condition of fish was as well. The fact that treatment of raw sewage is now a requirement in European law, and that present-day washing powders are phosphorous-free, will, in Dr Boddeke's view, keep a new gadoid outburst from bursting the nets of fishermen. Instead, he believes, because nitrates were not removed at the same time as phosphorous, there is a dangerous imbalance between nitrogen and phosphorous. This will produce the wrong type of algae. 'Only the northern North Sea, dominated by the influx of the Atlantic' he said mournfully, 'will be unaffected.' As I spoke to Dr Boddeke, a man of original cast of mind, he became philosophical: 'It is an interesting story. But . . . the whole world is full of interesting stories. No one is really thinking anymore.' The present-day waste of industry's phosphorous, which is treated as a dangerous contaminant, he regards as sheer madness.

Huge changes have occurred in the North Sea in the course of a century. The head of Europe's most prestigious fisheries research institution at Lowestoft, Dr John Pope, presented an international audience in Denmark with an amusing address in 1989. Veering away from the hot topics like multispecies analysis he presumed to profile the North Sea a hundred years ahead. His tone may have been tongue-in-cheek but his alternative scenarios show that his brain was very much engaged.

The first scenario was the North Sea as a playground. Northern Europe has detached itself from southern Europe and, allied to Scandinavia, is a prosperous high-tech community. Weather patterns had been interfered with to reverse higher sea levels and the Dogger Bank was partly an island. Northern fish had migrated north and had been replaced by southern species. Rampant conservation lobbyists had achieved the outlawing of wild fish catching, made more acceptable as the warmer North Sea was biologically poorer anyway. A few pot and line fishermen survived, as a tourist attraction. Fish products were synthesised and imported. The chief rôle of the North Sea was as a playground for sport fishermen and to this end dramatic fish like sharks were encouraged by a variety of programmes eliminating their competitors, including deploying species-specific diseases, pheromone seduction of males away from spawning females, and species-specific pesticides.

Dr Pope's second scenario pictures the North Sea as a foreign exchange earner. Northern Europe had collapsed and become a 'banana prawn republic'. The North Sea had still warmed, but now there is a preponderance of pelagics. Shrimp fisheries had been targeted as potentially earning vital foreign currency. Their success was helped by artificial recruitment. The former commercial fisheries were depressed to nothing. Submerged cities such as London were perfect artificial reefs, but economies were so low that their utilisation was licensed out to foreigners. Most research had ceased and the small offshore fishery was licensed and subjected to a limited fishing effort rule; licences were auctioned or purveyed illicitly.

The next scenario with which Dr Pope teased his illustrious audience was one which was first mentioned as an idea long ago – the North Sea as a giant fish farm. Global warming had proved a chimera. Europe was affluent, the Union a success. Everything was now cage-reared except salmon, presently cage-reared, but in the twenty first century ranched, or hatchery-reared and then released into the sea. Cage areas were akin to areas of farmland. All operations were highly controlled with spy-eyes in cages and a robot workforce. Three large fishing companies operated the big industrial fishery which was destined to feed farmed fish, and which was subject only to light research, populations being monitored by acoustics.

Finally he gave a social desolation scenario. The climate was as now. North Europe was a backwater with chronic unemployment and home to the smokestack industries, progress having moved to the Pacific Rim. The fertile North Sea in Dr Pope's 'larder scenario' was used as an industrial dump and a food source. All fish, regardless of species, were protein fodder converted into unedifying pastes. Fishing restrictions were as now, plus effort limitation. Half the biomass was caught each year, which appeared to maximise protein production.

Dr Pope rounded off by saying that whatever outcome proves closest to the truth, managers will try to create a North Sea mix of fish species which suits a particular purpose. That, effectively, is what is happening now. The biggest predator of fish in the North Sea, the grey seal, is left out of management because environmentalists have persuaded public opinion that they are too important to be killed. The same applies to whales, the common species of which are now, in the eyes of many resource managers, an under-utilised commodity. At this time conservationists are forcing upon fish-trading companies a move away from what are termed 'unsustainable' fisheries. Fish-trading companies are also looking in that direction for good commercial reasons. When thousands of processing workers lost their jobs overnight in Newfoundland the cost to the multinationals was dear. Suddenly their plants had no supply. It is instructive that the first tough action on sustainability –even though it may be too ambitious – has been taken by a company and not by a government. Overfishing on a continual basis will hurt everyone; commercially-aware businesses see this earliest and have the capability to act. In the shorter term the least 'clean' fisheries, such as the giant beam trawlers raking the grounds of the southern North Sea, may need to adapt or find new markets.

In a telling and wide-ranging address given to the World Fisheries Congress in Brisbane, Dr Serge Garcia of the FAO Fisheries Department accused fishery managers in general of dodging the difficult decisions. He said: 'Instead of addressing the structural causes of overfishing, facing the problems of fishing rights and resource allocation, they used less controversial and less effective technical measures such as gear regulations, closed areas and seasons. The TAC approach failed to limit capacity and effort as intended, aggravating the situation in some cases and leading to fisheries lasting from a few weeks to a few minutes a year . . .' Perceiving the overfishing of the North Sea in the 1950s as having been exported to all the world's oceans, he sees fisheries now, worth approximately four times what they were twenty-five years ago, as being driven by trade and market forces rather than governments. On this score he is concerned that as ingeniously manufactured fish products can replace those from another species (echoing John Pope), the need for biological diversity may get left behind. He highlights the disparity between the large profits often notched up by the processing and marketing sectors in contrast to the fishing sector which, taken as a whole, labours to break even.

The solution to the ills of fishing and fisheries as propounded by Dr Garcia refers repeatedly to the Code of Conduct for Responsible Fisheries, devised by the FAO in 1995. Some of the Code's main principles are: the recognition that states, in this case the EU, have a duty of care, or moral obligation to look after the resource; a user-pays principle, something accepted, for example, in EU 'third country' agreements but seldom mooted for EU nationals in home-waters; a scientific approach based on precaution, or conservative TACs; environmental considerations; more involvement of fishermen themselves in management actions (a particular favourite of Dr Garcia who has spent much time in fiishing boats and values fishermen's experience and input); and compatibility of management measures across the whole range of the stock. This last scrubs out 200-mile limits as logically indefensible, and moves towards Brian Tobin's definition of the 'contiguous zone'; it could as easily be called mutual management by common sense. Knowing that much of this menu, no matter how many countries, in however many different protocols, have signed up to it in principle, is a tall order, Dr Garcia concludes his framework proposals for sustainable fisheries by referring to the need for 'a long-term perspective ensuring a step-wise approach to fisheries optimisation'. His paper, entitled 'Fisheries Management and Sustainability', and co-written by R.J.R. Grainger, is long and profound, and is one of the best catch-all definitions of modern fisheries problems and possible solutions attempted by anyone.

He does not specify the problems of Europe. It would have made his presentation much longer. In 1996 the European Commission called for an over-the-board reduction in fishing power of forty per cent. Naturally, those with vested interests sent up a cry of terminal woe, and immediately began chiselling away at the figure. The EU has many grave problem areas, some of which I have not touched on. Subsidy cushioning has rampaged. Between 1983 and 1990 financial supports rocketed from US$80 million to US$580 million, a bewildering figure given the acknowledged overcapacity and that twenty per cent of this money was for building new, or modernising old, fishing boats. Not only will the pain be great if there is a fishery collapse, the embarrassment will also be great. New technology continues to outpace restrictions placed on fishing, making vulnerable the often called-for move from TAC to effort limitation. If more fish can be taken with less effort the level of effort has to be continuously attacked. The science has lacunae in key areas. Top scientists freely admit that the vital link in their stock assessment calculations, the matter of recruitment, is inadequately understood. Recruitment is the golden goal. The multispecies models, obviously a *sine qua non* if an all-embracing scientific approach is to stand as the basis of management, seem to be so complex and difficult as almost to have jerked scientists backwards. Yet the more that is found out, the greater the interaction between species is discovered to be. The American scientist, Nicholas J. Bax, has characterised single-species analyses as 'gross caricatures of complex natural systems'. Above all, the situation where overcapitalised and bitterly competitive fleets, concertinaed together under the CFP, regard doing each other down as part and parcel of their survival, cannot last. It is not due to do so.

Officials in DG XIV have been working on an 'EU Fisheries Policy', to be introduced in 2002. The presence of the plan was betrayed by an obscure item buried in the 1994 Treaty of Accession by which Finland joined the EU. Under the plan the CFP would go and so would national quotas. In fact everything 'national' would go. From 1 January 2003 the only fishermen allowed on the sea would carry a 'Special Fishing Permit' issued by Brussels, identifying what fish could be caught, where, and when. Countries rewarded with most permits would be those, like Spain, who had behaved best in reducing their fleet size. It would be decided in Brussels not in member states who would be allowed to fish in 'Union' waters. British fishermen, apparently, might end up with an allocation of less than ten per cent of the catch. Politicians in Britain, and maybe in other countries too, will be hoping devoutly that by then the number of surviving fishermen will be too few to matter. If they are wrong, fisheries have the potential to split the EU.

Chapter 6

Japan

If markets are not entirely driven from the top, they are certainly lifted by the top. That is why the most important single place in world fisheries is the fish market in Tokyo. Named the Tokyo Central Wholesale Market it is known as Tsukiji (pronounced skee-jee) after the neighbourhood where it stands. Tsukiji is a phenomenon, in any context.

It was described to me by Petur Bjornsson, Managing Director of Isberg, a fish trading company based in Hull.

> 'The Tokyo fish market is quite incredible. A huge number of people – 45,000 workers and traders, 25,000 visitors – are milling about in a very cramped space. 2,500 tons of fish, worth $22–$23 million, changes hands each day. The market is open round the clock, but the auctions start at 5.30 am. and end at 9 am. Approximately a third of the fish is fresh, a third frozen, and a third dried or conserved in some way. The auctioneers are frantic and passionate, shouting and hopping about. The handling is in hand-wagons; little is mechanised. The high tension and electric atmosphere are almost tangible.'

Petur Bjornsson sees it through the eyes of a major trader. He finds the Japanese fish auction fascinating because it shows at work a quite different philosophy about fish (whales in eighteenth-century Japan used to be buried with a memorial stone, the same way as deceased people; today in Tsukiji time is still taken for prayers for the dead fish), and a different philosophy about business. Petur Bjornsson sells mostly shrimps to Japan; it is the second biggest commodity after tuna. It says something of the fish trade, which may surprise outsiders, that as someone who has sold hundreds of millions of pounds worth of fish he has never sought a bank guarantee from another trader. To the Japanese, fish, and all details relating to fish, are, if not a religion, a central part of life. Not only is Tsukiji the biggest wholesale market in the world but the Japanese eat more fish than anyone (bar the Icelanders) and, maybe inconsequentially, tend to live longer. One of the surprises for Petur Bjornsson at Tsukiji was the absence of automation in the processing industry. For a country where large sums of money are bound up in fish it strikes a west European forcibly that processing plants are largely manually operated. This is partly a deliberate policy to maintain employment. Of late processing has been moving to China, where a day's labour costs what one hour's does in Japan.

High unemployment is anathema in Japan; so is high fish. Foreigners to Tsukiji have remarked on the absence of a fishy smell. The fish are so fresh and move so fast to the consumer that they have no time to smell. Fish from round the globe is flown into Japan, some of it alive, in a vaster migration than the seas have ever witnessed. Freshness is an absolute premium, and the travel routes for the merchandise are tightly synchronised. Petur Bjornsson said that in the case of cod the freshest product could be worth seventy per cent more. People talk fish. In Japanese restaurants where 'You spend the earth' a British executive told me, hastily adding, 'on expenses', they discuss every aspect of the fish being prepared with blinding dexterity in front of you. Freshness is the first part; then, where is the specimen from? Is the fat content right? Whether today's obsession with freshness is a backlash from the older days when makuri – herring stored until it had begun to ferment – was a local speciality, I am unable to ascertain. Even with salted or cured fish prices differ widely between different consignments.

The prize of prizes is bluefin tuna. Mr Haramo, a Japanese fish buyer based in Rotterdam, said that fresh bluefin (if supplies were scarce and if it came from cold waters, where fat content is higher), could sell for 20,000 yen (about £100) per kilo. His best tuna came from Spanish boats in May and June, and out of the American port of Boston between January and March. Frozen tuna is worth 3,000–5,000 yen per kilo, well under half. Bluefin are the largest of the mackerel family, famed as hard-fighting, fast-moving, schooling sportfish, accustomed to making migrations of thousands of miles in pursuit of shoals of foodfish, and they can weigh up to 1,800 pounds. A single bluefin could be worth $100,000 once in Tokyo. If the superlatives about bluefin have not been exhausted, it was reported in the fishing press that in 1994 the biggest super-seiners and longliners targeting bluefin tuna were grossing $500 million a year fishing round the American-owned Samoan Islands. No wonder bluefin are becoming so rare.

Tuna in the Tokyo market meet a strange fate. Being captured in distant oceans, most are frozen. Lying in rows on the floor, like so many mummies, they exhale a chill mist from refrigeration. They are similar-sized, numbered, whitened by freezing from their natural brilliant blue and grey, without tails. They are hauled into position by workers with long-handled hooks. Bidders are looking for subtle distinctions in the tuna, texture, colour, fat content, all playing a part. Despite the phenomenal prices paid, four hundred tuna are sold in an hour's hectic auctioneering.

What happens in the end to this most-sought of the great fishes? One possibility is the end-use described to me by Jon Reynir Magnusson. He was a guest at a tuna dinner in Japan in a hotel in 1973. Two actors in chain mail entered the private room armed with those legendary weapons of steelcraft perfection, five-foot-long tuna swords. They performed a slow and elaborate dance round the fish, which was in the centre of the room placed on a wooden board about a foot thick. Suddenly and without warning the two ritualists swivelled and began to chop the fish, which was some eight inches thick, into perfectly even slices, raising the swords high over their heads and bringing them whacking down with huge force and pin-point accuracy. Jon Reynir Magnusson described it as like a circus performance, brilliantly theatrical and demonstrating astonishing skills. The fish was then eaten raw. Such ritual could only derive from a culture of concentrated and ancient fish-eating.

Another delicacy in Japan is roe. The roe from many fish – herrings, Pacific salmon, sea urchins, even the roe of the small pelagic fish, capelin, mostly caught by Iceland, is sought. The detail into which Japanese buyers of capelin roe delve is typical of the precision developed by centuries of raw fish consumption. As with herring roe, the Japanese are only interested in roe at a certain stage of development. Too early or too late, and the roe is useless. They send buyers to purchase the capelin roe and supervise proceedings as it is boxed. The roe is then frozen in blocks and Japanese boats circle Iceland picking up their egregious commodity at different ports. Complying with all the details of the consumer prescriptions means that prices need to be high to make the whole performance worthwhile. It is a rare sort of business.

In 1992 I visited a Japanese fish processing factory in the midst of wilderness on the banks of the Alsek River at Dry Bay (so-named because the rain is incessant), Alaska. The presence of a processing plant in the midst of such desolate and wild country, with nothing attached to it but a track and a runway for aircraft, is bizarre. Inside the rudimentary building all the workers had been ferried in for the short autumnal salmon netting season. The mouth of the Alsek was netted by licensed Americans. The Japanese bought the often considerable catch. Fish were decapitated by a V-shaped slicing machine, and roe was extracted and handled separately. The roe was placed in wooden boxes, the salmon in bigger ones. Three times a day an old propellor plane landed, loaded, and flew the fish and roe to Juneau, further south on the Alaskan Panhandle. From Juneau the cargo of American salmon travelled immediately by jet to Japan. When we enquired about the prices of the roe and the salmon we were told the boxes of eggs were worth more than the far larger boxes of fish. In fact, despite the awesome logistics, the operation was lucrative. Again, the extraordinary Japanese appetite for fish had created an entire trade and processing operation in the most out-of-the-way spot. It happens in many places in the world, and now problems have arisen with American sportfishermen who are caught in the crossfire of the ethically-driven movement insisting on returning caught fish to the water, and the financial incentives to race big sportfish back to port to get them on a plane bound for Tokyo.

Many people question the effect of a nation such as Japan, which eats more than a tenth of the global fish catch (a third of world trade by value), on stocks of fish and on conservation of fish. Half of the Japanese enormous fish consumption (70 kilos per person per year) is imported. Known too for their dogged determination to continue whaling in defiance of much of world opinion, the Japanese are completely alone in their orientation to other states vis-à-vis fish. Japan's fish culture raises many issues. It is hard to deprecate a state whose citizens so appreciate the qualities of fish, and discern so minutely between one individual fish and another, and eat them raw, coming oneself from a culture in which a fish is only that, another fish. The Japanese, whose fishing boats are so immaculate, do not fit the folklore of tatty, marauding pirates. Furthermore, attitudes to fish have defined and developed a whole culture and style of resource management. The Japanese system of management cannot be ignored; indeed, many aspects of it can tutor the west, and are being looked at with rising interest. Its principles may be alien to westerners, but its lineage is much longer, its complexity greater, and it is integrated into national life in a way which western fisheries management most certainly is not.

Napoleon said 'Geography determines all.' Japan is an island system in a sea of fish, or habitat for fish. The first centralised Japanese state was established about 1,500 years ago and tributes were paid from the bounties of the sea. When Buddhism penetrated Japan in the sixth century it brought taboos against the consumption of four-legged animals. Since then fish has been the Japanese people's main source of protein. The study of fish grew alongside; surveys of its own fisheries had almost been completed by 1610. Eventually sardines and herrings were to be used as invaluable fertiliser to boost agriculture, and oil from whales was to become an insecticide. In 1691 the Matsumae domain forbade anyone to let off a gun within earshot of the sea during spawning time, lest the startled herrings failed to come close enough to the shore to be caught. Lighting bonfires was illegal for the same reasons. Japan's was from the beginning a fishing culture with a difference, perhaps the definitive fishing culture.

A local writer describes a Japanese scene between 1780 and 1800:

> 'Everyone in the Wajinchi fished. For three or four days at a time, several times each spring, local residents – whether they were fishers, farmers, merchants, or even samurai – abandoned all other work and heeded the call of 'Kuki' – Herring run! They were beckoned by the huge shoals of herring that travelled from the Sea of Okhotsk down the west coast of Hokkaido to spawn. In late March or early April the first shoals approached the shore, their arrival heralded by flocks of seagulls and gams of whales. During a large run the sea turned white and sticky from milt; the beach was littered with eggs, milt, and the bodies of fish washed up onto shore; and the water was so crowded with herring that a pole could almost stand unsupported. Anyone not too old, too young, or too sick joined in the business of hauling fish with nets, baskets, or bare hands . . .'

There was a complex array of taboos, rituals and religious performances connected with the fishery. Some present-day interpreters regard the purpose of a variety of the taboos – for instance on women going out fishing – as means to ensure pressure on the resource never became too great. An inordinate number of festival and therefore non-fishing days; taboos against individuals fishing who had recent association with birth or death; Japan's notoriously violent weather and the incidence of typhoons; and the obligation of fishermen to provide labour for the feudal overlord, drawing them away from fishing; all these combined to reduce the pressure on marine resources. Also Japan's seas are fecund in a variety of species; a single species was less likely to be targeted to perilously low levels when others were available which also commanded a good market. The Japanese never had the narrow herring/cod/haddock culture that, for example, Britain had. They ate almost all that swam, developed recipes even for the highly toxic blowfish, and consumed even the seaweed in which the fish sought haven.

As the debates about resource use and resource ownership have become more probing in the west, a school of anthropologists, social historians and ethnologists have been focusing attention on Japan, where management systems have evolved quite differently.

The area of particular interest is Japan's coastal fishery, responsible for only a quarter of all fish production, but a proportion worth about 41 per cent of the total because of the higher-value species which throng the coast. Today 80 per cent of fishermen still work in the inshore fishery. It was for the inshore fishery that in 1949, in a Japan shattered by war and under American occupation, a Fisheries Law was passed which finally transferred fishing rights in a formal document into the personal ownership of individual members of the Fisheries Co-operative Associations (FCAs). Sea was treated as an extension of land, equally divisible, equally accessible to management. The 1949 Act was the culmination of an attitude to marine resources which was deeply bedded in the history of the country from earliest tradition. Basing their hypothesis on the known fact that tribute was paid in marine commodities, and items such as sharkfin, sea cucumber and abalone were traded between Japan and Ming China, and that such systems required for their smooth operation a continuous and stable supply, the authors Kenneth Ruddle and Tomoya Akimichi hazard that sea tenure systems existed in Japan 2,000 years ago. However much earlier it originated, certainly sea tenure as a landlord concept was established sometime early in the Edo Period, or feudal era, which lasted from 1603–1867. The feudal overlord owned the sea as he did the land. Access to it was rented from him, in return for seafoods or in return for labour. The sea was a closed territory from about the time, as that noted authority on Japanese fisheries, Dr Arne Kalland, has observed, that European theoreticians were declaring that the sea should be open to all.

Japan was an organised, hierarchical, and also isolated society. From the mid-seventeenth century for over two hundred years Japan had almost no external contact; even Japanese sailors shipwrecked abroad had difficulty being re-admitted to their mother country. Self-regulation and self-reliance became almost an art form. Coastal villages were classified, either as farming villages, in which case the houses faced landward, or as fishing villages, where they faced the sea. The sea offlying a land territory formed an extension of the land and the right to exploit its resources, through dues paid to the landlord, belonged to the individuals in the community as a common right. How far the fishing territory extended depended on its productivity and the volume of pressure on it. The poorer sea areas were larger and stretched further, perhaps ten kilometres out to sea. Householders automatically had sea-harvest rights, whereas tenants had to pay a rental fee. Historians stress that circumstances and catches varied considerably between villages and the precise systems were crafted to suit particular circumstances. This organisational flexibility, and detailed compartmentalising of the fishery areas, persists still.

Within a shared fishing territory the use of a particular fishing method, or the harvest of a particular species, might be reserved for the inhabitants of a single village. Other species were available to neighbouring fishermen provided they belonged to the same FCA. Whilst simple fishing methods – hook and line, spearfishing, cutting seaweed or gathering shellfish – were open to all in the fishing village, small-scale netting was a category of its own, embracing gill-nets, seines and drive-in fixed nets (where fish were herded). These were licensed and subject to agreement by the village leadership. Heavier and more extractive technologies, capable of capturing tuna, whales and shoals of sardines, were treated separately. Because the large nets, usually beach seines, could only be afforded by better-off folk, commonly merchants, levies were charged by the feudal overlord for their deployment. One fixed principle was that despite the high capital risk of fishing with large nets, the local village should be a part beneficiary. Ten per cent of the catch was a typical dividend, split amongst the households in the village. These big nets needed more water both to work effectively, and to justify themselves, so sometimes special licences were issued, in tandem with foreshore rights for net-drying, on the basis that such rights, for that sort of net, were exclusive; and the fee was often set by the value of the catch. The keenest debates frequently centred on which fishing village owned which bay. Ownership of the beach was a vital handle on the resource. The typically pragmatic Japanese solution was often to make the bay a common territory for several villages. This system must have been one of the earliest examples of a move towards the privatisation of a fishing-ground. Built into it, however, were checks and balances to prevent the sprouting up of monopolists. Alongside this was an early awareness that the resource was finite, needed nurture, and should be protected by the identification of sanctuaries. As Dr Kalland has pointed out, this contrasted with the views of European theorists, like the sixteenth century Dutch jurist Hugo Grotius, who declared that fish were an inexhaustible commodity.

Licences were issued within a tight network of regulations to protect the underlying resource. Villages could swap their fishing rights with those of neighbours reciprocally, but the restrictions imposed by villages on their neighbours and on themselves were prodigious. Seasons, numbers of boats and nets were tightly controlled. Bottom gill-nets which harvest benthos, and mist-fine nets, were prohibited as too damaging. Night fishing with torches was only seasonally permitted, as was seaweed gathering for the reason that in the spawning season some fish's eggs adhere to seaweed. When the feudal period ended in 1868 the entire system was scrapped, and traditional fishing rights were declared null. There was a national upheaval, chaos quickly ensued, and in short order the tried and tested methods were reintroduced. Local fiefdoms took the place of the landlords, with the rights of sea tenure, and a system based on fishing villages and listed and regulated fishing tenants, retained.

Dr. Kalland makes an array of points in support of the sea tenure fishing system for inshore waters. Flexibility in the system reduced inter-village conflict; if fish disappeared from one fishing area they were accommodated, under certain terms, as guests next door. The co-existence of fishing sectors of different sizes led to efficient exploitation of the stocks. The key to the system's smooth working was flexibility. The small boat operators might work for the large netting operations as the seasons moved round. Different fishing technologies rotated round the year, affording the fisherman a permanent means of earning money. When all else failed there was the simple hunter-gathering, hook and lining, or spearfishing. This robust early fisheries management had, and still has, many of the characteristics fashionably espoused today in the western term 'wise use'. Domination by outside lawmakers was avoided. Most importantly, because only a small number of coastal villages were given their rights of sea tenure, in harness with the refined regulations, the resource was not swamped.

The first national fisheries law had been enacted in 1901. It consolidated the old system. The guilds of fishermen from fishing villages were made into Fisheries Associations, responsible as before for establishing fishing zones, ensuring restrictions on gear and fishing method were obeyed, and setting seasonal limits. In 1910 their constitution was amended to include co-operative marketing, but their main function of administering fishing in coastal waters remained, and is still in force today. Post-war Japanese theory, under American guidance, attempted to eradicate the last traces of feudalism and democratise the inshore fishery absolutely. Leasing fishing rights was abolished. A local fisherman's property right in seawater cannot be loaned, or transferred to others, nor be used as security for a mortgage. It is curious to consider the twin sustainability and limitation of the Japanese fisherman's sovereign fishing rights as against the value of the quotas for a fisherman in modern Iceland. Japan is the country breaking slowly out of the shell of a millennium of feudalism; Iceland is the modern social democracy without social hangovers, with a fishing policy forged by the dictates of the market-place. Where there is a gulf of difference is in the powers that have settled over time on the FCAs. For example, if an association wishes to block a coastal development, is uninterested in the compensation payments on offer (and these can be big), wishing rather to preserve the inalienable rights of its members, then even central government is helpless to overrule it. This is an unusual degree of local control in contemporary society, and one of the reasons the system has persisted for for so long is the social conformity of Japanese people. Community norms have a force unenvisageable in the west, so the possibilities of warring factions, which would be an everyday event in a similar structure in the west, is remote. Where there are irritations in the Japanese system they tend to be massaged away, at a level settled by locals.

Not only have Japanese coastal fishermen had their rights for a very long time. They still match economically the performance of other country-based workers such as farmers. Then there is one mouthwatering fact, sufficient to make western fishery managers stiffen with envy. Although Japan has slipped from its pre-eminent position in the 1980s as the world's largest fishing nation, it is the pelagic and distantwater fleet's catch that has failed to keep pace. The remarkable fact is that the catch of Japan's domestic fishery sector has remained stable since 1925, at two and a half to three million tons. This figure is roughly equivalent to the whole North Sea catch when it was being over-fished. Not only is it a very large figure on a sustained basis, but over this period the stocks of almost every other nation have gone through horrendous convulsions. Dr Kalland stresses that the explanations are highly complex, and only fully understandable in terms of Japanese culture. The principal reason, however, is co-operative management on a local basis. I asked him if it was the professionalism of the approach that had served Japanese fisheries so well. He used a different word: 'It is the pragmatism of their approach that is so important. They have turned a rigid system into a usable one. It may be technically rigid, in actuality it is feasible. And inshore fishermen are supported by prices which are especially high for very fresh local fish.' The powerful home market for fish must be a major contributory factor.

Many in the West question not only the science of the scientists, but their position in the structure. They have become, it is said, academic and detached, lost, in critics' eyes, in abstruse logarithms and predictions of stocks which are narrowly-based and expressed in abracadabra. In Japan scientists enter the scene from a different angle. The main scientific input comes from the state and the prefecture, or province. Scientists are spread through the structure, not superimposed on it. Some Japanese scientists work at the village level, others man the prestigious research academies (The most famous is the Tokyo University of Fisheries, over one hundred years old, with nearly two thousand fisheries science students, and a world leader in fisheries education and research). In parallel with state or prefecture scientists are those engaged by the fishing co-operatives. Generally not permanent, they are hired to address specific issues. Again, they can participate in the structure at any level, from state to village. Scientists have a strong though not decisive voice in framing quotas and drafting fishing licences, but the distribution of these is done at local, co-operative level. The scientists never supplant local fishing interests in the regulatory system. Neither do national politicians.

A culture so long acclimatised to regulating the take from a wild resource has, naturally, developed ways to effectively police the harvesting. The resolution of conflicts in fisheries is approached from the outset in an entirely Japanese way. Although a fishery police and a Coastguard do exist the vast majority of conflicts are resolved informally, at local level, after the issues have been turned over and examined from every angle. For this the West has no equivalent. Solving a fishery dispute on the quay at Peterhead by informal means would mean something entirely different!

In essence a fishery offence, be it gill-netting during the closed season, diving for coral without a licence, or using arsenic as a pesticide, is an offence against someone else's private property. There is no question of the sea being open to all (indeed, the small percentage of coastal dwellers actually within the fisheries co-operative associations is thought a reason for the resource having held up well), nor of anyone other than licensees having rights there.

The sense of community is also quite different. Community norms are considered important and are adhered to. The ultimate local sanction, social banishment, is described by Messrs Ruddle and Akimichi as 'real and horrifying'. They go on to say that the less proximate the body pronouncing judgement the less it means to the Japanese for whom local groupings are strongest. So the condemnation of a neighbouring community carries less weight, that of society as a whole less still, and that of foreigners is regarded as the subject of jokes.

Under the American Occupation fisheries regulations, normally so closely followed, and despite having been adjusted in a completely organic way, were frequently ignored. 'This fundamental characteristic' say Ruddle and Akimichi 'seems to have been well understood by fisheries administrators in Japan, who, in erecting the basic legislative framework, made sure to leave the details of implementation and administration firmly in the hands of the local community via the FCA.'

Woven into the formal strictures of the FCA are various eccentricities which say something about the Japanese sense of fair play and egalitarian values. For example, there operates a sacrosanct principle called first comer's rights. When the first day of the season commences at the best fishing-spots, fresh from being rested, catches may outperform others by three or four times. The first fisherman to work the spot is considered to 'own' it until he packs and goes. This working could be for several weeks with a fixed net, or for a few minutes if, for example, he is spearing cuttlefish. If a conflict does occur community pressure is brought to bear on the offender, which is almost invariably sufficient. First comer rights are enshrined in the formal regulations, meaning that as a last resort the Coastguard is called.

Fishing spots regarded as being personal to one individual, maybe because he discovered them, are also respected. A fisherman may have a low-tide and a high-tide spot during the spawning season. Fishermen as a rule are recognised as having fishing-spots near where they first learnt to fish. The prized fishing spots, often amongst dense coral or in seagrass beds, have faster replenishment rates and can be visited frequently. Along with the satisfaction of bringing home a good catch a fisherman who has secured such a good spot also gains kudos in the community. He may even enter the revered class of master fishermen. Tremendous lengths are gone to by fishermen with hot fishing spots to conceal their location from others, appearing to be focusing their energy elsewhere if anyone else approaches, and practising all the devious and understandable deceptions that are common to all classes and sizes of fishing boat, in all countries, to keep private a cherished and personalised source of limitless dividends.

Maximising sea dividends is something the Japanese have been familiar with for centuries. They started to enhance fisheries by building what westerners call 'artificial reefs', and what the Japanese term 'fish apartments', over 150 years ago. These were in the early days just agglomerations of dumped material on a promising part of the seabed which acted as a honeycomb for rapid colonisation by organisms which, in time, provided the basis for a whole fish ecosystem. Crustaceans and sea urchins were the first colonists, which brought in small fish, which attracted big fish. Originally junk metal and old cars were used;

latterly specially engineered tetrablocks, replete with suitable-sized cavities, were deployed. Once more the social discipline surrounding Japanese fisheries ensured that the primary obstacle in the west to artificial reefs – that they are ransacked by all and sundry, before the fish community has time fully to develop – was avoided. The FCAs also improve fishing-grounds, for example by creating artificial shallows for fish at spawning.

Japanese management of the inshore fishery is, by any measure, extremely thought-provoking. Here is that often spoken-of ideal, local management in action. Even though the prefecture and state are present at FCA meetings, there is an overall preponderance of fishermen present. In the EU fishermen go along to hear what politicians and scientists have to tell them whereas in Japan the FCA comes to a decision itself, noting the contributions of others. The administration of fisheries is therefore good – except in cases where fishermen do not see themselves as beneficiaries of management decisions, for example, with some flatfish, which are only within the FCA grounds seasonally – and the inshore fisheries have a solid, accepted, time-tested, legal standing. Management is strict, and fishermen who have violated the regulations repeatedly have their licences confiscated (a contrast to the CFP where multiple offenders continue to fish quite unhindered). The regulations devised by FCAs are often extremely detailed and complex, small compartments offshore being allocated for different and precise individual uses, but because the fishermen expected to follow the regulations know their own fishing-grounds well, and because Japanese people will work indefatigably to achieve consensus, and incorporate into strategy even the smallest, most personalised request, the precision and detail assayed by management is understood and respected.

It is said that within the FCA meetings, although there are occasional emotional outbursts, and sub-meetings are convened to soothe ruffled feathers, the usual conclusion is a communal collation accompanied by ample rice wine, celebrating a moderately acceptable compromise. With so much depending on this cultural abhorrence of schism, and rejection of autocratic top-down instruction, it is hard to see fishermen's associations working in the Japanese way elsewhere. In no other country are inshore fisheries in the same place in the social structure. Centuries ago some European countries had systems of local self-governance which might be comparable, but that time is long past. In Japan the old sea tenure concept was adhered to, and it has served the country well. Worth mentioning in light of the contemporary debate on fishing systems is a point which idealists tempted to leap on the Japanese example should note: the limited entry within sea tenure was not primarily for conservation purposes. Aimed-for sea tenure benefits were the maintenance of public order, the survival of fishing and the provision of food, and the supply of reciprocal labour.

Some feel the end is now nigh for this ancient system. A critical, possibly a decisive, factor is that the perception of being a fisherman has changed. Fishermen were always of low social stature, but the worldly, career-orientated young Japanese now expect much more. It is industry, and manufacturing, that has made Japan a world power in the late twentieth century, not fishing. The number of active fishermen, around 400,000, is decreasing. Young male fishermen find it hard to pair off and many are single. Seventy per cent of Japanese fishermen are over forty-nine years old, a lopsidedness which detracts from efficiency and speed on fishing boats. Dr. Kalland described fishing to me as 'a dying occupation'. 'In ten or fifteen years' he went on, 'some areas may have no fishermen left.' He thought the financial insecurity of fishing and lack of any financial safety-net in the event of injury were damaging. For deepwater fishermen at least there is no need to reside on the local shoreline; they are more mobile and can live in Tokyo. It is now prestigious to work for a big corporation and smart, when going out, to eat meat rather than fish.

This is not a sudden unfashionableness. Ever since the Second World War forces have been at work militating against the old system's survival. Post-war Japan needed to industrialise, and to do it fast. The coastal belt suffered very rapid change. Spawning grounds and tidal feeding grounds for fish were often damaged by land reclamation, digging out shipping lanes and building harbours. The offshore seabed was affected by effluent from pulp mills and heavy industries. Oil-ship accidents became numerous, with consequent inshore pollution. In one area, Matsushima, oyster culture was completely destroyed for a year by effluents from nearby processing plants. Mutations appeared in some fish. Although catches remained stable, or rose, the pollution-resistant species such as clams, anchovies, sand lance, flatfish, cuttlefish, sea cucumbers, and certain seaweeds, did well at the expense of the more sensitive types such as lobsters, congers and red sea bream.

Japan had no comprehensive legislation on environmental impacts. Fishing associations fighting developments therefore had no set of rules against which to judge the plans of developers. Their exact property rights came under scrutiny in decade after decade of legal fights between developers and local action groups. 'Access to a natural environment' laid down the Japanese Bar Association, refining the legal status of fishermen, 'must be completely separated from the ownership of the real estate.' So fishermen owned the right to take the catch; they did not own the sea itself. One of the beneficial consequences of this torrid period was the development of strong anti-pollution law. Various marine localities have climbed back from environmental desolation. Another benefit was a change of view in many fishing communities. Development took them out of the backwater, into the mainstream. For many maritime Japanese, caught between variable fish catches and fluctuating prices, and regular, more sociable, industrial work, made more tolerable by sweeteners proffered by large corporations for loss of fishing opportunities, was a step up. Not every fishing community succumbed to the blandishments of the men in blue suits, but many did. Some heroic defence actions were fought. Two thirds of the FCA members had to approve development projects if they were to proceed. Again, the situation is uniquely Japanese. The fishing-grounds were being fought for, not by environmentalists, nor pressure groups, nor by the bulk of locals: they were being fought for by the fishermen for whom they had so long provided a living.

Meanwhile, Japan did not wait and hang on industry's progress. Fishing itself was turned into a huge industry, but not fishing the coast. That might have overstretched the sacred national stocks. The gigantic fishing fleet that the Japanese built up after the War fished in other people's waters. Eventually Japan was fishing the waters of over fifty countries, and its catch became the largest of any. In fisheries the name of Japan became associated with an aggressive, government-driven, expansionary policy based on fishery access agreements structured around joint ventures, training arrangements with host states, and infrastructure assistance. Particularly regarding the over-exploitation of free-ranging oceanic tuna, Japan came to be seen as a country associated with high-tech fish-catching effort, willing and able to go anywhere and uncomplicated by restraint. Japanese fishing operations were global, fleets from the Land of The Rising Sun swiftly coursing the oceans from off south-west Africa to Greenland, and from the fog-wreathed Aleutian Islands off Alaska to the Caribbean, and south to the squid grounds at the other end of the world round the Falklands. Above all, the Japanese concentrated on getting the maximum penetration around its own islands, and for over a hundred years protracted territorial disputes with Russia and China. So great was Japanese enmeshment in the Russian fisheries that the fish salting and canning plants on Russian territory in Kamchatka and northern Sakhalin became 'little Japans', precursors of the processing plant I saw in Dry Bay, Alaska. Between 1908 and the Russian Revolution in 1917, in the space of just nine years, Japan had multiplied its catch in the north-west Pacific by an astonishing five times. In the 1930s Japanese vessels penetrated the home seas of Taiwan, Korea, and Manchuria.

The seven years of Occupation after the Second World War was an interesting period in Japanese fisheries. Aided and encouraged by America, which wanted to avoid shipping food to Japan from its own reserves, Japan built up and clawed back the fishing abilities which had been destroyed in wartime. Ships, docks, processing plant, and ship repair yards, both on mainland Japan and also on the coasts of all its neighbours, had been lost in war. The American Occupation officials calculated that, assuming the average Japanese person needed 30 kilos of fish per year, 2.2 million tons edible weight of fish were required to sustain the vanquished state. The Japanese, with American aid, rebuilt their fishery with alacrity. Their catch rose from 2 million tons in 1946 to 2.5 million tons in 1948. The Americans worked closely with Japanese fishery scientists. Conservation measures were accepted as a *quid pro quo* for the enlargement of the fishing area. When they saw the extreme democratic workings of the FCAs, American fishery administrators approved. By the time the American Occupation ended Japan had regained its fishing position with extraordinary speed: in 1952 with a catch of 4.8 million tons, and 19 per cent of the world catch, Japan was once more the leading fishing nation.

Fears about Japan's excessive fishing zeal were re-awakened. America and Canada were especially concerned about the effect of Japanese drift-netters on the very large, high-value, salmon fishery off Alaska, America's most significant single fishery. When Korea announced a 60-mile fishing limit over its fertile fishing-grounds in 1952 it took the capture of 146 Japanese fishing boats to persuade Japan that the declaration of the territorial limit was of serious intent! With Korea, China and the USSR pushing outward with their own fisheries, Japan adopted the policy for which her name stands today – as the principal fishing nation in the big oceans beyond all territorial limits – although not beyond all disputes. Because of the fraught history of relationships with her neighbours over fishing, Japan to this day has no agreed EEZ. Still in dispute with Korea over the Takeshima Islands (Tok-do in Korean), with China over the Senkaku Islands (Diaoyu in Chinese), and with Russia over the Kurile Islands in the north, Japan's partial fisheries zone lacks agreed demarcation lines in the East China Sea and the Western Sea of Japan. Neither inshore grounds nor oceanic grounds, these waters are termed offshore fishing-grounds and, unsurprisingly, given the confused state of regulations, they are severely over-fished. With oceanic catches falling and the inshore fishery static it is logical that, in a rich country, imports would escalate mightily. So they have, which has produced the phenomenon of Tsukiji, presenting under its low ceilings a cornucopia of the seas riches, caught from anywhere between the north and south poles.

To talk of Japan only in terms of its inshore fisheries is akin to talking of whaling in terms of scrimshaw. The inshore fishery is interesting for its antiquity, complexity, and the fact that in management terms, it works. However, it is little known. Japan, rather, is the name the world's fish fear. It is the country that has caught more fish in the twentieth century than any other, in more places, with the keenest and most dynamic, some would say the most unscrupulous, pugnacity. The Japanese state has looked at the world's oceans as a whole and set about harvesting them with a single-mindedness which was unprecedented. Space prohibits examining the effects of this everywhere Japanese fleets have pulled their nets and reeled in their lines, but the North Pacific is a venue where interface with the Japanese fleet has been experienced by nations equally strong and stronger, and where a half century of negotiation and wrangling within an all-embracing organisation has ended in a retreat by Japan, under treaty, from its distantwater fishery, and a new, if tenuous, equilibrium.

The North Pacific became known through the middle of the twentieth century as the locus of a rare thing: the halibut fishery there was a classic case of successful international management by strict conservation measures. Japanese penetration of the North Pacific therefore had been preceded by a case history which established the credibility of fishery scientists and fishery management where two sovereign states agreed to act together to exploit valued stocks sustainably. Briefly, the hunt for halibut, a large, toothsome, high-value, bottom-dwelling fish which could be caught year-round, began on an industrial scale in 1888. The fishery was prosecuted off British Columbia and southeastern Alaska by Canada and the USA. It peaked around 1912 at only a modest catch level of 30,000 tons, then fell, which stimulated concern for its future. The first Canadian/American convention agreeing jointly to manage the halibut was ratified in 1923.

Herring curing yards in the 1880s at Shaltigoe, Wick, Scotland. *(Courtesy of the Wick Society)*

Wick Harbour around 1900. On the right are 'Fifies', which sailed with a crew of seven, drying their sails at a weekend from their 38 foot masts; to the left are shops and curers' offices, part of Thomas Telford's great design. *(Courtesy of the Wick Society)*

This pelagic trawler, *Research*, was built in Norway at a cost of £9 million, for a partnership of eight fishermen from the small Shetland island of Whalsay which, because fishing is the only industry, is the most fishing-dependent community in the UK. *(James Nicolson)*

RIGHT: This 1993 shot by Phil Lockley shows 'Chunky' disentangling pollock and ling from a five-inch mesh wreck-net whilst aboard the 50 foot *Britannia 5*, skippered by Freddie Turner, registered in Portloe, Cornwall, UK. *(Phil Lockley)*

Fishing boat in heavy weather off Iceland's Vestmannaeyjar Islands. *(Sigurgeir Jónasson)*

Sorting fish for freezing aboard a Grimsby-based deep sea trawler in the Barents Sea in 1969. *(Dr Richard McCormick)*

LEFT: Redfish being pulled up the ramp of an Icelandic trawler fishing the Reykjanes Ridge, south of Iceland. *(Morgunbladid)*

This 'shot' (haul) of herring was of 300–400 tons and called 'quite small'. Quentin Bates, the photographer and formerly a fisherman, describes this sort of purse seining for herring as 'very much hunting, sneaking up on a shoal and picking the right moment to shoot the net'; if unsuccessful and the fish dive, not a single fish is captured. *(Quentin Bates)*

Atlanto-Scandian herring destined for conversion into fish-meal, caught in international waters off Iceland in 1997, are being pumped into the hold of the top-catching Icelandic purse seiner *Sigurdur*. *(Quentin Bates)*

The famously tough Captain George Gilley of the Bonin Islands, photographed in 1887 when he had to fend off Cape Prince of Wales natives from his trading brig the *William H. Allen* whilst in the Bering Strait, Thirteen Eskimos died in the fight.

(Old Dartmouth Historical Soc. – New Bedford Whaling Museum)

Bundles of whalebone in New Bedford, Massachusetts, around 1860.

(Old Dartmouth Historical Soc. – New Bedford Whaling Museum)

Opposite: Whalebone around 1883 stacked in the San Francisco yard of the Arctic Oil Works, the first company to catch and process whales from the same base. *(Old Dartmouth Historical Soc. – New Bedford Whaling Museum)*

This photo portrait taken by Foto Pico is of an Azorean whaler named Manuel Pereira de Lemos. *(Foto Pico)*

American photographer William Neufeld's 1958 shot of an Azorean fisherman 'throwing the lance' from the open rowing-boat, hunting for sperm whales as these mid-Atlantic islanders preferred, 'face-to-face'. *(William Neufeld)*

The Azorean light-tackle whale fishery in action. *(William Neufeld)*

Tim the ship's cook, in order to get help with the washing-up, is sorting a catch of monkfish, John Dories, megrims and sole aboard the Newlyn-based beam trawler *Silver Harvester* when fishing England's Western approaches. The catch will go to Spain or France, and the crew of six will be at sea five to seven days. *(Quentin Bates)*

The reason that British saltwater angler Steven Collis's knees are bending is that this blue trevally, caught off Christmas Island in the Pacific, is a rod-landed record. Though Steven believes firmly in catch and release, 'The guide was determined to keep this world record, and the fish was duly despatched before I could speak'.

Barracuda about two feet long are schooling, which is usually a defence mechanism or for reproduction. Photograph by Jeff Rotman in warm waters in Walindi Bay off New Guinea. *(Jeff Rotman)*

This 600–700 pound bluefin tuna is meeting its end on Sicilian gaffs at the 'Mattanza' (slaughter), an ancient fishery in which migrating tuna are snared in a strategically-positioned maze of nets in the Mediterranean. *(Jeff Rottman)*

These Atlantic salmon have moved 35 kilometres up the Dartmouth river on the Gaspé Peninsula in eastern Canada and were photographed in autumn, about six to seven weeks prior to spawning. *(Gilbert von Ryckevorsel)*

Fishery conventions between two countries were unusual in the 1920s, but uniquely in this case trust was placed in the hands of an independent research body. This, the International Fisheries Commission, gradually won its laurels. Vessels had to be licensed but their numbers could not be limited; fishing seasons, closed-off areas, gear specifications, and quotas well below historic catches, were introduced. Until the 1950s only sailboats were permitted; no engine power was allowed in the fishery at all. The fishery had been preserved, by edict, in the style of the 1900s. It was an uncommon case. Fishermen themselves were consulted and asked for their views. Improving catches, added to familiarisation through dialogue, helped scientists gain the confidence of fishermen. Within a few years quotas were being caught in five months instead of the nine it took previously. The stock was responding well to good governance. The scene onto which the Japanese arrived in the 1950s was remarkable for one other thing. Sharply in contrast to its own form of regulation in the inshore fishery, in the halibut commission's fishing area there was no control of fishing effort. Stemming from the Canadian and American feeling that everyone had a right to harvest the open sea, all attempts to limit fishing effort had been resisted. This led, inevitably as the market value of the halibut increased, to a greater number of boats fishing more frantically in a shorter and shorter period. Eventually the season became restricted to a matter of days, or even hours. Suddenly empty seawater would be covered in an armada of fishing boats, which would then vanish altogether. The consequences of market gluts, excessive capitalisation for boats (ultimately about 4,000) which were mostly idle, and prolonged unemployment for the fishermen involved, all features in due course of the North Pacific salmon fishery, were accepted. The cardinal *sine qua non* remained in place: anyone who bought a licence could fish.

The postscript is that the halibut stocks are still in good shape. Maybe inevitably, open access has been scuppered. Management by quota, stretched year-round, has replaced the frenetic and dangerous 'derby' days. Fresh halibut is available in the stores. But that is another story.

The involvement of the Japanese in management started to be discussed after the Second World War. As described, it was an aim of the Marshall Plan to resurrect Japan and restore some degree of seafood sufficiency. Pre-war Japanese fishing pressure around the American territorial sea limit, then three miles, had aroused concern in both Canada and America. However, the success of management in the halibut fishery, allied to more general post-war political and trade needs, fuelled the desire for a convention in the North Pacific. Peace in fisheries was to be part of a wider peace. The North Pacific was one of the world's great fishing zones possessing, mainly domiciled on the North American side, the world's biggest salmon fishery, and also containing very large stocks of herring, pollack and crab. Additionally, particularly for North America, a large number of those in employment were connected to the fishery.

On the other side the Japanese had much greater fishing power. Already by the time the North Pacific convention was ratified in 1953, with almost miraculous speed, Japan's world-beating fishing power had been restored. America and Canada had traditionally used the north-east Pacific fishery at a low-tech level. The sail-boat halibut fishery was emblematic. The salmon were fished in small boats as they came close to shore, preparatory to swimming up their natal rivers. The Japanese fishing fleet was the opposite, potent and ultra-modern. The distantwater type of fishery which Japan was starting to deploy, and for which she was to become well-known, was what was called a 'mothership' operation. The mothership was generally a large vessel converted from other uses into a fish-processing factory and supply base for a fleet of catcher boats. At first the 30-plus catcher boats were around 90 feet long manned by a crew of 21 men; as time passed they grew in size and power. Some acted as scouts for the rest. Eventually they were each to set 15 kilometres of floating gill-nets in the upper surface layer each day. The fish-extracting capacity of these gill-net fleets became awesome. It has been calculated that when the fleet in its heyday had some 700 catcher boats attached to different motherships, they were retrieving about 10,000 kilometres of net daily. The British solo mariner, Toby Bromley, told an audience at London's Royal Geographic Society that he had sailed past 430 miles of consecutive Japanese gill-nets off the fog-bound Aleutian chain. It was interesting to note that although this vast operation was taking place in American waters, when he informed the US Coastguard, they had no idea any fishing was taking place at all. In addition to the mothership operation the Japanese also had a fleet which unloaded at various ports in northern Japan. At its peak this consisted of 200 vessels. Veritably these Japanese fishing boats were sieving the oceans, or one ocean.

It may seem strange that a small nation, on the losing side in a world war, competing in the same waters with two of the world's big industrial nations, both victorious in war, should have so overshadowed them technologically. That is to forget that to the western USA and west Canada fisheries were only of local importance. They did not have a vast market for fish, and their outstanding food export was grain. They had no particularly strong tradition or history in the fishery; indeed, Alaska, where the salmon mostly homes, had until 1867 belonged to Russia. The tradition, as far as it went, for salmon and halibut, was that they had been fished by native peoples. Conversely, conservation of the Alaskan fisheries had widespread public support in America; it had become almost a symbol of American assertion of proprietary national resource rights. This was the background history to the formation of the International Pacific Fisheries Commission (INPFC).

The INPFC is interesting not only because it demonstrates Japan deploying opposite attitudes to fish to those enshrined in its inshore fisheries philosophy, but also in its often tortuous and tortured trajectory it has demonstrated and unveiled many of the features peculiar to an attempted share-out of a wild resource, a resource whose mobile constituents respect no international maritime boundaries but only the dictates of instinct. This instinct is moved by the need for food, their responsiveness to the changing characteristics of the ocean, and the need to procreate. The INPFC has been described, tentatively, by some of those whose hard work and expertise held it together, as a success.

Conversely an Atlantic salmon fisherman who participated in the Bristol Bay fishery in Alaska in the 1960s and 1970s told me he thought for Japanese vessels to catch half-grown salmon on the high seas was a crime, and that no one in the fishery with him felt any differently. The legislator view and the view of those subject to legislation is, maybe inevitably, different; the legislators had in mind wider political objectives with Japan. Tracking the INPFC, however, produces a crop of revelations.

First, the institution. It was composed of representatives from America, Canada and Japan. Russia, the fourth state involved in the North Pacific, remained outside. It was distinguished from the start by the high quality of its committee members, and their long service in the Commission is a form of testimony itself. Many had big reputations when they joined. During the prolonged, often heated dialogues, friendships were forged and understandings reached. Elmer Rasmuson, a native Alaskan who served as chairman of the Commission in five separate years and became famed as a peacemaker and catalyst, wrote of it: '. . . much is to be learned from the record of tempering of self-interest and political power applied to the ocean fisheries through scientific research, international debate, and decisions within mutual interests.' Lastly, the Commission's scientists made strides in the advancement of knowledge on fish of the North Pacific, particularly salmon. Their publications were highly reputed. The picture which materialised of migration routes for the five species of Pacific salmon was one of great complexity and quite different from what had hitherto been assumed, with hundreds of genetically-different stocks each following their own migratory routes criss-crossing those of others, and it served to underline that in fisheries the ownership of the resource is often a tangle; that incidental by-catches are in some situations unavoidable; that the ocean is a dynamic theatre in which geographically-determined resolutions are due to get short shrift; and that fisheries management, when away from the coasts, is seldom easy.

Japan needed new fishing-grounds after the Second World War because the salmon run off Kamchatka on the far eastern tip of Soviet Russia, the arena of eternal Japanese/Soviet wrangling, started to fail. The big Japanese salmon fishing companies pressed their government to research fishing possibilities off the Alaskan coast, particularly in Bristol Bay, the huge, wide-armed bay into which numbers of migrating salmon ran into the big rivers in almost biblical quantities. In some years scientists reckoned more than sixty million salmon assembled in the bay prior to running home rivers. When Japanese investigations commenced, Americans shuddered. The territorial limit, as said, was three miles. Japan's record of plunder preceded her. Before the World War reached the Pacific in 1941, concerns about the fish resource in the Bering Sea, which divided Alaska and eastern Russia in the far north, were well stirred.

From the time of the 1953 International Convention for the High Seas Fisheries off the North Pacific Ocean, Japan was seeking new fishing-grounds. The USSR wanted to push Japanese fishing effort eastwards to America, and the USA wanted to push it west. As negotiations about the wording of the convention got underway Japan's open ocean fishing effort accelerated. In 1958 a fleet of sixteen motherships fed by 460 catcher vessels, a truly awesome fishing force, began netting in the western Pacific and the western Bering Sea.

The first stumbling-block proved to be the contentious principle, which INPFC members had finally signed up to, of 'abstention'. This was the idea that a stock was of such importance, at certain vulnerable stages, that it must be the subject of total protection. It was designed to protect coastal fishermen from the ravages on stocks which could be wrought by distantwater fleets. Japanese scientists argued with American and Canadian scientists over whether any stocks deserved 'abstention' status. The argument settled on salmon, to which the principle of abstention was especially applicable because of their anadromous lifestyle, and attempts to define the word 'stock'. Halibut management in the same area had been comparatively simple. Different halibut populations remained physically separate, and the fish was long-lived, tending to stabilise the population. Salmon, so the Canadian scientists were discovering, were mixed in the North Pacific to such an extent that a single net could haul fish of three or four different stocks. Salmon populations were strongly cyclical. How then could you define stock health? One fact, energetically deployed by the Americans and the Canadians, was clear: numbers of salmon were way below their peak. In the 1930s catches were around 770,000 tons, in the 1950s the figure was 400,000–500,000 tons. However, the Japanese, no newcomers to fishery science, exploited the holes in the presentations, saying the graphs inadequately portrayed the true health of stocks.

In years when American scientists had underestimated the size of returning runs the Japanese pressed home their discomfiture. The resource was being wasted, they argued. The 1965 sockeye salmon run in Bristol Bay was of record size and far bigger than the American industry could properly harvest. The concept of wasting an annual fish harvest was more applicable to Pacific salmon, which all die after spawning, than to Atlantic salmon, some ten per cent of which survive to spawn again. It was already understood that runs of spawning salmon could only optimise their freshwater habitats to a certain point. In any river there was a finite amount of habitat to support young salmon. The principle that you could cream off salmon surplus to restocking needs was accepted. In this way the Japanese justified their claim to harvest the sockeye in Bristol Bay, even when the fish were immature. Many Americans thought to harvest mixed stocks in the sea was a nonsense and that salmon should be taken when in rivers, so that escapement of an exact proportion of the run could be achieved. However, Japanese delegates in the Commission were being pushed from behind, with politicians of parties not in government calling for complete freedom of access for Japanese fishermen anywhere on the high seas. Consensus, given these seemingly mutually-exclusive arguments, would prove a difficult trick.

The Japanese used a remarkable logic. They said it was unfair of America to monopolise fish stocks outside the three-mile limit which should belong to everyone. They claimed that only multinational management adequately protected stocks. The fact that their own inshore fishery zones often went out to sea, with their detailed management prescriptions, much further than this, exposed a hypocrisy not uncommon in fishery disputes. And, of course, the Japanese cherished monopolistic management in offshore waters too. On the other hand, the Japanese were scrupulously careful to control their catches of sockeye outside Bristol Bay with minute precision. Observers from other countries were allowed on Japanese vessels. Between 1965 and 1970, when the runs fluctuated wildly between nine million and sixty million salmon, Japanese exploitation percentages only moved in the narrow band of 7.8 to 11.5. Their fishing expertise and discipline were noted with a mixture of admiration and trepidation.

A new set of developments, occurring worldwide amongst coastal states, overtook events in the 1970s. At a meeting in Caracas in 1974 most coastal states agreed to finalise the process of nations edging their jurisdictions seawards by establishing 12-mile territorial limits and a 200-mile fishery conservation zone. Nations with good fishing to the 200-mile limit could now control access. Distantwater fleets would require permission to fish within these zones. Only the high seas, in the middle of big oceans, remained all and anyone's territory – for the time being. As 95 per cent of the world's fish stocks swim within 200 miles of land, the open ocean fleets were being pushed into a tighter and tighter corner.

The INPFC's Convention was amended in 1979. A whole new set of ground rules had come into play. Bilateral agreements with nations wanting to fish in another nation's home-waters could now be negotiated. America gained protection of the Alaskan salmon, but not completely: INPFC research had shown they mingled with Asian-homing salmon inside Russia's new 200-mile zone. Japan's reaction to the new order in fishing was typical: she rapidly expanded her salmon-ranching programme which would boost a fishery mostly contained in home-waters, and she redoubled her efforts at the negotiating-table in search of bilateral agreements. These resulted in generous access to America's huge stocks of groundfish – cod, pollack, sablefish etc. – in return for restraint on salmon.

The actions of Japanese diplomacy in the INPFC shows several things. The fact that a state has great fish-catching power, and the skills with which to deploy it, is argument in itself. The world's biggest fisherman could not be ignored. Japan's formidable manufacturing output and the need for other countries to import her technological goods ensured her position at the INPFC's negotiating table even though, it might have been argued, they had few fishing-only reasons to be there. In the eastern North Pacific, for example, she had no long-established right-of-use. It was in the western North Pacific that her huge fish catch had largely been accumulated. Japan's presence, in effect, was justified by her neediness, and the absence of any international law excluding her.

The history of the Commission shows too that participation in a convention at which other powerful nations were present required acknowledging the principle of fish conservation. For the Japanese fishery management by stock conservation was an old-established concept, *vide* the inshore fishery. Conservation of stocks which, after 1977, were established as belonging to other people, was a new thing. It would never have become so rooted had not the INPFC's scientists produced high-class research which attracted universal recognition. From the start, science was the premise for action, an order of play which fishery managers in other parts of the world noted. Furthermore, the scientific evaluation of stocks was used by businessmen, several of whom had positions in the Commission, as the basis for operational and market planning. The three member states were, by value, among the world's largest traders of fish. It became accepted that science had a critical rôle to play in fishery development, from all points of view.

Meanwhile Japanese inroads into the North Pacific fisheries had been halted and turned back on the other front, the west. In 1956 Japan had co-signed the Japan–Soviet Fisheries Commission (JSFC). Although the JSFC set rules on herring and king crab, the principal subject was salmon. From the start the Soviets had insisted, and had been in the dominating position to insist, that the Commission would set catch quotas for the Japanese whilst the Soviets would set their own. The principle of Soviet-origin salmon being regulated mainly by the home state had been implicit.

A new phase opened in 1977. The Japanese agreed, in return for their fishing inside the Soviet EEZ, to pay a 'fisheries co-operation fee', or user fee. The user fee was sometimes in the form of technical assistance, for example, helping the Russians to build hatcheries. The Soviets had a fixed aim. It came out in the open in 1988: they said fishing for Russian salmon outside their EEZ must cease after 1992. During the 1980s America too, on the eastern seaboard, had been demanding reductions in Japan's high seas salmon fishery. The trend against Japanese participation was inescapable; the doors were closing fast. In 1990 the once-feared mothership fleet caught only one per cent of all Japan's salmon landings for the year. Canada, the USA, Russia, and Japan became signatories to a convention under the North Pacific Anadromous Fish Commission (NPAFC) in 1992/93. This declared a total ban on high seas salmon fishing. It heralded the end of an era for Japan.

The convention's turning-point document is worded with great care. It bestowed ultimate authority for migratory fish unequivocally on home states, where the fish spawned. Considerations for any other user states – in practice, Japan – did exist, but they were so framed as to disempower the odd one out. States with no history of participation in the fishery were simply denied entry. These types of prescriptive statements were way beyond anything that had been attempted under the old JSFC and INPFC. One of the factors which had toughened attitudes in the USA was that the low catches of salmon in the 1980s were being linked, suppositionally not scientifically, to the high seas drift-net fisheries started up by Japan, South Korea and Taiwan in the early 1980s. This type of fishery had achieved notoriety in the media under the catchphrase 'walls of death', characterised as mangling and drowning vast swathes of

166

marine life, both fish and mammal. A different high seas ban on drift-nets, also powerfully affecting Japan, this time by the UN, of which more later, took place in 1992 too. Raw nerves had been scraped in 1990 when a fleet of North Korean-flagged ships which had been illegally catching Soviet salmon was apprehended by the Americans. Further enquiry revealed that the boats were crewed and owned by their convention associates, the Japanese.

The new convention, although salmon-focused, referred too late to species 'ecologically-related' to salmonids. All along the focus on salmon only, as if they lived independently of other marine life, was likely to prove the weakness of American and Russian efforts to protect their most valuable North Pacific fish. Subsequent American and Canadian research tracking back a hundred years shows that atmospheric pressure fluctuations, related to the atmospheric influence of the far-flung Aleutian Islands, coincides closely with sea surface temperature movements in the North Pacific, and in turn with salmon abundance. Weather cycles were affecting the young salmon food supply to a critical degree. Scientists showed how low pressure periods in the Aleutian Islands corresponded historically with the abundance of copepods, the dominant form of zooplankton, which in turn stimulated production of all marine species. Salmon-only views risk looking simplistic and not working. Scientists advocate strategies for larger aggregations of stocks as well as for the inclusion of industrial stocks.

The 1993 Convention moved from the science-orientated thrust of the INPFC towards a drive for enforcement. High seas catching of Pacific salmon was to be targeted even if it was incidental. Furthermore, innocence had to be proved by the accused, rather than vice versa. The new convention gave itself authority which has been described as 'quasi-judicial'. It had been possible because in the international community there was nobody who was going to gainsay these powerful states. Canada had an external motive to approve the new reach for wider powers, stretching further out. Canada realised it would not be long before it had a fight on its hands on the other coast, on the Grand Banks. The productive part of the Banks crept out further than the 200-mile EEZ, and was being heavily overfished.

Enforcement in the North Pacific was rigorous. Fishermen there have told me that no one gave a thought to breaking the rules. Many boats had observers on board to ensure fair play. Each state was allowed to do as it thought fit with its own surveillance vessels. Interestingly, reported vessels sightings included all drift-netters. However, the UN moratorium on high seas drift-netting was voluntary. Two points arise: the convention's members saw fit to watch all high seas drift-netters; secondly, it was assumed they were, intentionally or otherwise, catching salmon.

The future, for the convention, may face the problem of stock enhancement programmes. Japan has responded to the decline of its distantwater operations – fifty per cent of the fleet was decommissioned between 1991 and 1993 – by boosting farmed salmon production. Its vast chum salmon hatchery programmes started as far back as 1887; as usual Japan's thinking had shown an impressive head start on everyone else.

Most of these salmon hatcheries ranched salmon, releasing them to sea. By 1892 Japan was releasing sixteen million fish, a remarkably advanced achievement for the time. A century later the figure for chum salmon alone had attained a phenomenal two billion. It gives pause for thought. The scale of the operation puts anything else into the shade. None of this progress had been made with the benefits of scientific or cultural cross-fertilisation; it had been a solo effort. Things the Japanese had been doing in the nineteenth century were appearing in magazine articles and spoken of in wonderment in the West a century afterwards.

The question arises of the carrying capacity of the North Pacific as more and more hatchery-reared fish are released into it. All those fish are relying on the same ocean to sustain their growth. What worries scientists is that sizes of adult salmon have dropped significantly. So, are there too many? The Japanese have seen the carrying capacity question coming, and are nervous that someone might recommend the curbing of their hatchery production to make way for Russian, Canadian and Alaskan hatchery fish. Speaking of the future in the North Pacific, Yasuhiro Ueno, of Japan's National Research Institute of the Far Seas Fisheries, wrote: 'It is obvious that studies on the carrying capacity for salmonids (will be a) serious political issue.' If ownership of anadromous fish is to be vested, as a guiding fisheries principle, in the states where the fish had their origin, how are those states to re-coup exclusively their own stocks in the sea? At face value the problem looks insuperable. Logic suggests that once the principle is accepted that the ecology which supports anadromous fish in rivers in their early life makes them the property of the home state, then the ecology which supports them later, as adults in the ocean, similarly makes them the property of whomever's copepods, herring, and so on they are fattening off. Whose they are depends on where they are. Hatchery programmes where the adults' feeding ground and natal state are different conflict with this logic.

These sort of questions, described in a thesis by Yvonne L. de Resnier of the University of Washington, in a paper entitled 'Evolving Principles of International Fisheries Law and the North Pacific Anadromous Fish Commission', go further. Problems could stem from the realisation that even stocks 200 miles out are thoroughly mixed. What of the principle of homing salmon sovereignty then? The 200-mile line is arbitrary, a line on a map, not the walls of an aquarium. What happens to the incidental catch of the valued migratory salmon if boundaries are pulled in?

Setting up salmon as the linchpin stock most vigorously to be defended may have its complications and its inconsistencies. Japan's presence, with a large and affluent population, a high-protein diet nearly half of which is comprised of fish, still the world's most significant and advanced fishing nation, will exploit them. Or try to. Meanwhile, unamazingly, America's catches of the major commercial species, sockeye, have done well. In the absence of Japan the sockeye catch soared to a recent record of forty million individual fish in 1993, and climbed to forty-four million in 1995. The 1996 run then broke all records, although there was a correction in 1997. Whatever arguments arise, they will circle matters of some substance.

Looking back over the history of Japan's presence in the North Pacific it is easy to see the accumulation of pressures against this great fishing state culminating at the end of the second millennium with her virtual expulsion from foreign waters. It did not stop there. Japan's role on the high seas, as mentioned, was also attacked. International action against Japan's North Pacific drift-net fishery for squid, along, to a lesser extent, with the drift-net fisheries of Korea and Taiwan, eradicated a major fishing action in which Japan deployed over 450 boats. The ban was at the behest of the UN, and therefore lacked legal force; in place of legal force there were threats of trade sanctions. This particular event, orchestrated by America, in turn being pressurised by Alaskan salmon interests and a phalanx of heavy-duty environmental bodies, was of a different type to the high seas salmon fishery ban. The banning of all high seas pelagic drift-netting was decided in the absence of any reputable scientific data, for an open ocean not an anadromous fish, without any examination of alternative ways to reduce the offending by-catch, in disregard of the economic consequences or economic alternatives for the expelled fleets and fishermen, in a politically-driven process riddled with inconsistencies and contradictions, not least of which was the obvious fact that distantwater fleets were to take the brunt of public distaste for drift-nets while the much thornier problem of coastal states conducting similar fisheries within their EEZs remained untouched.

No fishery crisis in the North Pacific was at hand. It was an example of the normally laudable precautionary principle being used in extreme fashion. Although conservationists jumped for joy at the decision, a detailed study of the way it was approached exposes a number of shortcomings, one of which is the role of the UN itself, not a body vested directly with responsibility for fisheries. America paid a quarter of the whole UN budget. In the high seas drift-net ban America might be said to have been getting its money's worth. These entanglements, and the lack of a science-based rationale, have meant that many see the drift-net squid fishing ban as discrediting an already precarious international organisation.

The North Pacific squid fishery in 1990 landed about 300,000 tons, seventeen per cent of the world supply. In the late 1980s, as mentioned, high seas drift-net fishing had been targeted by the media as a Number One horror story. By-catch included the highly sensitive subject of salmon and the equally livewire subject of marine mammals and turtles. The largest by-catch was actually of blue sharks, which carried less weight; blue shark numbers exceed sixty million, and rose through the drift-netting period. In addition there were 'drop-outs', uncounted drift-net victims which fell away as the net was pulled in, and escapees which died later. 'Ghost nets' made the headlines, being nets lost by fishing boats which continued to entangle living creatures as they drifted from ocean current to ocean current. The media never focused properly on lost fishing gear continuing to fish. In fact it is in trap fisheries that most gear goes missing, American studies showing up to thirty per cent of gear falling by the wayside each year, as compared to a figure of one per cent for gill-nets, rising higher for bottom trawls which are inevitably more prone to snagging. The drift-nets, in 10-kilometre sections joined to a maximum length of 50 kilometres, were long, which led to media phrases such as 'marine strip-mining'. Some commentators said drift-nets left biological deserts in their wake, hyperbole at odds with the record North Pacific salmon runs of the mid-1990s.

Those who feel that the 1992 high seas drift-net ban set an unfortunate precedent, and they include some of America's top fishery scientists, object to the sketchy and partial evidence submitted, in a small sample, by the observers placed on fishing boats to report on the by-catch. Simply, the data was inadequate for the contemplation of a global fishery ban. Differences between the three differently-conducted fisheries of Japan, Korea and Taiwan, which could have formed the basis for regulations in the fishery, were ignored; so too was any consideration of time/area restrictions to improve matters. By abolishing the fishery the chance of any proper monitoring of its effects was surrendered. Drifting gill-nets are employed worldwide. They have been in use for over a thousand years in self-replenishing fisheries. In some circumstances they are highly species-selective. It is not the type of fishery which determines the volume of discards – fish traps and long lines can entail huge by-catches – it is where they are fished, how, and for what. These big fisheries from the East were an opportunity to look at gill-nets and squid fisheries in detail. Neither are going to disappear. They are minor villains in by-catch. In the North Pacific squid fishery Taiwanese drift-netters were observed with a commendably low 9 per cent by-catch, Koreans with a 5–6 per cent rate, Japan with a 36–39 per cent by-catch rate. One point is that exploring reasons for the discrepancy could have improved methods of cleaner fishing, another is that the worst shrimp trawls reject over 90 per cent of what is landed.

Shrimp fisheries out-devastate any other fishery and represent a third of by-catch globally. In the Bering Sea king crab fishery, in a lamentable example of discards within one species, nine illegal-size crabs are discarded for each legal one retained. West Atlantic shrimp trawls are worse: twelve are rejected for each one kept. In the leading study of global by-catch by Lee Alverson, John Pope, Mark Freeberg and Steven Murawski, squid drift-nets actually clock in as having one of the lowest by-catch of any fishing method, squid drift-net fishermen over the world averaging a nine per cent by-catch compared to an all-species rate of 35 per cent.

There were far better grounds for outlawing other fisheries. But moving at UN level against, say, India's 150,000 coastal drift-net vessels, which also have a by-catch of cetaceans, might have proved difficult, to say the least. Within America's EEZ few American fisheries will have as low a by-catch rate as those recorded in the Korean and Taiwanese drift-nets. One cannot help feeling that the rapacious Oriental was a softer political target than subsistence fishermen off America, or anywhere else.

Japan is on the retreat, back into her own fisheries borders. There the development of inshore fisheries, with artificial reefs and marine ranching systems, has been taken further than anywhere. For ten years the Japanese have been experimenting with artificially-created upwellings using ramps on the seabed to push nutrient-rich water to the surface, simulating the natural upwellings which have the best fish densities. Results are encouraging. They are involved with experiments creating artificial sea-bottoms, the utilisation of deep water layers, the fertilisation of the sea, and offshore fish farms, at a level of exploration which is mere science fiction to other fishing states. The number of juvenile fish being ranched throughout Japan has risen to a stupefying fifteen billion, using over eighty species. Wild fish capture, ranching, and marine aquaculture have been treated complementarily; they regard the divisions as an artificial impediment. The marine and fisheries laboratory complex opened in 1997 in the Nagasaki Prefecture cost a cool US $86 million and in terms of scale and range of activity re-defines the meaning of high-tech fishery science. It is designed to be able to rear a wide range of fish, such as yellowtails, bartails and abalone, to fry stage, in huge numbers, is complete with a sophisticated seafood processing centre containing extruders, dryers, steamers, packaging etc., is programmed to study fish diseases and recurring environmental phenomena like red tides, and has a worldwide computerised information system linked to its network of national fisheries organisations. What can be performed on the whole eleven-acre site will make fishery managers in other countries feel faint with envy.

America is starting up the sort of marine re-stocking which Japan initiated a century earlier. Dr David Garrod, director for ten years of the august fisheries laboratory at Lowestoft, recently advocated as a solution to Common Fisheries Policy ills the adoption in Europe of a management system incorporating protected areas, and inshore fisheries under local control, along lines quite similar to the ancient Japanese model. If Japan does not occupy her customary lead-position on the matter of avoiding by-catch it is because the Japanese historically ate everything they caught. By-catch was meaningless. If Japan has behaved with arrogance on the world fishing stage it was an arrogance born of wide knowledge and understanding in fisheries, not ignorance. Other countries must have seemed to be floundering in the Dark Ages, at which time, returning to where we started, sensible marine tenure systems in Japan were already in embryo.

Chapter 7

Sharks and Shark Fins

The world has many horror stories. Fisheries are not a horror story: they are the history of man trying to make accommodations with an urgently needed resource against the backdrop of economics and the considerations of politics. As in other natural resource businesses there are one or two abuses which are completely unjustifiable, explicable only in terms of unbalanced and unbridled exploitation. One such is the trade in shark fins, where the fins are removed from living sharks and the extraneous body, unable to swim any more, rolling helplessly, is kicked back overboard to be devoured by its fellows.

The purpose of the sharkfin trade is to provide people in Taiwan, Singapore, Macao, and China, and Chinese residing abroad elsewhere, with fins for making into soup. Japan has acted as the main country for trans-shipment of fins to China. Sharkfin soup is an Oriental speciality of ancient origins known to date back over 2,000 years to the Han dynasty in China. Because catching sharks so long ago was difficult, fins were rarely secured, and sharkfin soup was a luxury enjoyed by the élite. If the chef's preparation erred, he could be decapitated. Then around 400 years ago Chinese herbalists declared sharkfins had medicinal properties. The mystique of these cartilaginous extremities of the world's most primitive family of fishes was enhanced again.

Today there are some twenty-five ways in which sharkfin soup is eaten. To make soup the fins are boiled and softened and the skin is removed leaving the finger-like structure of cartilage. The bony cartilage is extracted and what is left is the raw material which is so prized. When boiled up it resembles rice noodles or fine spaghetti. To get to the gourmet delight there is a lot of preparatory work. The highest valued fin is in powdered form and drunk as an aphrodisiac. Sid Cook, of the Milwaukie consultancy Argus-Mariner, reporting from Singapore in 1996, said shark dishes cost between $130 and $215, and had to be ordered in advance to allow for re-hydration of the fins. The fins which will do for culinary use are the dorsal, tail, and lateral fins; the 'clasper' fins of male sharks, modified pectoral fin extensions serving as sex organs, are unsuitable.

Sharks have had uses for humankind of many sorts for a long time. Their famously sharp teeth have been employed for carpentry, armour and weaponry. Their skin makes excellent leather, and its plated, scaly surface used to be used as a form of sandpaper. The cartilage in the fins was used to make gelatin. Sharks' liver oil is utilised in cosmetics for cold creams. At one time, owing to its abnormally low freezing point, sharks' liver oil was used in aircraft hydraulics. Time was when the streetlamps of Dublin were fuelled with a mixture of basking shark and rapeseed oil. Today shark jaws, if they are big enough, bristling with curved, wicked-looking teeth, can sell for thousands of dollars as curios.

This is the age of added value, one reason wasted shark carcasses are so dis-

172

tasteful; liver oil has new uses. Capsules made from the liver of a small, deep-water shark found off Australia are sold in Tasmania as capable of improving the human immune system. Investigations into this product show its development has had more to do with bumping up the value of by-catch in orange roughy and other deepwater fisheries than enhancing anyone's health. Other shark liver capsules have a longer lineage – the Japanese call theirs 'marine gold' – and their range of benefits would be impressive were they supported by proofs from clinical trials which had won general acceptance, which they are not. Another shark by-product which has dubious efficacy is the use of shark cartilage as an arthritis and cancer cure. Again, the product is sold in pills. Shark cartilage, so the hucksters say, can cut off the blood supply to tumours, causing them to wither and die. The background to this utilisation of shark is a torrid scientific controversy, the researchers who first conducted tests saying that extrapolating from their findings a cure for people with cancer is deceitful. If it were truly efficacious presumably there would not be a shark left. Finally shark meat is food; but it goes off fast and this has militated against its food value. Fins were the most readily marketed part of the sharks and propelled the first 'directed' shark fisheries, setting out specifically to capture sharks.

Excluding the sport fishery for sharks – not only is shark fishing a worldwide sport in which the fish is often released but in some Third World countries, such as the Seychelles, the fishermen put a fish-head on a hook and play sharks for as long as possible for relaxation – the capture of sharks today is by and large not directed. Certainly tuna boats will deliberately bait lines to target sharks if there are no tuna showing, but principally sharks are caught as a by-catch, mostly in the tuna fishery, typically by Japanese and Taiwanese long liners. (Europeans, though, need not feel shark-finning is a distant, distasteful event: it happens off the Portuguese-owned Azores, as a by-product of the swordfish fishery. Selective use of shark carcasses is practised in the multinational deep-water fishery west of Scotland in which shark livers are cut out for the cosmetic industry.) The sharkfin trade is a strange one, ancillary to directed fisheries, and such is the value of the end-product of dried fin that it has attracted the rough-est operators. In the week of putting pen to paper on this chapter two Taiwanese gangsters were gunned down in Capetown harbour, and currently the question of how to control the gang warfare which has grown up around the sharkfin trade is one of the hottest subjects in South Africa.

Tuna is, pound for pound, the world's most valuable fish. The fishing fleets which target tuna are high-tech, fast and mobile, and they are fed daily messages, routed by satellite, about the ocean surface temperature movements which are critical to the passage of tuna. Tuna long liners pay out a staggering sixty kilometres of line in each set, and the number of lines could total a hundred. High-speed winches can retrieve this vast apparatus at startling speeds, speeds often physically perilous for the crew. Three fleets of tuna catch-ers at the time of writing in winter 1996/97 were working around the tip of South Africa, one on the South Atlantic Convergence, one in the south-east Atlantic, and one in the south-west Indian Ocean. The tuna are blast-frozen to minus 160 degrees centigrade and accumulated in huge holds. When the holds are full, or it is becoming too expensive to fill the last corners, the fleets repair to Tokyo Bay. Electronic scoreboards tally daily fluctuations in the price of tuna and when it is judged correct they dock and unload.

Sharks come into the picture in a curious way. The fins are a bonus for the crew. They have the critical advantage of taking up little precious space, and it saves paying crews extra wages. To catch sharks either the long lines are baited with pilchards, or whatever is locally available, or they are caught coincidentally as the tuna are being winched aboard. The long lines for tuna are usually set, depending on location and temperature, at depths between 80 and 400 metres. Sharks are frequently near the surface. As the struggling tuna is ratcheted past them they bite at it, risking getting hooked too. Encouraged by a South African insider, a small London film-making company called Cicada made a film on the sharkfin trade basing their investigations on the African centre of the activity, Capetown. The film has been described as accurate and the material is fascinating. John Dollar, the film-maker contracted by Cicada Films, managed to get aboard a Taiwanese vessel (an impossibility on a Japanese ship) and film the hapless sharks being turned on meat hooks as the fins were ripped from them, then rolled back through a special gap in the gunwale, disabled whilst still living. The scene was graphic and revolting. The fins are stored separately. As the ship heads for Capetown the fins may be dried in the rigging, ready for sale. Wind-drying of fins used to take place in special yards in Capetown docks but with all the gangland violence this is now done in the concealment of closed sheds.

John Dollar sailed on boats with sharkfin operations and also got access, presumably at some personal danger, to stores and warehouses. In one he saw the fins from an estimated 9,000 sharks, and 82 crates of fins weighing 800 kilograms each. Shark weights being up to 5 per cent fin he reckoned 8 tons of fins equalled 160 tons of wasted shark. Small boats fishing off South Africa can catch 200 small sharks a day on one long line, and the number of sharks passing through Capetown as counted by their fins is reckoned at several million a year. This fits, approximately, with the figure of 70 million, the World Wildlife Fund's (WWF) calculation for the annual mortality of all shark species. An American scientist speaking on the film estimates that five years from now there will be no sharks to fish – perhaps an unduly dour prognosis, but an indication of the stock damage being done. The struggle between competing Hong Kong Chinese and local Taiwanese crime syndicates for the sharkfin trade started only about four years ago. In the midst of the high-profile public brouhaha about armed mafiosi patrolling the Capetown docks one government minister said there was very little that could be done about it. South Africa, however, has awakened to the problem and public feeling is hot. The capture of great white sharks, which were demonised by the film *Jaws*, has been made illegal in South African territorial waters. One or two troublemakers still have to be killed, but anyone possessing a great white's fins with no body to match and no licence commits an offence. In 1996, Australia, another great shark water, followed this lead by protecting two species of shark. South Africa, a major shark state, issues around a hundred licences to oriental tuna boats for fishing in its own waters. Ninety per cent of the sharks landed are either the large mako sharks, in which case the whole carcass is eaten, or the smaller oceanic blue sharks, whose carcass is high in ammonia and of low value, and consequently often thrown back.

Politics is changing the sharkfin trade. South Africa is looking for better relationships with mainland China and breaking diplomatic ties with Taiwan, with which Red China is still nominally at war. China may have no bluewater fleet to match Taiwan's, as yet, but South Africa's shark stocks are fading; new waters may be a necessity for the Taiwanese tuna fleet in any case.

The sharkfin trade is set to continue. Korea has become a big player. Japan and France are expanding their longlining in the Indian Ocean. The value of fins in winter 1996/97 is falling, meaning the crew either do not get a decent bonus, or they fish harder, in a more directed way, at hooking sharks.

Already the sharkfin trade has profoundly unbalanced localised shark populations. Increasing affluence in the upper echelons of Chinese society has pumped up fin demand. During the era of Red China under Chairman Mao sharkfin demand was flat: Mao thought sharkfin recipes smacked of the old order and of privilege. At the same time tuna and swordfish stocks elsewhere were diminishing. In the Atlantic and the Gulf of Mexico shark fishing was compatible with the tuna and swordfish gear. More fins met rising demand. Then the shark populations of the Arabian Sea, off central America, in the southern half of the Caribbean, and off Nigeria and Mexico, began dwindling. America rose to fill the gap. The fins of sharks swimming American waters were superior to those in tropical fisheries. The National Marine Fisheries Service focused on sharks as an under-utilised species and encouraged their capture. The value of shark fins traded on the world market, according to the UN's Food and Agriculture Organisation in Rome, rose between 1980 and 1989 by over three times, while the weight of fins doubled. In 1987 the US Department of Commerce was swamped with enquiries, some individual trade demands exceeding the projected shark catch in American waters for a whole year. Buyers traditionally discriminated between the fins from different sharks, but as prices spiralled in 1988 shark species which were previously of no fin value – blue sharks, makos, and threshers – were caught for processing into a sharkfin hodge-podge and sold in boxes to the bottom end of the market. All sorts of adventurers had rushed into the market, just at the point when numbers of sharks were falling and, ominously, the average size of those landed was diminishing too. By 1989 the Americans had realised their mistake and were busily drafting recovery plans for the very species they had wanted killed a decade earlier.

China was still the main end-user, but often this fact was cloaked because the whole business was run by hoodlums and trans-shipments were confidential. China anyway will not divulge statistics on sharks to the FAO; indeed, all fishery statistics, from the world's biggest producer of fish, are surrendered only in generalised terms. However, the fact that China has made public its intention to develop its distantwater fleet, which has risen from nothing to the sixth largest in the world since 1985, coupled with the fact that the biggest change in the methodology and balance of the Chinese fishery is a doubling of the gill-netting and longlining sector (from 8 to 16 per cent of total fishing activity during the period 1979 to 1993), bodes ill for sharks. With Hong Kong reverted to Chinese ownership the possibility for state-sponsored nationals to partici-pate in a more major way in the sharkfin trade looks strong.

Sharks are unsuited to an intensified fishery. Attempts at sustainable exploitation of sharks have always failed. The directed porbeagle shark fishery of the 1960s is an example of one that, seemingly inevitably, went over the top. Another example was the Taiwanese gill-net fishery for sharks off northern and north-western Australia. The Austalians, who declared their territorial limits and brought shark fisheries into management in 1979, suspected overfishing when they began to participate in the shark fishery themselves. Their scientists found out that the Taiwanese were obliged to double the length of their nets to maintain the same weight of catch. Even so, the main shark species being tar-geted were getting smaller and smaller. Pushed by conservationists who objected to the by-catch of dolphins enmeshed in the nets, the Australian government introduced a length limit on the gill-nets, shortening them from the customary sixteen kilometres to two and a half. This rendered the Taiwanese operation uneconomic and they departed. It is estimated by Sid Cook, along with Leonard Campagno of Chondros, an Oregon-based publishing organisa-tion dedicated to exposing the danger posed to the world's sharks, that of all the directed shark fisheries initiated in the twentieth century, ninety per cent have run out of stocks.

Sharks are vertebrates, and dissimilar to other fish. They are primitive and unchanged in structure having simple gill-slits, skins covered in small thornlike scales, and skeletons of a cartilage which is flexible. They do not spray the environment with their seed and cast off millions of eggs, rather, they repro-duce slowly (for a mako shark the gestation period is two years, one of the longest in nature), and invest considerable time in their young. They grow slowly and mature late. Many sharks have young numbered in only single figures. Acting in the absence of man as an apex predator they have no, or few, natural enemies. Sharks are an ancient fish which along with jellyfish, squids and horseshoe crabs, emerged in their present form 350 million years ago. Evolution determined that shark survival rested on a stable environment, free from competition. Fishermen targeting sharks changes this. Spread over about 370 species (one family is the beautifully-named requiem shark), they occur over most of the world's oceans. Around a hundred species are presently exploited. However, in most places sharks have never been a natural resource which could lend itself to a directed fishery; or to being a heavy by-catch. Scientists have calculated that stocks of the western Atlantic sandbar shark were reduced by 85–90 per cent in just ten years of fishing off America and Mexico. As with large land predators it is possible to deliberately eliminate sharks altogether from localised areas; they are, after all, easy to attract and easy to catch. Once down, shark populations have only the ability to increase incrementally, not explosively.

The rehabilitation of sharks is underway. The public at large is being persuaded by those conversant with these diverse, large, strong and fast creatures that behind the menacing appearance and natural aggression is a fascinating animal. The fact is being quoted that bees kill more people round the world than sharks, putting their menace into perspective. In this context the practice of 'finning' looks horribly unacceptable. As debate about marine resources, and access to them, waxes, the idea of wasting a large carcass for the sake of some bony apppendage, in the name of increased sexual performance or arcane, fetichistic recipes is outmoded, out of balance and indefensible. The wider matter of the maximum utilisation of all fisheries products will become the focus of attention in the twenty-first century. It will be less acceptable to waste parts of a fish, and, perhaps, more acceptable to see abandoned resources, such as sea mammals, also as potential wasted.

The WWF meantime has sponsored an international investigation into the trade in shark fins, cartilage, meat, skin and oil. The sights that Sid Cook witnessed in Thailand where tiny, as he called them 'popsicle' sharks, only a month old, with signs of their umbilical linkage still visible, having their pathetically small fins cut off, should be beamed far and wide now that we have the communication means to do it. Then those sitting down in the twenty-first century to sharkfin soup will pause, once because they will know what they are doing is regarded over much of the globe as abhorrent, and again because, the price of sharkfin having rocketed with shortages, the bill in prospect is sky-high.

The directed fishery for sharks for their fins would suffer some injury if properly exposed. However, regulating the fin trade would have no effect on the greatest proportion of sharks killed, those hooked up or enmeshed as a by-catch in other fisheries. Getting turtle excluding devices established in shrimp fisheries took the Americans a lot of effort; they had to ban the import of shrimps from fisheries in which turtles were still being ensnared. Turtles are a lot rarer than sharks, and closer to human sympathy. I have seen them paddling kookily past me on a Pacific atoll. It would not have needed a lot to make me indignant about their capture, especially if the bulk of the body was not to be used. I did not feel the same about the black-tipped reef sharks in the same domain. Sharks are the helpless victims of their own success as predators. They may be primitive, but our attitude to them should not be.

Howard Stock, a London-based sub-sea photographer who has spent much of his life below the waves, told me many reported 'attacks' by sharks were really the shark deliberately bumping underwater swimmers, usually on the shoulder, so that an olfactory device in their noses could tell them whether or not this visitor was customary prey. He said the hard thing was to get near sharks, which are shy. Some of those who do get attacked are spear-divers, carrying bleeding fish attached to their belts. 'Sharks' he said wistfully, perhaps remembering his treasured encounters, 'need a reprieve.'

Chapter 8

The Balance: Seals

'Maria will make me a pair of sealskin trousers, which it is essential to have on long winter journeys. When it is especially cold I will be wearing my Jaeger flannel trousers and then sealskin trousers above them with windproof trousers over both the other pairs. I have now sufficient clothes for my body and head; warm thick polo sweaters, a heavy Jaeger flannel coat, and a windproof blouse which also has a hood to place over the top of my Balaclava helmet. In addition I am having two pairs of sealskin gloves made, lined with puppy skin, which is the warmest you can have; these will have two thumbs so that when one side is wet, the other can be used. As regards footgear, I shall have two pairs of camiker, the outers being made of sealskin, the sole itself of the thickest type of skin, and the insides of the foot of dogskin, and the leg part of sealskin.' (Letter from Andrew Croft, *Sledge*)

'Evidence exists which suggests that protectionist non-governmental organisations have influenced scientific research in the field of marine mammal/fisheries interactions over the past two decades . . . The scientific record has been substantially distorted and impoverished as a result of this.' (From the Killilea Resolution, adopted by the European Commission in 1994).

'If the large-scale pup hunt were discontinued, the major focus of opposition to seal hunting in general would be removed, and polls show relatively low levels of opposition to other aspects of the hunt. There is, in fact, strong positive support for continuance of subsistence hunting to provide food and clothing, particularly by Inuit, and opposition to the hunt as a means to provide cash applies less to Inuit and local communities than to groups engaged in large-scale hunting.

Whatever policies are adopted towards seal hunting, acceptance by the public is likely to be improved by increased knowledge of all aspects involved, including the status of stocks, the nature of the hunt, and the significance of the hunt to people who undertake it. The polls have shown clearly the general level of public knowledge of all these aspects is extremely low.' (From the Canadian Royal Commission on Seals and Sealing in Canada, published in 1986, after interviews in Canada, USA, Britain, France, Germany, and Norway).

William Conway said: 'From a place where people were surrounded by wild animals, the world has become a place where wild animals are surrounded by people.' The ocean remains the exception. However, the attitudes developed from the human experience Conway described have been stuck onto marine creatures as if they were on land. Ecology and balance are not implicated in attitudes to sea creatures over the world. Instead there is hypocrisy, illogicality, and the supine acceptance of fashionable attitudes called political correctness.

Sharks, presumably, because they still arouse primeval fear in human beings, are hard to champion. The repellent realities of the sharkfin trade are, by and large, tolerated. Seals, which look helpless and vulnerable when on land, as they hitch themselves forward on attenuated flippers, whose young resemble doe-eyed dogs dressed in white fur, arouse instincts of human protectiveness. They have been dubbed saltwater labradors. Yet in the environment they inhabit seals are apex predators like sharks, often devoid of natural enemies. Whereas sharks feed only off a part of the fish life ranging their habitat, seals feed with less inhibition on everything they can catch. Partially because of the cuddliness factor, which is particularly photogenic, seals have taken on a sancrosanct identity among large sections of the population in the northern hemisphere countries they principally offlie, Canada, Britain, the USA and Norway. Precise figures are not known but organisations such as Greenpeace have raised millions of pounds in Britain alone for seal protection.

Public attitudes to seals have moved away from simple protection. It no longer matters, in the view of many fisheries people, if scientists prove that seal numbers have exceeded the carrying capacity of the environment to support their numbers in equilibrium with everything else. This is evident from the experiences of wildlife managers off the coast of California. Here sealion numbers have steadily risen since protection was afforded by the Marine Mammals Protection Act of 1972. From 3,500 in 1947 the Californian sealion population has reached around 120,000; the rate of growth remains unslowed at ten per cent, an extremely high rate of increase for any wild animal.

The impact on the coastal environment is considerable. Californian sealions are very large carnivores, up to eight feet long and, at 650 pounds, three or four times the weight of a man. The job of wildlife rangers has moved into the realm of protecting the public from their own over-sentimentalised foolishness. Simpering citizens are running up and stroking and cuddling sealions at risk to themselves.

Fishermen, meantime, are tearing their hair out. Participants in the seals' environment as a co-predator, any sentimentality has been stripped away. Netsmen have to fight the big mammals off as they haul their nets. 'Lions', as they are called locally, have jumped onto boats and bitten fishermen. The experiences reported by saltwater anglers in a new 'interaction reporting' system are astonishing. Five out of six fish hooked by anglers are lost to attacking sealions as they are being reeled back into the boat. In another incident suggesting sealions are far too numerous, a herd of 1,500 young males took over a whole dockyard in Monterey, closing down launching pads for leisure and fishing boats.

The anomalous protection of sealions while other fish stocks off California, where there is a rich fishery, are subjected to strict management measures, reached a head when the impact of sealions on a remnant run of Californian steelhead began to endanger the very survival of this highly-prized sport fish. The steelhead is a sea-running anadromous fish, either rainbow trout or a char, scientists remaining split, which is famous among sport anglers for its supreme fighting qualities and habit of 'tailwalking' over the water when it has been hooked. What made this story of particular interest is that steelhead stocks are themselves on a tenuous lifeline. Once common in Californian rivers, they are now rare, or gone; even in British Columbia, Washington, and Oregon steelhead have rapidly lost their niche. Tooth and claw scars made by sealions were found on fifty per cent of steelhead examined during 1995/96 on the San Lorenzo River in California, and on sixty-five per cent of steelhead examined in 1991/92 on the Big Creek Hatchery in Oregon. These figures, naturally, only refer to the fish which successfully escaped being eaten at sea or in the bay. On the Puntledge River in British Columbia in Canada in 1993 only twenty summer steelhead managed to run the gauntlet of the sealions at the rivermouth, along with three hundred chinook salmon. Biologists desribed, it as 'a salmon-killing zone'. Here was an endangered fish, gathering in the bay to run the home river, being hunted down in public view by a vastly-bigger and manifestly very numerous carnivore. Environmentalists took up a stance opposing any interference.

In true American fashion a panel was set up to judge the thorny issue. The first solution – a classic idiocy – was transportation. The sealions, strong swimmers, immediately returned. Then the government's National Marine Fisheries Service (NMFS) recommended capturing the sealions and administering lethal injections. The Humane Society sued them. American Vice President Al Gore stepped into the high-profile debate and got the sealions removed to a zoo in Florida. Throughout what should have been a sensible wildlife decision-making process there was no indication that the majority of the public saw things the steelhead way. Protection of the sealions, at any cost to natural diversity or to other living creatures, had become an unchallengeable principle.

An administrator with the NMFS, whose job is only to administer the law, remarked to me sadly: 'Marine mammals seemingly will always win out in these cases. You allow a harvest in similar situations in the terrestrial environment. But sea mammals are different. They have an allure to the public that defies reason.' The story of the sealions and the steelheads took an extra political twist when native Muckleshoot Indians from Washington State, exercising their right of exemption from the game and conservation laws, on grounds of subsistence hunting, declared that they intended to protect migratory fish if no one else would and sealions would be shot with guns, crossbows and harpoons.

Preservationist sentiments that defy reason nonetheless have reasons. One of the features of this Californian tale is that it illustrates the generally overwrought sympathy for all forms of seals. Except in eastern Canada, where a return to management with a cull of 240,000 harp seals was authorised in 1996, and where the low human population is, or predominantly used to be, engaged in some form or other in fisheries, and in Russia for different reasons, all northern hemisphere countries have human populations whose zeal for more and more seals as yet knows no bounds. The reasons must lie far back in the human psyche (in his book *Seal Cull* John Lister-Kaye suggests human affinity with seals is connected to seals' taxonomic relationship with dogs, man's oldest friend); in the feeling that man has dislocated much of nature and must not wreak further havoc on the most visible marine biota; and also in history.

Some biologists have observed that during the Stone Age seals must have been a mainstay support for human communities. In seal rookeries on the rocks only a club, the most primitive weapon, is needed to kill adults and pups, providing essential skins, fat and meat. Seal remains have been found in the earliest human settlements in Britain, which are round the coast. The Icelandic and Norse sagas of the ninth and tenth centuries record a strong reliance on seals. For the first time, in a specific year, a sealing expedition is on record for the Hebridean island of Haskeir, visited by Outer Hebridean islanders in 1549. References to sealing after that multiply until, striking a new note, John and Charles Sobieski Stuart in their *Lays of the Deer Forest* written in 1848, report that seal numbers on Haskeir have gone down, owing to 'long and continual slaughter'. The first preservation order in Britain, except by local landowners (Sir John Orde halted the Haskeir expeditions, it is thought on humane grounds, in 1858), was the 1914 Grey Seal Protection Act, which afforded grey seals a close season. It was suggested that seal numbers might have fallen as low as 5,000, a figure often quoted, but retrospectively, in the light of problems met with accuracy in contemporary seal counts, this was probably a serious underestimate.

Britain never had a commercial seal involvement to compare with the American one. It was principally the Americans who prosecuted the immense seal, sealion, and sea elephant harvest in the South Atlantic. This was no less vast for being long ago. By 1830 the fur seal was very rare off southern South America, in the Falklands, and in Antarctica. A corvette called the *Aspasia* killed 57,000 fur seals in 1800 and a staggering 122,000 the next year, the skins mostly going to China. The carnage must have been gruesome; big bulls were shot with a musket through the mouth.

At a later date the human harvest for seals around the North Pacific was vast too. In 1870 when a firm called the Alaska Commercial Company leased the northern Bering Sea Commander Islands from Russia, and the Pribilof Islands, recently sold to America by Russia, from the US government, the trade was prodigious. Prior to this the Russians had killed 70,000–85,000 pups a year on the Pribilofs, but the population under this pressure had crashed. The Americans applied a quota of 100,000 on the Pribilof Islands northern fur seals and narrowed the cull to only young males. Northern fur seals bred only on these two island chains. The US government rent was $50,000 a year with a $2 royalty on each skin shipped out. The value of all the northern fur seal skins sold in the first twenty-three years of the lease has been computed at $33 million. By today's values this was a colossal and apparently sustainable trade, levied from one animal. Then in the 1880s the seals began to be hunted whilst still at sea, British and American schooners employing native Indian hunters to spear them from small boats. Abandonment of constraints hurt the trade. A considerable controversy flared up between America and Britain and in 1891 a commission of experts was set up to investigate the effects of the fishery on the stocks of seals. This must be one of the earliest bilateral commissions set up to research stocks of marine creatures, and like the later American and Canadian halibut commission, its locus was the Bering Sea. Vigorous sealing also took place in the North Atlantic, and in the Russian White Sea, and in the southern oceans of the southern hemisphere; seal populations in some places were exterminated. In these days the skins were used for leather, sealskin boots being a standard fishermen's item in the nineteenth century, and oil for burning in lamps was distilled from the fat. Europeans stripped off the skins in the seal rookeries, but the true seal culture native people removed and put to use every part of the animal. Greenlanders used seal intestines to make waterproof clothing, a vital vestment. When they killed a seal the harpoon wound was sewn up again for transportation in order to hold in the blood, which was even more valued than the meat. For long durations rich seal meat constituted the entire Eskimo diet; life alternating between privation and abundance, they were famous for how much of it they could consume at a sitting. Stretched skins covered the light kayaks in which they prosecuted the dangerous activity of hunting the seals, their sealskin overcoats buttoning over the seat in the kayak-top making the whole thing watertight. The stitches only penetrated halfway into the skin, so there were no leaking apertures; fat was rubbed over the whole to complete the envelope. Early Arctic explorers recorded an indifference to the fragrance of Eskimo women, who covered themselves in seal fat to keep out the cold.

Whether sympathy for seals arose out of their over-exploitation is an interesting question. Messianic seal protectionism may have something to do with the over-exploitation in the eighteenth century of the Pacific coast sea otter, taken to a knife-edge of survival, and the actual extermination, mentioned earlier, of another marine mammal, the Steller's sea cow, which was consumed into oblivion by otter hunters with little choice of diet. By the turn of the twentieth century idiosyncratic attitudes to marine mammals were starting to take shape. F.G. Aflalo, in his book on sportfishing *Sunset Playgrounds* published in 1909, reports of California too much public sympathy for seals for any punitive actions against them to have been contemplated. The fact that seals could be tamed and responded to different people like household pets must have humanised them from the start. Their liking for musical sounds enhanced this. It seems that underwater film showing them hunting and biting the bellies out of salmon would probably fail to reverse this: for most coastal visitors seals are seen at their most harmless on land, or positioned upright in the water like bottles, their wide eyes unblinkingly meeting their viewers'.

Given that human sympathy with seals is so strong it is interesting to explore further what informs it, as attitudes to seals have started to typify developing attitudes in the western world to a whole range of wild creatures, including fish. An organisation now exists, aimed at the French coastal dwellers who dig and rake their tidal waters for tasty crustacea, championing the rights of crabs, winkles, and so forth. In *Seal Cull* John Lister-Kaye conducted identically-structured interviews with seven individuals just after the 1978 occasion in

Britain when a seal cull authorised by the government on the Orkneys had to be called off because of a rising tide of public protest. Women clutching babies to their breasts had run screaming over the beach shouting 'Nazis' at the Norwegian professionals hired to reduce the seal population.

His interviews were principally with eminent marine biogists, ecologists, and fisheries scientists. Included in the list was a director of the Greenpeace Foundation; he maintained a reply-mode of reiterated, brief, unexplained opposition to the cull. A wide range of questions was asked and in general the interviewees agreed that the seal cull had regrettably been hyped up by the media, contributing to a scene in which balanced judgements lost out. The answers to note, however, addressed the question of seals' impact on fish stocks. It was partly in defence of fishermen, after all, that the government had called the cull. The interviewees responded, in mildly differing ways, almost with unanimity. They thought that with the possible exception of the inshore fishermen, seals were irrelevant to the catches of the principal commercial species. Simultaneously, most regretted, from a biological view, that the seal cull had been called off. Collating these answers then, this bunch of distinguished people thought the cull should take place because the stock of seals was out of balance and needed clipping. Its interference with the jobs and livelihoods of fishermen was a non-factor.

The author of *Seal Cull*, one of the rare books dedicated to the interface between seal biology and public feeling about seals, is himself a biologist and also a noted conservationist. Stepping aside from the impact on fisheries side of the argument, because of the lack of good scientific evidence, he laments the termination of the cull, its proponents cowed by uninformed sentiment. He attributes rampant feelings about the Orkney cull in particular to the Canadian one of the much more numerous harp seals, as a coda adding that the wave of public horror at the eastern Canada cull on the ice-fringed rocks went into reverse at the end of 1975 when the public began to realise how profitable seal culling had been for the conservationists battening onto the issue as a fund-raiser. He then points to the gulf of difference in the British and Canadian situations. In Britain the grey seal numbers are known not unknown; the proposed cull would only have checked population growth, not cut into the stock; and the cull would have had no financial benefit for anyone. It was, in short, a management measure for pure conservation reasons. He concludes by saying that the conservation bodies which ranged up in opposition to the cull, given all this, lost face. He presciently predicts, writing in 1979, that the task of conservation bodies, producing responsible recommendations for the management of wildlife, would in the future be more difficult.

The situation near the turn of the millennium has changed considerably. For a start, the official scientific research body responsible for advising the government on seal numbers, the Sea Mammal Research Unit (SMRU), appears either to have changed its science or its orientation. In 1978 when the SMRU recommended a cull in Orkney the British grey seal population was 70,000. It is now, in the view of SMRU, even after an adjustment to methods of censusing (using a multiplier figure from pup counts to adults which had the effect of reducing earlier base-lines retrospectively), much higher at 110,000. Yet the SMRU no longer recommends a cull. Asked when seal numbers would reach a level to justify a cull, the SMRU chief of the moment said no such time could be envisaged. In other words SMRU never imagine again having to go through the baptism by fire of prescribing a cull. It is thought by the Scottish Fishermen's Federation (SFF), in whose area the vast bulk of British grey seals live, and by most people working round Scottish coasts, that the SMRU has been highjacked from the top by pro-seal sentiments or by those who, in the course of study, have 'gone native' on seals, and that they now manufacture evidence to suit their case. The polarising of attitudes on seals has been exacerbated as fish stocks have fallen and seal stocks risen. SMRU methods of censusing, by air when seals come onto Scottish coastlines between September and November, are by the admisson of its own scientists, unsuitable for counting in some locations such as in Shetland, where the narrow geos and inlets cannot properly be viewed from above, or where seals congregate in caves. Scottish fishermen attempt their own seal population estimates, arriving at a figure for the total of 140,000–150,000. The SMRU view is that seal numbers are growing by 5–8 per cent; fishermen say that the true figure is 10 per cent, even 20 per cent in places like the Monach Islands where seals have only recently started the process of colonisation.

It is on diet that SMRU scientists and fishermen really fall out. A licensed drift-netsman from the Northumberland coast described the ongoing battle with seals in his nets. He fights them off with every haul. He dubbed seals 'Chinese cormorants without the string.' 'When a fish touches the net it is a race between you and him. The seals have learnt to fish the nets better than we have.' Hauling a net of salmon and seatrout off Northumberland has become something of a trial; and the seals seem to get ever more numerous and more reckless.

Seals robbing static gear is not confined to drift-netters of salmon. A government report commissioned in Orkney found, as have Canadian pot fishermen, that seals have learned how to fiddle the clasps of lobster pots; pots are being robbed of their bait. Of each ten fishermen interviewed eight reported some damage to creels and bait-robbing, and six said the interference was significant.

Such signs of desperation, or hunger in seal populations, would not be quite so irritating were it not that SMRU scientists have actually disputed that seals ordinarily eat salmon at all. The fact that salmon are routinely witnessed being chased by seals, which often now come several miles up rivers to continue the feast, matters not a jot. This material is, SMRU scientists have claimed, anecdotal evidence, not science. Sub-aqua film showing salmon being chased and bitten and killed by seals is also inadmissible as evidence. Scientists revert obstinately, some would say blindly, to their accustomed methodologies, and deny what is known to all and indeed cited in standard textbooks on seals.

SMRU diet research methods are based on findings of otoliths, or the small ear-bones in salmon heads, in seal droppings. No otoliths mean no salmon in the diet. SMRU scientists calculate that seals in the North Sea feed principally on sandeels, next on cod, and next on ling. Other fisheries scientists will say that salmon otoliths are fully digested in seals stomachs, which is why they are not present in droppings.

Professor Basil Parrish, for one, has attacked SMRU contentions powerfully. As a former Secretary General of ICES, his views merit attention. In a paper submitted to the SFF in 1993 he said their data were characterised by 'inaccuracy and bias'. Seals frequently did not eat the hard parts of fish like cod and salmon, so otolith measurements were irrelevant. The predominance of sandeels in seal diet he regarded as misleading; it could have been due to sandeel otoliths found in seal droppings originating in the cod eaten by seals. Professor Parrish estimated that seal diet was not less than 80 per cent formed by commercially useful species including cod, salmonids and sandeels. Choosing from the mid-range of estimates for seals' daily consumption of fish he said, assuming an average intake of 15 pounds, British seals ate 2.4 tons of fish a year each. He cautioned that seals actually kill more than they eat, so the impact on fisheries might be greater. The amount of fish killed, in his view, was 250,000 tons, four fifths of which were commercially useful species. Reckoning that valuable gadoids constituted at least half of the total, he than made the arresting point that if adult salmon were as little as one per cent of seal diet, it exceeded the whole salmon fishery of Britain and Ireland combined! The scale of the problem was echoed in 1997 in the report of the Scottish Salmon Strategy Task Force. Also assuming the conservative one per cent of diet figure this government-commissioned document said Scottish seals ate 400,000 salmon a year against a total catch in the wild, nets and rods, of only 168,000 fish. Professor Parrish concluded in his paper, 'Since grey seals and fishermen prey together at the top of the food chain, the case for imposing restrictive control is a compelling one.' Such a view, from such an authority, makes the SMRU stance look, at best, peculiar; as peculiar perhaps as the SMRU 1997 updated stance on the salmon content of seal diet which is that it is impossible to come up with a reliable estimate, and as peculiar as the fact that it is ten years, ten years of rising concern, since the last salmon-specific work on seal diet was conducted. The late Professor Parrish, according to one scientist working in the field 'was ignored for political reasons'.

A European Commission report contracted to the Irish Sea Fisheries Board, published in 1997 and entitled The Physical Interactions Between Grey Seals and Fishing Gear, is set to cause a considerable stir in fishing communities. Examining the effects of seals on western Irish inshore gill-nets set for cod and hake the authors found that, apart from all the fish the grey seals removed entirely from nets, 96 per cent of damaged fish retained in the nets had been subject to seal attack. It says losses to fishermen were 'of considerable commercial importance'. It went on, as Professor Parrish had, to condemn the otolith reconstruction method as an inappropriate and misleading research tool used in isolation, in effect saying that the years of work spent nosing around in seal faeces had been misdirected. In its recommendations the report points to 'the need to integrate the management of marine mammals and fisheries.' It questions whether grey seals, at their historically very high population level, should be protected at all. In keeping with so many reports in a developing trend, the Irish one calls for studies looking for the interactions between stocks and species, and flags the need 'to explore the dynamics of large marine ecosystems'.

The SMRU position on salmon in seal diet may soon in any case be overturned by those in the western Atlantic. Canadian researchers are looking at a dietary analysis which focuses on fatty acids. These are in unique combinations in each fish species. The fatty acids remain unchanged passing from the stomach into the blubber; ditto in seals' milk. Another analytical method being developed hopes to identify dietary components by examining proteins and bony parts through electrophoresis. Work is ongoing with these techniques. In due course salmon farmers who have had their nets torn by seals, and their fish's bellies ripped out, and the netsmen fighting seals off as the nets are pulled into their boats, may be reassured that they were not hallucinating all along.

In other ways British seals' behaviour has changed during the 1990s. Now seals follow fishing boats when they leave harbour to set the nets. They have begun eating fish, such as gurnard, which never used to be part of diet. They frequently consume all of what they catch leaving only the fish's head, whereas previously they were known for one bite, in the soft underbelly, and then moving on.

Seals have moved on, and upwards into the public eye. In Scottish firths they can be seen all over the sandbars and adorning the top of every rock poking through the mud. On Stroma Island in the Pentland Firth, separating Orkney from mainland Scotland, I have seen what I took to be a spit of rock projecting into the sea until I realised it was a mass of seal bodies lying about each other. The skerry of seal flesh suddenly heaved and lumbered and was no longer there. It does not take a scientist to calculate that so many carnivores in one place must be bashing hell out of the already over-pressured stocks of commercial fish species. The Californians have the extra cross to bear, in the last convulsions of the imbalance, that shellfish beds have had to be closed, poisoned by faecal contamination from the sealions.

Such anomalies cannot go forever unremarked and beyond being questioned. Fishermen have felt for a long time that the very live issue of seals for them never enters the purview of politicians. Most politicians see the issue as dynamite, but unstable dynamite, not a political weapon that could readily be deployed on the enemy. However, there are political people with wider, non-party, even non-national, constituencies. Eight years ago the European Bureau for Conservation and Development was founded, the brainchild of a Canadian whose country had, after the EU ban on all seal products, lost a staple industry for its most vulnerable peoples. The social cost in Canada, to seal-dependent native communities, had been horrendous.

The Brussels-based Bureau tries to promote the sustainable use of natural renewable resources. It believes in multispecies management without sacred no-go areas unless their populations are defined, by solid science, as threatened. Seals and whales should, in this organisation's view, be managed along with everything else. They represent a huge unutilised resource. Over-protection can be as harmful as over-exploitation. Its Greek spokeswoman, Despina Symons, took their case a step further: 'Unless you protect the environment to satisfy human needs, the policies will fail.' It is certainly possible to think of many case studies which could be adduced demonstrating failures.

Despina Symons believes the sheer common sense of the principle of multi-species sustainable use is gaining ground. Already, in the realm of seal politics, this is so. Links have been forged between the SFF and parallel organisations in Norway and Denmark. Salmon farmers from several countries have found a common cause. Indications of changing attitudes are not only perceptible in fishermen's groups. A 1996 poll in Ireland revealed that if grey seal numbers were shown to be causing job losses, a majority of people would accept the need for culling. If some sea mammal scientists felt in the past their work was almost regarded as being heretical – the 1994 Killilea Resolution written and accepted by the European Commission referred to the intimidation of scientists engaged in establishing the effect of sea mammals on fisheries, and the possibility of careers being imperilled by such researchers – there are clear indications that the truth can no longer be ignored, bent or suppressed. The logically absurd proposition that fisheries should be run principally for the benefit of a bur-geoning multitude of sea mammals is due to be challenged. It has no place in a fisheries policy which purports to be scientifically-based, in which hard choices are often made involving fishermen's lifestyles and livelihoods, in an environment with increasingly taut conflicts, where the battle for space and for the valuable resources is becoming more and more acute.

Chapter 9

Sport Angling

'Like all good fishermen, they were boys at heart.' (Zane Grey, *Tales of the Anglers' El Dorado: New Zealand*)

'You pace the boat to the angler. The guys can get an adrenalin burn-out. You have to watch them carefully.' (Joseph G. France, record-holding Atlantic blue marlin skipper in Faial, the Azores)

'I'm not a fisherman. I do it for the hunt. I enjoy putting my boat together.' (Joseph G. France)

One of the most attractive ways to use a resource is enjoy it, then leave it behind. Sport fishing has this unbeatable, quite unassailable virtue: it does not harm the stock. Either the numbers caught and retained are so low as to be irrelevant or, in the USA where sport fishing has been taken further than anywhere else, the fish are played and then put back. In sport fishing meccas like the Florida Keys, fishing guides report that almost nothing is retained by anglers anymore; even potential record-breaking specimens are being returned unmeasured. Indeed, returning fish has taken on an almost mystical significance. An American salmon angler buttonholed me in a Helsinki hotel, preparatory to a fishing trip on Russia's Kola Peninsula, and staring hard into my eyes said, 'Hitting that fish on the head would be like killing my mother.'

As an unimpeachable way to exploit the fish resource sport fishing has few equals – except perhaps reef viewing through a glass-bottomed boat. Over this simple voyeurism sport fishing has the advantage that its adherents actually get to grips with the fish in its environment, selecting a lure which will attract it, then playing it on tackle expressly designed to communicate the power of the fish through the length of the rod to the angler, and finally appreciate its splendour at close range. Simply the frisson of being in physical contact with big fish, shining, unblinking, disturbingly different from their captors, is one of sport fishing's inscrutable attractions.

The socio-economic characteristics of sport fishing are highly favourable. It provides relaxation in the open air on the earth's least-pressurised surface. It can be performed by anyone of any age regardless of handicap. Even a blind man can be rigged out and put in the fishing-chair to wait for the bite. It provides, if a sample of the catch is retained, top-notch food for the table, fresher than it could be procured any other way. Its environmental impacts are almost zero; for those who fish off their own flat feet from the rocks or the beach, literally zero. Sport fishermen are the best and most numerous early warning reporters of anything peculiar affecting the environment they use. Sport fishing can be done at any level, and only in countries where the beach or shoreline is privately-owned is it necessary that there need be any monetary exactions at all. Where there are, they support jobs. With their catch and release habits sport fishers are in a perfect position to tag fish at the same time. American scientists have fulsomely acknowledged the role of 'the sports' in accumulating data for vital stock assessments. Specialists in rigging up lines of baits, calculating the colours, sizes, and sequences in the rig of each constituent part, and at what depth, speed, and location to fish them, have contributed to scientific knowledge about fish behaviour. For reasons like these, plus the one that saltwater angling can be as challenging, thrilling, dramatic, and soul-refreshing as any other activity, its popularity is growing.

Furthermore, saltwater angling has had a major effect on fish stock conservation; several species have been saved by the angling lobby. All in all, armed with these points, it is possible to face the fact, without incredulity, that in America the value of the saltwater sport fishery taken as a whole hugely exceeds that of the entire American commercial fishing industry. Excluding indirect spending – travel and hotels – sport fishing is worth $72 billion a year and provides 1.3 million people with jobs. By numbers the most popular sport in which people actually participate (an alleged 60 million practitioners), sport fishing has also been called 'the salvation of American fishery resources'.

The ethos of American sport fishing has another dimension. Game fish, like land-living game, except in some cases under special licence, cannot be sold. This was laid down as one of the early provisos of American hunting lore, a wise recognition that over time it would militate in favour of species protection. It applies equally to fish and fowl and acts as a brake on over-exploitation. In any debate about the partitioning of resources game anglers open operations on the moral high ground.

This is reinforced by another circumstance, surprising to non-Americans, even surprising to some Americans. Particularly in states on the southern seaboard there is a considerable amount of subsistence fishing described as sport fishing. Subsistence fishing normally refers, in both the USA and Canada, to fishing by native Indians or Inuits. This can be more an ideal than a reality, and native people are unrestricted in their technology. I have seen native Indians armed with supermarket shopping trolleys capturing spawning salmon in the heart of Vancouver! The subsistence fishers of the American south, however, have to use sport fishing gear and follow regulations. For America's coastal poor, fish is free food. Many who realise this – and there are worries about the quality of fish eaten in inner cities – nonetheless would not want to see the present state of affairs imperilled. Getting to sea and putting food on the table is an important social safety valve in needy and depressed communities.

America is indubitably the place to examine saltwater angling. It is wedged firmly into, and has helped define, American culture, a point realised and enlarged by the artist Winslow Homer in his paintings of the lonely fisherman with sinews stretched in a small boat on a zesty sea. To American literature

Ernest Hemingway contributed *The Old Man and The Sea*, which told a truth to be found, buried in all the regalia which often bedeck American sport vessels heading for blue water: that sea fishing is the act of a free man, that the fish is the last subject of hunting conquest to which no stigma is attached, that the whole performance defines man and nature in the biggest frame in a stark, uncomplicated, and eternally irresistible way for individuals whose spirit still harbours something a little wild. Herman Melville told this tale too in *Moby Dick*, but there the aim is deeper, the scope more majestic, the metaphysical message more bleak and frightening.

Sport angling is not a new thing. It starts with taking pleasure in catching fish. The language of commercial fishermen in the nineteenth century shows that in those days, when skippers had to have a 'nose' for the fish, when actually bringing them aboard was entirely manual (as in many Third World fisheries it remains), when there were no echo-sounders or ship-to-shore communications, commercial fishing was itself often enjoyed as a sport, but a sport to which gruellingly hard work was attached. Sea fishing purely for fun and recreation and 'the pot' started over 150 years ago.

A recreational fisherman then was a versatile fellow. The British sport angler Lambton J. H. Young, in *Sea Fishing as a Sport*, pulled aboard his 25-ton yacht all manner of devices invented to remove marine species from their accustomed habitat, including small trawls, primitive bottom-nets, a launce rake for sifting out sand-dwellers, a dredge, 'lines of every kind', crab, lobster, and prawn pots. Lastly, he would take a gun with which he shot duck. 'Let it be laid down as a golden rule that you never go in your boat without your gun,' he wrote. One fondly ponders on what such an assemblage would do to the eyebrows of the contemporary fishery police officer on fishing grounds covered by gear regulations! Even between the World Wars lines were of linen or silk, rods of bamboo, and fishing reels with little click drags barely adapted from those used by fly fishermen. Tangling with more energised saltwater fish, these simple reels of the early days frequently exploded. Rods shattered, lines parted at the slightest mistake, and 'hanging in there' was often an unaccomplishable art.

Today's sport fishing boat is a different story altogether, armed with racks of rods, powerful engines capable of reaching the hot spots early in the morning ahead of the crowd, and electronic screens on which fish appear as moving blobs. Rods fishing different baits, at varying depths, fan out from the back of the boat to rake the water. Lures may be held in the surface-film by kites flown alongside the boat, or alternatively by balloons. Dreamers about really big fish use reels made of aircraft-grade aluminium. The latest trend is fishing from inflatables, feeling the water pound the boat-skin, often for fish – tuna, barracuda, bonito – that are nearly the boat's length. These inflatables are no toys; tough and fast they also accommodate the customary console of electronic aids, and a raised rail at the back holds an array of rods rigged to catch separate species.

It is interesting given the credentials of sport fishing that, in America, sport fishermen are starting to perceive themselves just as endangered as many

American fish stocks have become through overfishing. The last two or three years have seen a turnaround in attitude by sport fishermen. Today they are on the warpath to defend their sport, tirelessly critical of the commercial fishing sector's cavalier attitude to stocks and, in the shrimp fishery, of the horrendous level of by-catch, and tub-thumping about the role of sport fishermen as fish conservationists. They have of necessity become, perhaps in fashion with all those doing traditional things out-of-doors, active and vehement defenders of their rôle and their integrity. One of the problems is that, to an outside eye, American fisheries appear to be managed archaically.

Despite being a huge trader in fish, fisheries have never been part of the larger framework in the economy of the USA. Rather, the fishery resource is treated as a political pawn. One of America's top fishery scientists, Dr Michael Sissenwine, told me that fisheries were turned over and weighed up by politicians to see how most votes could be extracted. A supporter of the ITQ system as set up in New Zealand, where open access policies were abolished in favour of established fishermen getting vested user rights, Dr Sissenwine believes Americans fear the evolution of the domestic fishing industry from small-scale operations to large-scale ones. To an extent, perhaps, they see all skippers as Melville's Captain Ahab, every vessel a Pequod.

Even today not all commercial species are subject to quota. Rather, fishing effort is controlled. This has failed to address over-capitalisation. It leads to the absurdities of fishing 'derbies', referred to already off Alaska for migratory salmon. The New England area was famed for the preponderant influence of numerous small operators. A large number of often low-income fishermen were dependent on the fishery. New England stuck out its neck and attempted to manage solely by controlling the sizes of net meshes. In once-glorious fishing-grounds the collapse was fast. In the New England fishery at the time of writing the failure to tackle over-capitalisation may mean a groundfish season, already cut to eighty-eight days, going down to around thirty. The blame for the manifest failures of management on parts of the American seaboard is often laid at the door of the management structures.

The instruments settled on by the Americans to achieve full representation in national fisheries were Fishery Management Councils. There are eight, distributed around coastlines, collectively managing US fisheries in the federal zone beyond three miles and out to 200 miles. The water within three miles (sometimes more) is state-run, states often co-operating with each other. Mike Sissenwine thinks the open processes on management councils, with wide representation, has merits. He calls the system 'participatory'. Inevitably there are many positive aspects . . . and many stalemates. It could not be called 'smooth-running'. Management decisions are described customarily by American scientists as 'risk-prone'.

The beef of sport fishers is that they are not sufficiently represented on the powerful federal councils. The framework is weighted to favour the commercial sector, despite its smaller economic rôle in American fisheries. Quotas are therefore routinely too high. Other council members include officials from each state, scientists, and statisticians. Councils take account of the social effects of their decisions, which means fishery measures are too soft on fishermen and too hard on stocks. Writing in the journal *Fisheries*, Mike Sissenwine and a member of his staff, Andrew Rosenberg, put it succinctly: 'Under the pressure from the fishing industry, in the face of uncertain scientific information, risk-prone fisheries management decisions (bias towards over-utilisation rather than conservation) have been the norm. On average, risk-prone decisions are followed by further resource declines, more economic stress, more pressure not to reduce catches, and ultimately severe biological, economic, and social consequences.'

The voice of sport fishers lacking representation on the councils is theoretically safeguarded by the National Marine Fisheries Service (NMFS), a part which many sport anglers feel with considerable heat, the federal service is altogether failing to fulfil. The NMFS is the ultimate arbiter of fishery regulations. It vets and can alter the recommendations of councils. Sport fishermen feel the NMFS lacks independence and can be pushed by politicians, frequently at the behest of big businesses associated with fish processing, food retailing, or the larger single-ownership fishing boat fleets. With the exception of one fishery council, that on the south-west Atlantic, where sport fishing is hugely popular and a headline industry, the bias of fishery management favours commercial fishermen.

One of the reasons for this is that in the eyes of America at large, a country mostly far from any coast, the commercial fisherman is seen to epitomise the American Dream, elevating himself by hard work, facing the elements in a primal struggle to conquer territory. While obviously not an accurate representation of the average professional fisherman, this image has a kernel of truth. On some American coasts, for example in New England, immigrants still unversed in the English language form a hard core of fishermen. Legislators are reluctant to come down hard on them. Behind this embedded sympathy for the common man is the march of American history, in which the New World set out to slough off the husk of the Old World. In Britain the Crown over time arrogated most fishing rights to itself, even in estuaries and sea-bays; the New World wished to be different, prizing the principle of an open access fishery. Also, Americans are used to importing fish for the table; only some ten per cent of marine fish eaten are caught in home-waters. Despite the huge potential of its seawater to nurture fish, with the wide range of sea temperatures and circulating sea currents, America, although a very large catcher of walleye pollack and salmon off Alaska and of the low-grade herring, menhaden, in the south, has never regarded itself as a leading fishing state nor has it been regarded as such by others. It is a strange thing, as the world's largest fish exporter, but commercial fishing has tended to be seen as a sunset industry and, as in some other sunset industries, many of the stayers-on are subsidised. This itself blunts the impetus of change. Now it has become obvious that expectations of good marine management have been let down.

The fishery management council members need political backing to be appointed. Sport fishers feel fish processors and the large industrial fishing interests push forward their own people. The councils hold around ten meetings a year which last for three days, a long time away from the helm for a sport fishing outfitter or indeed a fishing recreationist. The big industries can afford place-men. In the words of one long-time sport fishing outfitter, 'Fishery councils bury you in bullshit; they're only casually interested in the real issues.' This view is common.

Certainly if American fisheries are to be judged by their management of stocks, some species with high profiles as sport targets have drawn terrible and tragic trajectories. Sharkfin hunters decimated American sharks from the late 1970s until, with stocks shattered, the practice of de-finning was prohibited. Sharks today have been described as 'recreationally extinct'; a fisherman from New Jersey told me that only fifteen years ago he was getting fifteen sharks a day in a ten-day period. The charismatic broadbill swordfish, like a saltwater unicorn with its elongated flattened jaw, capable of growing up to 1,300 pounds, was fished almost to extinction in America's waters by longliners. Today's swordfish have a pitiful average weight – around 40 pounds – pitiful when you consider they have to attain 250 pounds to be capable of spawning; and taking into account that thirty years ago the average weight of swordfish landed was 266 pounds.

The legendary bluefin tuna, a key target for sport fishermen in the western Atlantic since the 1870s, was fished down by commercial pressure to an estimated eighteen per cent of its pristine population. This was the fish physically endowed with a unique circulatory system giving it extraordinary metabolic rates. A bluefin exhibits volcanic energy when experienced attached through the mouth to a short fishing-rod. One outfitter said, 'It's hard for a 200-pound man to get his ass whipped by a small tuna of 50 pounds. This was the fish that brought you to your knees!' With bluefins capable of taking off in searing runs at over 60 mph, one can see how. Grotesquely, in retrospect, America's Bureau of Commercial Fisheries encouraged fishermen in the 1950s to seine for bluefins to develop the product for putting in cans. When the punishment inflicted on the bluefins became undeniable, and the Atlantic Tunas Convention Act was passed (1976), representatives from the canneries were given seats on the board! Americans were going to stick to their hallowed management principle of everyone having a say through thick and thin. Prior to the bluefin stock really collapsing was the period when the large bluefins capable of replenishing stocks, and differing from the smaller shoaling bluefins in being either solitary or in small schools, until then not targeted, were hunted down singly and air-freighted to Tokyo to meet the vaulting market prices there.

Today, bluefin, which once comprised eighty per cent of America's western Atlantic catch of ocean fish constitute under ten per cent, a tremendous tumble. Certainly there is plenty the matter with stocks of what could be called America's charismatics. Although these big fish are highly migratory, and capable of crossing and re-crossing oceans, therefore harvestable in many different fishery jurisdictions, American fishery managers showed a lead only in novel ways to make money from the crop. The USA always was, and remains, a major catcher and consumer of tuna; together with Japan, America consumes seventy per cent of the world's tuna landings. Conservation was but a whisper lost in the clamour for exploitation. Recommended quota cuts were themselves routinely slashed by the toothless regulatory body, the International Commission for the Conservation of Atlantic Tuna. It is salutary to remember that when South American states in the 1950s began declaring 200-mile territorial limits to protect stocks, the USA took a stance as an opponent of extended jurisdictions in the ocean which was at variance with its equal eagerness to protect its own politically sensitive and regionally important coastal fisheries. When under the Magnusson Act America did finally declare a 200-mile limit the only fish expressly to be omitted from exclusive home-state control within 200 miles was tuna. This was a political fix to retain American access to tuna in other people's waters. It revealed the strength of the tuna fishing lobby's foothold in government. For in the 1970s the San Diego based distantwater fleet of some 130 vessels controlled about seventy five per cent of the valuable tuna resource in the eastern tropical Pacific. Proper management of tuna would have to wait. For the 'sports' management of tuna had communicated grim warnings, forebodings which became acute in the 1990s. By the calculations of its own top marine scientists, of the 156 fish populations in American waters for which resource assessments have been done, forty five per cent are classified as over-utilised. Saltwater sportsmen could not look to federal bodies with much confidence that they would protect the resource.

If the problem has been weak management the best way to demonstrate it is to take a fish whose stocks can be measured with scientific certainty, and for that you need an anadromous fish or one that returns to freshwaters to spawn. In both America and Canada counting anadromous fish has been taken seriously; in many rivers they are counted both in main stems, and in their upper tributaries, where one man sits with a clicker on a plank extending into the river where the fish pass through a funnel device while his buddy guards his back from bears with a pump-gun. On the Taku River, which emerges at Juneau, Alaska, large river-side paddlewheels slosh migrating salmon into holding boxes, where tagging followed by records of re-capture form the basis for stock assessments.

Unfortunately the supreme North American migratory fish, the king salmon, or chinook to west coasters, does not perfectly fit the bill and is harder to count than smaller salmon. However, the quarrel over the chinooks has its own value as a tale, with features that are tragically symptomatic. Chinooks spawn in large gravel often in large rivers and in turbid water, idly pushing 12-inch rocks around to make way for a redd. They evade river-edge counters opting for strong, mid-river currents. This window of uncertainty about spawning numbers, in the way of things in fishery politics, is fiercely exploited by those wishing to prise open the fishery to more exploitation.

Anyone looking at the plight of chinooks squarely sees the greatest salmon

194

in a parlous state. Professor Carl Walters of the Fisheries Centre at the University of British Columbia in Vancouver reads it this way. Focusing on the Strait of Georgia, between Vancouver Island and mainland British Columbia, once the greatest chinook location in the world, he calculates that the wild stock has virtually collapsed. In biologists' terms this means there is no reproduction. Chinook numbers are ten per cent of what they were in the 1970s; this is five per cent of the 'virgin' stock. Chinooks are now a mere two per cent of all west coast salmon. To anyone quibbling with this Professor Walters points out that one of the biggest sport fishing clubs recently caught only thirty tyees, the famed bigger kings of over thirty pounds, in a whole year. Scientists have sent up their smoke signals of warning for a decade, largely unheeded.

It is hard to exaggerate the losses which go with this state of affairs. To west coast sport fishermen chinooks are the fish of all fish, the monsters capable of topping a hundred pounds. In a television programme made by Eric Malling for Canadian CTV tributes to them poured forth from witnesses struggling to out-do each other: 'The spirit of the whole Pacific north-west lies in its chinooks'; playing a chinook is to one angler 'about the best fifteen minutes in life'; concluding the eulogies, Mr Malling remarks, 'the king salmon: it is more than a fish.' That is why a vast sport fishery hangs on the kings, and even when four out of five fishermen were returning home empty-handed after a week of angling, the boats out hunting kings in saltwater were still all fishing. Sport fishermen may ideally want densities of fish, but when the resource is low and uneconomic to harvest commercially, the sport fisherman, at any rate the chinook fisherman, is still there, the triumph of hope over experience. The biggest of the British Columbian fishing camps, deploying fifty sport fishing boats, cost $16 million to create, and claims to pump $6 million a year into the economy of Seattle, where most of the goods and services are purchased. Chinooks mean business.

In Eric Malling's film the Icelandic salmon conservationist, Orri Vigfússon, about whom more later, is invited to sit in on one of the meetings at which the chinook user groups and government representatives try to hack out a policy of restraint. Orri Vigfússon, with a track record of extraordinary success at brokering international deals between competing fishing interests, is supposed to try and knock heads together to settle a portioning of the resource.

The ensuing scene is an illustration of how not to conduct the management of a natural resource. There is an unbelievable medley of organisations involved in the capture of the few remaining chinooks. There are fishermen's unions, trollers' associations, gill-netters, commercial vessel owners' associations, salmon scientists, fishing camp chiefs, federal people, representatives from the provincial government, fishery managers, and native Indians whose take from the fishery is unrestricted by virtue of their status as native peoples requiring subsistence food. The only users not represented are those who take the chinooks illegally. No one, seemingly, will budge an inch; all want the others to take the brunt of the restraint; tempers flare. Plainly, the number of consultees is absurd. The fishery councils of the USA may be unbalanced in their representation, but Canada's effort to operate a system involving everyone seems inherently incapable of reaching tough decisions. The Malling film showed a resource being scooped at by everyone, and in many ways that rowdy scenario was the paradigm of much fisheries management, but with the cameras present. The fact that there were licensed netsmen and trollers enjoying provincial government subsidies to patrol the same water with their gill-nets and hook-bearing lines as sport fishermen trying to attract the big fish with lures seems nonsensical; the fact that other salmon-netters targeting sockeye catch incidental chinooks is tragic; the fact that native Indians can yank the autumn chinooks from their redds in uncontrolled numbers, when they are stale, to fetch a mere 40 cents a pound, pitiful. Yet this pathetic and lethal muddle is occurring in a province where sport fishermen number 400,000 and where even today's depleted sport fishing, minus most of its chinooks and steelhead, is worth $1 billion annually. To cap it all, in 1994 those catching chinooks for a commercial return received approximately half the price they got in 1979. The poor fish, as a slab of protein, had become almost worthless.

The position of the government spokesman was simple, but unhelpful. He would introduce regulations on the basis of mutual agreement. An Indian then made the contribution that he would go back to fishing with spears if all the white Canadians returned to Europe. A troll fisherman parried with the confusing though quite believable statement, given the plummeting prices and recalling the comment of the Canadian fishing writer Roderick Haig-Brown, that trolling was most akin to sport fishing, that fishing this way for kings he did mostly for fun. It made good television if regarded as knockabout comedy, in the abstract. The result of this strangely-arranged meeting was perhaps the only one possible: all user groups agreed to halve their allocation equally for the following season. No one, however, really believed it would be enough. A closing remark was: 'Memories may soon be all that's left of the big chinook fishery.' Judging by the catch landed from a single sweep of the net in British Columbia in 1932, collecting some 15,000 chinooks, which were winched onto dry land with difficulty, the memories will get very faint before, if ever, they are re-created.

One cause of the latest round of troubles for chinooks was climatic. The whole ecology on which chinooks subsist was altered by the El Niño events of 1991, 1992 and 1993. Warmed-up oceanic currents coming from the southern tip of South America meant three consecutive brood year losses for a fish which matures for breeding between the ages of three and six. The predators of young chinooks, chub mackerel and hake, increased in number and feeding in the sea declined. Survival of young chinooks at sea was discovered to be wildly variable, two different chinook stocks recording figures of 0.1 per cent and 15 per cent. Historic harvesting rates were obviously inappropriate. It seems appalling that in rooms packed with the angry exponents of narrow self-interest the fate of this once-mighty fish moves towards a decision; yet it is not atypical in fisheries.

What makes things worse is the management of chinooks when in American waters is under the aegis of the fishery councils. Trying to find home rivers the chinooks travel northwards up the coast through American waters, then Canadian, then American again when they reach Alaska. Chinooks are shared by America and Canada. A USA/Canada treaty on Pacific salmon dating from 1985 exists, but the two countries are, and always have been, uneasy bedfellows. Historically American catches of chinooks have been over twice Canada's. In at least seven rivers the spawning chinooks swimming eastwards cross national boundaries on the journey to the redds. Both countries want safe passage for the fish when in the other's bailiwick and accuse their neighbour of not providing it. The American response to the chinook crisis has been the simplest one: to hatch and release them. On the mighty Columbia River sixty per cent of autumn chinooks are hatchery-bred, eighty per cent of springers. Everyone agrees that reared fish lack the adaptability, survival rates, and obviously the genetic range of wild fish. In the long term hatcheries can only augment runs, not replace them.

Sport anglers are in a mêlée of conflicting interests which even Eric Malling's programme could not encompass. Not only are chinooks a by-catch, in small numbers, in Japanese fisheries in the North Pacific, but some chinooks move as far as Japanese waters in the western Pacific before turning round and returning to Alaska and the Yukon River in Canada. Chinooks are highly mobile, one of the attributes which excites sport anglers. The fish you catch is an envoy from distant places, trailing the scent of faraway seas; it stirs the imagination.

In one fundamental respect bluefin tunas and billfishes such as swordfish, and chinooks, are the exception; the vast bulk of American catches are resident within their EEZ, therefore susceptible to good management. The litany of fish stock collapses round the American coasts – redfish off the southern States; sardines off the coast of Monterrey (maybe caused by environmental change); cod, haddock and flounders off New England – demonstrate the inadequacies of the past. Sport fishers sense a new dawn. The overwhelming arguments supporting their case are becoming more widely known. Previously considered by officialdom to be playing at fishing while real fishermen in real jobs put food on the table, the evolving view of the sports is as conservation-reliant, environmentally-sound, non-greedy participants with a huge amount of somewhat disorganised economic clout. 'There is a feeling about ' said John Brownlee, senior editor of *Saltwater Sportsman*, and a member of the South Atlantic Fisheries Management Council, 'that we have to decide as a society how we use the fish stocks.' Acknowledging that for many commercial fishermen going to sea is a way of life more than a way of amassing a fortune, he pushes forward, 'We need to stabilise the number of participants in the fishery'. He continued, 'We need to try and get an equilibrium. I believe in ITQs and limited entry and a few vested people. We must avoid the Catch 22 of prices rising as catches fall, dragging more and more people into the fishery.'

A Texas-based body to represent sport interests has been formed. The Coastal Conservation Association's (CCA) mission statement is 'To promote conservation and preservation of marine resources for the general public.'

David Cummins, the CCA secretary, says pushing its agenda of eliminating harmful fishing gears, protecting marshlands to help brackish marshland breeding fish to get to sea, getting game status for endangered fish etc., is easier against the loose-knit groups, for example, of shrimpers, than against the bigger fishing companies chasing large tonnages of, say, menhaden. Today in America you cannot buy domestically-caught billfish, an achievement CCA notched on its scoreboard when it got gamefish status for all the big charismatics. Recognising where the levers of power are, the CCA is ultimately aiming to get its people on the fishery management councils. As David Cummins says; 'It is essential and imperative and the only way we can change the system.'

One of the stimulants giving new confidence to sport fishermen is that here and there in the gathering gloom of stock depletion there have been sparkling stories of success, moratoria imposed on both commercial and recreational extractions which have resulted in bounce-backs from stocks. The fabled story in marine fish rehabilitation programmes is the revival of red drum in Texas. It is a remarkable tale and demonstrates to doom-mongers that with sensible policies in place, with good biology backed by political will, all things in fishery resuscitation are possible. The red drum rehabilitation in Texas was conducted principally because of the sheer weight of sport angling numbers and pressure.

The red drum, a member of the drum family (so called because its constituents can vibrate muscles in their air bladders so producing drumming noises), is an ordinary-looking fish, generally copper in colour, with a black spot on the tail. Also called simply the redfish, it is a good grower and after three years weighs 6–8 pounds and measures 22–24 inches, still a far cry from the all-time record which hit the scales at 83 pounds. The main population lives and spawns in the Gulf of Mexico, sexual maturity not being reached till the third or fourth year. Eggs drift on tidal currents into the bays, and young red drum stay near shore till age three when they progress into the Gulf. In certain weather and seasons they cleave again to the shoreline. When feeding in shallow water they stand on their noses, tails waving above the water like flags, similar in their ground-hoovering habits to another famed sport fish, the bonefish. Red drum prefer shallow coastal water and will move up rivers. All-importantly, red drum are also omnivorous. These characteristics make them particularly accessible to sport fishers with basic tackle.

Red drum were fished off the southern American states since 1700, when they were a delicacy amongst southern gentlemen landowners. After about two hundred years the range seems to have contracted and the fishery became focused on the Gulf of Mexico, in particular off Texas, Louisiana and Florida. For a hundred years red drum have been caught for sport as well as by commercial fishermen, the popularity of sport fishing really accelerating in the 1930s. At no time did red drum become a major food fishery or support a large commercial sector.

The recreational fishery for red drum was biggest in Texas, amounting to a third of the whole catch. Red drum were fished off spits, piers, jetties, docks, rocks, any projection into the sea, and also by wading and from boats. Boat fishermen in a larger space have historically notched up more fish. The key thing was that getting into action on the red drum was never a hard proposition. These fish were not spectacularly difficult to catch. They were numerous and not picky or choosy. When they ran in multitudes the water turned red and tackle in the way had to be withdrawn from the water.

Management of this popular fish started over a century ago. Size limits, top and bottom, were set in Texas. The avowed aim was conservation. The measures were ineffective, focusing on fish which were too small to sell anyway and so not much sought. There were efforts to protect adults when spawning, by outlawing netting in certain seasons. Gradually the regulation book got longer. In 1929 the most destructive engines, drag seines, were prohibited on the Texan coast. Worries about stocks were already pronounced. The trotlines favoured by Texan commercial fishermen for catching red drum were limited in numbers of hooks per line, number of lines, bait types, and seasons. Quotas were applied to the commercial sector.

These rules were made in an information vacuum; mostly they derived from management principles developed for inland lakes. No one knew how many red drum there were, how many were being caught, where they spawned, or at what age. Concern about the absence of rationale behind fishing restrictions came to a head in the 1970s, but identical criticisms had been levelled since the 1920s. Occasional researches had established that red drum were slow growers, and when they spawned, but it was not until management of the fishery was vested in the Texas Parks and Wildlife Department, along with industry and other public sector involvement, in the early 1970s, that scientific evaluations began to get up to date. By this time the fish was disappearing from usual haunts and getting smaller; catches had been falling since the 1960s. The 'sports' were upset.

This was bad news for Texan legislators. In 1978 there were 667,000 sport fishermen; this figure had climbed to nearly a million in 1983. Texans are forceful people and a million unhappy forceful people is an event which makes managers sit up. As there were well under a thousand commercial fishing licences there was never much doubt where the restorative actions would be directed.

The Parks and Wildlife staff wasted no time. They began a programme of wide and detailed interviews with game anglers to build up data on the fishery. Harvest surveys were started. They legislated that only natural baits, primarily pinfish (small and silver), could be used, which increased the proportion of larger red drum in the catches, and eased the plight of younger fish. Laboratory studies of red drum were commenced. Stocks were enhanced by the introduction of young fingerlings. These initiatives were implemented quickly.

By 1977 the Red Drum Conservation Act was law. It stipulated a maximum commercial harvest from each area of the coast, made reporting of sales mandatory, and opened the account with sport anglers, restricting them to catching ten a day, and restricting the number they could have in their possession to two. Improvements in the fishery failed quickly to materialise. Thus in 1981 red drum, along with spotted seatrout, a generically similar fish to red drum, and also a fêted beneficiary of Texan fishery conservation, were classed gamefish only; all sales were prohibited. This was an admission that managing commercial fishermen had not worked. Thirty-two resolutions passed between 1975 and 1981 had still not stopped overfishing. Part of this picture was illegal fishing; some folk did not like infringements on their customary liberties. Gill-nets were being detected and hauled up by the mile. The solution was to pass full authority for rescuing the resource to the Texas Parks and Wildlife Department Commission in 1983.

The Commission's research showed that gill-nets caught thirty per cent more fish when bays were artificially enhanced by stocking. Hatcheries were put up. The number of red drum larvae stocked in Texas bays in 1983 was one million, the number of fingerlings seven million. By 1993 these figures were 214 million and 32 million respectively. As they say, everything is big in Texas. Nearly a billion larvae had been planted out between 1983 and 1993, an effort on the Japanese scale, performed by one American state. The purpose was not munificence for its own sake. The wildlife people knew that stocking would help recovery from natural catastrophes; stocking was to enhance wild stocks, not obviate the need for them. Sure enough, catastrophes came, natural ones. There were severe freezes in 1983, 1984 and 1989, and a poisonous 'red tide' (red tides are now recognised as natural, not man-made events) in 1986; stocking helped reduce their impacts. Not content with the improving situation, and anxious about a purse seine fishery under federal rather than state control, which opened in the Gulf of Mexico in 1986, the regulators prohibited the use of all nets in 1988. Not only red drum was being helped; twenty-two other species of saltwater fish were protected by bag and size limits, and limits on how many could be possessed. Law enforcement was toughened, and in Texas law enforcement starts off tough.

The upturn in red drum availability has been dramatic, even astonishing. Sport fishing boats between 1993 and 1994 caught a record bag of 275,000. The largest ever Texan red drum (54 pounds) was landed in 1996. With bulging populations of red drum, a fly-rod fishery has arisen. Not only catches but scientifically-measured recruitment, rates of catch and availability of red drum are all rising. Numbers of fishing guides have doubled. Since 1981 the average size of a Texan red drum has more than doubled to over five pounds. Runs of the big autumn red drum, known as 'bulls', have been reinstated right through the red drum range in Texas, North Carolina, Florida and Louisiana, and these are weighing in between 20 and 50 pounds, big fish to hook off a beach. Anglers were permitted in 1994 to start retaining two oversize red drum, a move which deliberately encouraged the development of a trophy fishery, drawing attention away from younger immature fish. Red drum has become one of five major finfish landed from Texan bays which are collectively now worth over $2.5 billion to the state each year. This is a stiff figure to argue with.

Through this transformation the authorities have had wide public support;

and considerable corporate financial assistance. Looking back over the accomplishment of two decades, leading managers have attributed their success primarily to three actions: the net-ban, the passing and enforcing of strict sport fishing regulations, and hatchery production. Perhaps the legendary toughness of red drum, as contrasted to the related, but more difficult-to-manage spotted seatrout which has been less prone to recovery, was another factor. The Coastal Fishery Policy Director, Hal Osburn, wrote in an official letter in 1995: 'The biological, social and economic benefits to the State accrued by these recoveries have been undeniable.' Later in the letter there is a significant sentence: 'One of the secrets to this turnaround is the ability to control fishing mortality much more easily in a sport rod and reel fishery than in a commercial net fishery.' He goes on to say that the exclusion of nets has created the right climate in which to develop more hatcheries. Many Texan game anglers believe the resuscitation of other fish species will be hatchery-led.

Whilst saying on the one hand that the grip on healthy red drum stocks will never be let slip again, managers are stressing too that solutions to their own problems in Texas should not be regarded as panaceas. Different situations may need different solutions; in some other states, which adopted differently accented policies, red drum are back too, providing either easy targets and hard ones, depending on which way anglers choose to fish them and which fish they hone in on.

It is necessary to understand that the triumph of red drum is, from a wider perspective, circumscribed. Although indubitably that rare thing, a commercially successful sustainable use of a mainly wild population, the red drum case history is in many ways peculiar to itself. For a start, Texas is unusual in having a state territorial limit extending to sea not three miles but nine. The remit of the wildlife staff covered a much bigger area than it would in other states. Then, although red drum spawn beyond this limit, in federal waters, for much of the time they are concentrated, when the tides 'run red', near the coast. Also, the state officials had good relations with the federal people about red drum. Federal agencies worked with the wildlife department to protect red drum and the damaging 1986 seine net fishery for drum was rapidly closed down again. Texas lacked the tribal peoples' rights in fishing which bedevil fishery conservation policies in the north-west Pacific, and the state has no true subsistence fishermen whose need for food could distort policy perspectives. State authorities had, in short, the freedom to manage wisely.

There is a general governmental trend in the USA towards decentralisation. Fisheries have not escaped this drift. It has been observed by many that the management of the vast American EEZ by a small number of fishery management councils could be improved. There is talk of fishery regulation moving towards state-by-state management; it is echoed in Europe by initiatives from fishermen themselves, sick of being political pawns, favouring coastal zone management. Mike Sissenwine, talking from a mobile phone as he was about to enter another interminable fishery management council meeting, typically, had sharp points to make. The scale of the remits of fishery management councils was reasonable, in his view, relative to the scale of fish stocks. This has obvious application when considering all large resources in highly mobile charismatics, like tuna, marlin, swordfish and albacore. Even the much smaller red drum has a considerable range. Particularly for free-ranging fish, the more consensual the management over big areas the better.

Dr Sissenwine believes the evolution may be possible of a 'cascading process' from the large-scale to small, local-scale management, when the fishery science argues for it. Urging caution in too-dynamic change, he points out the legal framework is presently not in place for major steps in reform of fishery management.

America is presently in the throes of mighty upheaval and debate is hot over the introduction, so far very restricted, of ITQs. Maybe in America more than most countries, to take away what is perceived as an entrenched freedom, like the one to bear arms, is immensely difficult. Fighting the introduction of ITQs at every step are those who would be disenfranchised, as business consolidation took its natural course. The position of America in fisheries is peculiar: very advanced in the development of sport and game fishing but mired in an out-of-date open access mentality when it comes to many coastal and inshore fisheries.

America's rôle in the evolution of sport angling as a major use of the fishery resource is critical. Although there are other countries with large sport angling fraternities – Australia, South Africa, New Zealand – America is the home and originator of saltwater fishing, and the place where most is happening. The sport bandwagon is growing and the extreme bias towards commercial fishing on fishery management councils is being shaken by the sheer paucity of fish and the gathering drumbeats of complaint.

It is a defining time in American marine fisheries. Some experiences are discomfiting. On the North Atlantic coast, the famous fishery for the striped bass or 'striper', a large bass which can grow to over six feet long, was by the application of a total fishing ban resuscitated from chronic over-netting, only to be subjected to a management plan by the Atlantic States Marine Fisheries Commission (responsible for the inshore fishery) which sport anglers claim is far too generous to fishermen and far too punishing on the scarcely-revived stock. Under the plan for the 'restored' fishery, commercial quotas were tripled and size limits were fixed allowing the capture of females, most of which will not have reached the size for reproduction. The pressure for resumption of commercial exploitation, typically in so many stock recovery situations, was intense.

Sport fishers and marine conservationists are appalled. They said the evidence on which the so-called recovery was based was either rigged or scientifically flawed. They maintained that estimates for the number of stripers caught as a by-catch in the dogfish fishery were eleven times too low, and that the new measures are opening things up much too fast. They said the striped bass, with its freshwater spawning lifecycle, has always been a predominantly recreational species, citing the value in 1990 for the commercial fishery in the Chesapeake area of $1.1 million compared to $22.3 million for the recreational fishery in direct revenue alone. Stressing their conservation credentials, they maintained that this revenue is generated without significant stress on stocks; over ninety per cent of striper anglers now put their catch back. Striper survival after catch and release has been studied off Maryland; survival rates are exceptional, up to ninety-eight per cent. The striper, they said, remains vulnerable to water pollution; acid rain 'pulses' caused by storms were shown to have killed off entire stocks of young fish in some of their headwater rivers. In short, such reckless handouts to the commercials are rash.

The striper was once described as 'possibly the most adaptable species on earth.' This is because in 1879 and 1881 a total of 435 yearling stripers caught by seine nets took the train from the east coast, across America, to the unfamiliar climes of the colder waters of the Pacific off California. There they thrived. By 1899, a mere twenty years later, commercial netsmen from California landed 1,234,000 pounds of stripers. The sea is veritably a fecund place: which is more reason rather than less that it should be managed honestly and responsibly. The bottom line is if stripers take another beating, the sports will lobby for it to be classed a gamefish only.

Sports are conscious that their own public image must appear squeaky clean. It is a source of some sensitivity that the bread-and-butter target of east coast sports, the bluefish, at one time caught in greater quantities by sport fishermen than by nets, seems stuck in a disturbing population slough. Historically the sports have sometimes been looked at askance, principally because they appear to be more affluent. Flashy boats with racks of rods streaking from port at dawn do not convey an image of the underdog. On the other hand high-tech sport angling means a bigger dollar lobby, and, of course, the expensive outfits are merely more conspicuous not more numerous. Surveys show that forty per cent of American inshore anglers still have no boat. In order to dodge the accusation of causing fish unnecessary pain, and to play the catch and release card effectively, there is a movement away from too-light tackle in favour of strong gear with which hooked fish can be winched in quickly and let go with minimum stress. Sports have increasingly played up the virtues of the act of fishing, as opposed simply to the act of catching fish. Spokesmen say, 'We limit our catch not catch the limit.' Sport fishermen are usually not subject to quota as such, but daily bag limits equate to effort limitation. Perhaps rather dangerously, given that they are calling for strict treatment of the commercials, some sports are presenting themselves as the standard-bearers of open and truly equal access to fisheries. While this may contain a grain of truth, wise counsels for fishing American waters tend, in most situations and as an ideal, aim to include both recreational and commercial sectors in management. Management means restriction, on occasion even abstinence. Most sports recognise this. Surprised sports took heart from a vote in Florida in which a constitutional amendment limiting the use of nets at sea was approved by three quarters. The rise in conservation awareness should head uncommitted supporters their way. The fact that a few of the larger charismatics, like Pacific marlin, can still be sold by sports under special licence, is viewed by some with embarrassment and unease. The chorus of objection to this, from within sport fishing circles, is rising.

It remains irritating proof, however, of America's sustained embroilment with the commerce of fishing that it is in other smaller and poorer countries that completely debarring certain important sport species from all commercial exploitation has first occurred. New Zealand allows no commercial exploitation of any marlin species within its EEZ; several South American countries, for example, Venezuela and Guatemala, and up to a point Mexico, have legislated to protect sport species from commercial use in favour of the valuable tourism afforded by the charismatics. Naturally, America does not crave the sport angling tourists as these other poorer countries do; and highly vocal user interests are more influential and have more power.

'Fishery management' opined John Brownlee, 'has always been reactive; only recently has it become pro-active.' One of the proofs of this was the mid-1990s birth of the Recreation Fishing Alliance (RFA), an umbrella organisation for those with vested interests in practising and servicing sport fishing. With a more sport fishing emphasis than the Coastal Conservation Association it has set out to restore American fisheries, protect the jobs in sport fishing, and provide saltwater angling for the clients. Already the RFA has turned back a move to raise the moratorium on stripers in further-out federal waters and been prominent in the fight to keep bluefin tuna in the recreational sector. Its further agenda is to: identify some fish particularly important to sport fishermen and have them declared gamefish, thereby outlawing their sale; to better protect inshore spawning grounds; to build artificial reefs in places unsuited to commercial fishing; finally, to assail the over-exploitation of large pelagics such as sharks, billfish and tuna. The RFA's appearance has been called 'the single most important event in the arena of marine fisheries management to take place in the last twenty years.' In addition a 1996 Sustainable Fisheries Act produced some conservation measures aimed at consoling, even encouraging, the sports. The days when sports were simply chased off the water by angry commercial fishermen may be exiting.

Meanwhile sport fishing itself has, in some departments, evolved. Partly it is the shortage of fish which has caused saltwater angling to become more high-tech and more attentive to the particular characteristics of the target fish, partly it is a trend which has occurred anyway in freshwater angling and migrated across. As in all recreations which have their own dynamic, there has to be a novelty, something new to be tested out. So, some wild-eyed men are now struggling with large demersal species which they hook up on deepwater fly tackle. Fishing deepwater flies in tidal flows is no easy trick. Other sports can be part of the era of 'multi-lure magic', 'spreader bars', 'daisy chains', or any other of the rigs which the more thoughtful of the exponents of fish-attraction have come up with. By studying the characteristics of baitfish, the fish which anglers' lures are trying to imitate, some American innovators have developed entire theories to support their complicated fishing rigs. Colour, size, and the number of lures simulating baitfish in any one configuration, become critical considerations. The American author Fred Archer has taken this as far as constructing geometric rigs of imitation baitfish which pull closer together, imitating small fish responses to attack when being chased by a predator. He hones in on the challenge of getting a disinterested, non-feeding fish to strike. One type of rig is put out on one side of the boat, and a different spread, representing different prey, coiffured up with something to hint that they are vulnerable, is put on the other. In the process of building up his store of information on baitfish and the triggers prompting predator action, the likes of Fred Archer have increased knowledge of fish ecosystems. Mr Archer says: 'We do research and try to teach, and consult really good fishermen. Our purpose is educational, not informational.'

The trend toward refinement in teasing into action solitary fish is possibly connected to the movement away from offshore angling with heavy gear to inshore fishing with fly tackle. Clear water fishing for bonefish or tarpon actually targets a visible fish, to be galvanised into action, whatever its mood. Fly fishing has hit the mood of the times, artistic, delicate, as accessible to women as to men. By-products of the sports' intense investigations into their pastime come in unexpected shapes and sizes. For example, species-selective baitfish are being looked at in Norway as a possible tool in commercial fisheries management.

The importance of sea angling is growing. It offers what many seek – solitude and relative peace. As one sport angler remarked, 'There are less people out there, none if you leave early enough.' The sport has many features to commend it in a contemporary context and a few, such as over-generous licences to sell the catch, which besmirch it. Rod anglers saying they need to market the catch to pay for the cost of getting to sea have a frail case; other recreations cost money too. To what extent sport fishing holds promise as a use for the marine resource in general in the wider sense is hard to judge precisely. In very cold climes the appeal is more limited. The Barents Sea is unlikely ever to be smothered in craft manned by recreational fishermen. Sport fishing for crustacea sounds like a fantasy, unless you have visited those seaside resorts where the youngsters – along with a few grown men in a state of mental fixation – dangle hand-lines for the obliging crabs. Saltwater angling as a visitor activity requires infrastructure, but can pay good dividends. In the small harbour off Horta in Faial, a westerly island of the mid-Atlantic Azores, the one-hundred-day game angling season starting at the end of June grosses around US $100 million for the local economy from just the three principal sportfishing outfitters. American sport fishermen will pay $100,000 to get their own boats delivered to Horta in order to have a crack at the record-size Atlantic blue marlin and bluefin tuna. Anywhere that it catches on, and the list of sport fishing Third World countries is lengthening, will find that more careful management of marine gamefish and their environment pays off. From the fish-in-the-sea view, the more sea anglers the better.

Sport fishing's rôle in fisheries is, for most countries, complementary. Even the most hyperbolic advocates of aquaculture as a replacement source of seafish for human populations admit it could never do more than contribute to our food needs. A controlled harvest from the sea is infinitely the best method of putting fresh fish on the table. Hatchery production figures pale into insignificance when compared to wild fish reproductive performance. So in any foreseeable scenario commercial fishing's use of the fish in the sea will be the main one. Sport fishing is a useful and interesting ancillary use to this. In its capacity to assist researches by tagging fish which are then released, it can play a valuable part in fishery science.

The fact that it has completely supplanted in economic importance the inshore fishery in America is significant. Essentially its orbit will probably remain within range of shore for a charter or private boat in one day's motoring. Although a form of angling exists for most species of seafish, barring real deepwater species, the sports will tend more to the inshore coastal species and charismatics like tuna and billfish. Nonetheless it should be noted that the heart of sport fishing in waters largely devoid of charismatics, like those around Britain, is in humble species such as cod and mackerel and pollock. As pressure on terrestrial space intensifies more boats will push to sea, and more sea anglers will clamber onto the rocks and cast into the incoming waves, the strains and tribulations of working life slowly draining away while the sun's face describes its arc around the sky.

Chapter 10

Orri Vigfússon: An Icelandic Mission

'Science has given us knowledge – we KNOW the salmon is under threat. What science has not given us, and perhaps never will, is the WISDOM to stop the decline of a species we all cherish.' (Orri Vigfússon to the Livery Dinner at London's Fishmonger's Hall, March 1997)

'Politicians can always say, there is no true answer. They prefer a complicated solution'. (Dr Ricardo Santos, Department of Oceanography and Fisheries, University of the Azores).

The gradual replacement of the inshore commercial fishery in America by sport angling consisted in essence of the maximum value of the fish being realised. The sport fisher is looking for fish one by one; the commercial fisherman, crustacean gathering aside, is looking for a shoal. Naturally, far higher value accrues, fish by fish, to the one suspended from the angler's tackle. A philosophy of wise use examines where the maximum value for each fish can be gained. This entails too, for the commercial fisherman, catching the fish in its optimum condition, not, for example, post-spawning, when it has lost weight and condition. However, for fish to realise their optimum value, a fisherman has to abstain from catching them at an earlier date. One initiative started in 1989 took on board the value of this abstention from the harvest, tried to quantify it, and then reasoned that the abstaining fisherman should be compensated, paid not to fish. The initiative was extraordinary in more ways than one: it was private; it came from Iceland; it has been successful; it has resulted in a new understanding of the possibilities of resource management; and the individual who broke the fresh ground was himself from a family of commercial fishermen and had vested interests in commercial fishing.

The object of Orri Vigfússon's rescue mission was the most furiously championed fish of any. Many may support the protection of seals but the majority of supporters will never have been near one. The Atlantic salmon was a famous gamefish in Europe a hundred or more years ago. The books devoted to its praise run into hundreds. It arouses in its admirers a near-religious fervour. The beauty, flavour, and sporting qualities of the fish have made it a symbol of environmental health in the North Atlantic on both the west side and the east. Orri Vigfússon did not have to explain to anybody what the Atlantic salmon was.

Until the late 1950s Atlantic salmon were exploited by most salmon-homing

207

countries – the main ones were Canada, Russia, Iceland, Norway, Ireland and Britain – in home-waters, with both inshore net fisheries and rod fisheries. Most fluctuations in the stock size, and the timing of the runs, could be attributed to environmental changes. By the early 1960s two interceptory fisheries for Atlantic salmon had started up in the North Atlantic, both by states taking fish which were on passage, of stocks mixed up from all over, fish at the far extremities of their migratory range.

One was off Greenland where Atlantic salmon from both North America and Europe, in roughly equal proportions, spend the winter feeding off euphausiids, capelin and sandeels. Greenland is a country with a small population almost all of whom are fishermen or from fishing families. Shrimp is the principal catch and the cornerstone of the national economy. Greenlanders being fishermen it is unsurprising that they were already catching salmon in the early 1900s. However, it was in 1959 that some Greenlanders discovered salmon could be caught in fixed gill-nets in certain fjords. Salmon were valuable fish in the 1960s and it was not long before Danish, Faeroese and Norwegian fishermen had joined in, fishing now with floating drift-nets. Ice was a serious hazard and for this reason almost the whole catch was taken off west Greenland, although when fishing was possible east Greenland appeared to have as many fish. The nets, originally nylon, then monofilament, were fished fast, because of predatory seals and the vagaries of arctic weather, and usually less than two miles from the shore. The salmon harvest must have seemed like manna from heaven; opportunities for profitable employment on the narrow strip of Greenland's east coast were not exactly varied. Catches rose steadily from 60 tons in 1960 to 2,689 tons in 1971, a huge total when it is considered that Atlantic salmon is a comparatively rare fish, even in its favourite haunts. In the 1972 season ICES calculated that the Greenland drift-net fishery had removed 33 per cent of the whole stock present in west Greenland. By this time the Greenland fishery was attracting international attention. Fishing by outsiders was stopped by the Greenlanders between 1972 and 1975, and in 1976 a total allowable catch (TAC) was set of 1,190 tons.

The Greenlanders tried to regulate the catching of the TAC, introducing licensing for all salmon fishermen and provisions to prevent processing plants from buying from unlicensed fishermen. There were both free quotas, open to all, and regionally-allocated quotas. As time went by and tagging of fish gathered pace it became clear that the fish caught off Greenland were mostly of Canadian and Scottish origin, mostly females, and mostly older fish which would have spent two winters at sea by the time they re-entered home-waters.

Landings plummeted in 1983 and 1984 to 310 and 297 tons respectively, massively underperforming the TAC. Although they rose again, by 1989, the year Orri Vigfússon got into action, the catch was down to 337 tons on a TAC of 900 tons.

Until the 1950s the number of salmon caught in the sea each year off the Faeroes, the other *arriviste* salmon catcher, had not exceeded five individuals. The Faeroese fishery deployed long lines baited with sprats not nets, and took off later and more slowly. It gathered pace when Danish vessels began to participate in 1978, and the Faeroese deployment was itself forty-four boats when the catch of Atlantic salmon reached its peak in 1981. Over a thousand tons were landed. The Faeroese catch appeared, from limited tagging data, to consist of fish mostly from Sweden, Norway, England and Scotland; representatives from Canada, America and Russia were few. Irish fish showed up in the catch but only of small size, and below 60 centimetres length it was an offence to land them. The average weight of Faeroese fish landed was slightly higher than off Greenland, 7.7 pounds as contrasted to 6.6 pounds. Salmon probably occur off the Faeroes year-round and it seems while some are on course to return to natal rivers, others are headed north for offshore Greenland. In 1982 the Faeroes government agreed to apply a quota and in the first year it was set at 750 tons. Catches stayed up at around 500–800 tons until 1988, usually underperforming the quota, by which time regulations had limited both the number of boats and the fishing season.

By the late 1980s and early 1990s Atlantic salmon catches were starting to slide into a steady downward trend in all major salmon-homing states except Iceland. The reason Iceland's catch figures showed improvement not decline is because that prescient state had developed a large salmon ranching operation, accounting for two thirds of catches. Total reported catches from all countries, including the two interceptory salmon fisheries, had fallen from a mean of about 12,000 tons around 1973 to 4,000 tons in 1991. Prolonged debate has circled on the question of Greenland's and Faeroese relevance to this decline, some scientists arguing that natural cycles and fluctuations, triggered by the environment, have a far greater effect than man's depradations on the stock. Others query how the Greenland and Faeroese catch can fail to be significant, when a proportion of those fish would have avoided natural mortality on the way home, circumvented any drift-nets deployed by home states, kept their mouths shut when offered an enticing fishing fly by a hopeful angler, and survived to spawn. People even calculated to the last salmon how many might have been saved, or not saved, had the nets been removed, factoring in figures for natural mortality, netting stations, and so on.

Orri Vigfússon approached the matter from a different angle. He thought the only correct way to exact a controlled harvest from a migratory anadromous fish was in the river of origin. He abhors high seas or interceptory fisheries, hijacking stocks not only from different rivers and different countries, but even from different continents; and also offshore drift-nets such as those of England, Ireland and Wales, taking fish which are homing sometimes to different countries, in the manner of an opportunistic scavenger. He abhors too the in-river nets, ensnaring stocks which radio-tracking has shown might depart the river again and go up a neighbouring one. Principally he abhors nets because they catch fish at the lowest point in their value. The prices of Atlantic salmon in the world market have been in crisis in most years since 1989, when Orri Vigfússon began his one-man rescue mission for the species in its wild state. In this time the wild fish catch has become a tiny fraction of the production figure for farmed salmon and the success of farmed salmon has suppressed the value of the product, providing extra force for the argument that the value of the fish in its wild state must be maximised. The point of a wild salmon is no longer primarily for human consumption. It is far too valuable to meet its end in the net of anyone, be it a Northumbrian drift-netter in a small boat, or a Greenlander in his small boat.

The Vigfússon argument is as follows. Stocks in 1997 are at an all-time historic low. Probably not more than four to five million adult Atlantic salmon exist in the world, a small stock by the standards of a smallish marine fish. Around 150,000 are being clocked returning to natal rivers in all of North America. A proportion of this total is of hatchery-reared fish, of similar genetic make-up to each other and of weaker ability in the wild to negotiate safe passage back to breeding streams. He is acutely conscious that even these estimates, all through the 1990s, have turned out to be over-optimistic when numbers of mature fish have been counted reappearing in rivers.

Commercial netting had to stop: there was no biological view on which it could be justified. The fish in a net was worth, on the most inclusive and comprehensive figure, one fortieth of the same fish to an angler. The availability of that fish in the recreation arena vastly widened the range of people who could extract enjoyment from it. As a consequence a rod-caught salmon supported far more jobs.

Orri Vigfússon believed governments in this scenario would be incapable of acting effectively. The salmon was too politicised. Politicians, perhaps in a politically sensitive constituency, would always be looking over their shoulders to see who would be disenfranchised, and where. Also, in the 1990s, governments are prone to instigate more research as a way of dodging practical actions. Research looks virtuous, pleases the scientists to whom people listen, and is hard to quarrel with on its own terms, without going to the expense, and incurring the delay, of commissioning different scientific programmes. Governments would always look for loopholes in the freestanding logic, and research was a way to make an art of loopholing. Salmon scientists have proved, in his view, that proliferating uncertainties is a perfectible art. Governments also need to watch their budgets; hoofing someone out of the fishery costs not only votes, but money, in compensation or unemployment benefit. As Orri Vigfússon pointed out, it is easier to raise money for a natural resource once it is gone. He quotes the famously expensive reintroduction programme for the wolf in America.

The initiative therefore had to be private and it had to be quick. Also – a critical thing – it had to be fair. He comes from a family in Iceland that did well in the herring boom, until the herring which swarmed past the north coast vanished. He is familiar with the quick descent from feast to famine. The fishermen would need compensating. He looked at the case of the Atlantic salmon through the eyes of a Greenland drift-netter and a Faeroese long liner. The fish was fattening in their waters, eating their capelin, their shrimps. Its physical perfection was built there. Without this winter bounty the salmon could never gain condition to return and spawn. Their entry into the fishery might be recent, but it was valid. They had rights and any agreement not to exercise them had to be paid for. He would be a fisherman speaking to another fisherman, conducting a business transaction. There would have to be reliable data and market price information on which to base a deal.

Lastly, it was not enough to appear and say, 'Stop fishing; here is some money'. He investigated ways to open alternative industries. The disenfranchised fisherman would need a new livelihood. It was not to be a social assistance programme but it would help with ideas, training and technology. It would not leave these native fishermen bereft as Canadian sealers had been when their exports were banned and they were supposed to be solaced with extra dollops of social assistance. Orri Vigfússon is from northern Iceland, the tough, uninhabited side. He is fundamentally sympathetic to isolated communites and those living on the fringe of civilisation. He did not want ex-salmon fishermen feeling that they had been out-manoeuvred by international money and left high and dry. He wanted to establish a common interest, not bludgeon home only his own. He has said, 'The reality of today's problems is much more in the realm of international diplomacy, in buying and selling and negotiating with interest groups.' To get his 1993 buy-out with Greenland he travelled to and from that outpost fourteen times. The alternative ventures are in snow-crabs and the production of lumpfish caviare, the latter particularly suited to the inshore fishery which was catching salmon. Harvesting sea urchins for their roe was another possibility, but top prices were paid in Japan for sea urchins which reached them within 24 hours. Greenland's communities are far-flung and the timetable was unrealistic.

The fact that quotas had already been set and established before Mr Vigfússon arrived provided a basis from which to go forward. The harsh truths that Faeroese and Greenland fishermen had had to swallow when they had been unable to actually catch their quotas made things easier. The reality was staring everyone in the face: either the resource was at risk of dying out, or it was plumbing new depths in environment-triggered cycles which no one had seen before.

Orri Vigfússon had the motivation and the negotiating skills and being a fisherman from Iceland, a politically neutral country, he had the credibility. He now needed the money. When he started what was to be a mini-lifetime of beating a path to the doors of politicians, conservationists and fishermen, he represented nobody. He had neither backing, nor any organisation. Playing on his few contacts he managed to get an audience for a few minutes with some senior figures from North America's Atlantic Salmon Federation in Washington. This group included some powerful and influential persons. He spoke for six minutes, after which he was halted. A chequebook was produced and they underwrote his attempt to buy the long lines off the Faeroes to the tune of $50,000. This was despite the fact that, then, only one American-tagged fish had ever been recorded in the Faeroese fishery. The Americans' immediate, imaginative grasp of his mission is one of the things he looks back to with

greatest appreciation in all his catalogue of negotiations and appeal-meetings.
Customarily the response he met from government officials was, simply, 'It can't be done. What you ask is impossible.' The Greenlanders said this initially; the Irish said this when asked to control their notoriously unbridled salmon drift-net fishery off the west and north-west coast (subsequently he was asked to join their salmon task force as an adviser, and the Irish passed legislation regulating salmon drift-nets in 1996); the Canadians mimicked the rest. The British Minister responsible for fisheries refused to meet him, but at a later date capped the number of licences in the biggest salmon drift-net fishery off Northumberland with the aim of phasing it out over thirty years.

Mr Vigfússon strung out more nets to ensnare fishery ministers than his forbears ever did to catch herrings. When the European Union Commissioner for Fisheries made remarks in 1996 about fair decommissioning schemes he wrote to her saying, yes, but include in your proposals wild salmon, the most pressured stock of all. Before the appallingly low figures for the 1996 Northumbrian drift-net fishery became public, but as the last net was hauled at season's end, he wrote to the then British Fisheries Minister and, citing previous ministerial promises about re-thinking policies if the stock situation became critical, said, stocks *are* in crisis now, your move, proceeding to enumerate suggestions about what to do. He was indefatigable, sifting new lines of reasoning, varying his arguments, inventing graphic new illustrations of his points (one of these is a medium-sized cargo ship on the sea, capable in its hold of carrying the whole Atlantic salmon population supposed to exist today), cross-referencing one country's responses to another's, tirelessly exposing the absurdity of managing a migratory stock at sea when it could be managed in natal rivers at a greater benefit to more people for incalculably better economic returns. Like many successful negotiators he attempted to profile his political adversary prior to any meeting, and see how he could tie in his own arguments with presently laid-down policy aims.

Meanwhile, erratically, unpredictably, in dribs and drabs and sometimes large lumps, the money arrived, sometimes from governments but in the main from individuals. Methods of collection varied. In Iceland a committee was set up, fishing clubs were approached and also fishing tenants from abroad, the succulent centre of the business, were approached. Friendly corporations were tapped. Different structures were established, country by country. Some individuals wrote six-figure cheques, some sent cash in envelopes with no attached address. Cash-laden envelopes arrived from all over, addressed to Orri Vigfússon, North Atlantic Salmon Fund, Iceland. Nonetheless, sometimes Orri Vigfússon would enter negotiations without knowing what money would be needed for a further phased buy-out and end up having to guarantee any shortfalls from his own pocket. A more recent development is committees of fundraisers in donor countries arguing over the extent, in precise percentages, to which they were beneficiaries of the buy-out. Unlike in America where saving the Atlantic salmon was initially perceived as a worthy end in itself, despite the fact that there was only a very slight vested interest, in Europe some states and

North Atlantic Salmon Fund committees tended to view their potential gain through slit eyes, calculators at the ready, extruding explanations as to why someone else should pay more, themselves less. Notable exceptions were France and Spain who donated more money proportional to their catch than was raised in other European states. The governments which never contributed at all were those of Canada, Britain, Ireland, Denmark and Finland. Private individuals from these countries, maybe niggled by their national parsimony, were magnanimous. A single dinner in 1995 in Ireland raised $200,000, and Ireland is a small country.

The volunteer-staffed North Atlantic Salmon Fund (NASF) was founded in 1991. The vision of what was being attempted sometimes caught on, sometimes did not. Seldom did Orri Vigfússon enter a negotiation with any cushion of cash.The norm, rather, was that he would underwrite the deal, then try and raise the cash and get NASF branches galvanised into action, afterwards. It must have taken a lot of nerve – and conviction.

By the end of 1996, after six years, NASF had raised $5 million. A figure in excess of this had already been committed to keep interceptory fisheries off and to implement the programmes of alternative employment. Between 1991 and 1996 the North Atlantic Salmon Conservation Organisation (NASCO), the international body responsible for issuing quotas to high seas commercial fishers, had recommended quotas totalling 3,669 tons. NASF had successfully bought off 98 per cent of these quotas, representing a phenomenal achievement. An analysis had been done demonstrating that some 400,000 salmon had as a result, between 1992 and 1995, enjoyed an unimpeded run home to North American and European rivers. If such an escapement had been contrived in hatchery programmes, of younger one-sea-winter grilse, it was calculated the cost could have risen to $110 million. The British arm of NASF has calculated that each of the extra fish in home-waters has cost £4.50 in donations to NASF. Sporting agents calculating rod fishing rents use a figure of around £300 a fish likely to be caught on the basis of historic records. The economic logic of NASF accomplishments is inescapable.

The stock assessment for Atlantic salmon bothers Orri Vigfússon. If the exercise of explaining how it is arrived at is somewhat tedious, the point is partially made: the assessment is messy, yet on it much hangs. Scientists from ICES do the stock assessment. They attend an annual spring-time meeting and report their findings. The findings go to the salmon advisory council of ICES, which consist of many of the same people. The advisory council makes recommendations to NASCO; often identical national representatives are present. The figures are then sent to the governments of all salmon-producing nations. After their submissions, NASCO tries to finalise a stock assessment. But anyone can exercise a veto; including Greenland and the Faeroes, both of whom are present and both of whom naturally wish for as large a stock assessment, followed by as large a quota, as possible. For this is the quota which Mr Vigfússon is on the doorstep wanting to buy. At no point are these comings and goings independently scrutinised, by outside analysts.

213

This procedure has been called sharply into question by the Canadian-based Atlantic Salmon Federation, the Atlantic salmon's most authoritative body. The federation believes quotas have wilfully been set too high. Furthermore, Dr John M. Anderson, the federation's vice-president, writing in its journal, has pointed out that ICES scientists have chronically miscalculated salmon returns to North America. Forecasts for the six years up to 1995 have been around forty per cent higher for the target sample, of salmon which have spent two winters at sea, than numbers actually being recorded coming home. No wonder Orri Vigfússon is sceptical of, and worried by, the underlying processes providing data on the basis of which he must do business.

Meanwhile, sometimes inspired by the Icelander's example, sometimes scenting the change in perception of how migratory fisheries are being looked at by all and sundry, various supporting organisations were abetting the drive to save salmon. To underpin Mr Vigfússon's 1993 two-year settlement with Greenland, Canada bought out salmon netting licences in Newfoundland and later in Labrador. All netting for salmon in Icelandic waters ended in perpetuity in a key breakthrough in 1997, with the Icelandic government paying substantially towards the buy-out cost. In Britain the Atlantic Salmon Conservation Trust had used private funds to buy off a range of estuary, headland and in-river netting stations round the coast. Many statutory river boards, responsible in Scotland for managing stocks of migratory fish, had done the same. 1997 was the first time for 300 years that salmon would run the Tay, Britain's most prolific salmon river, without any netting encumbrances at all. In most Atlantic salmon homing countries where the runs had reduced to a trickle, like Spain, France, Germany, and Denmark, restoration programmes were underway. Many major rivers had made huge steps in the improvement of their water quality. Fish-passes were repaired, salmon ladders re-built, both to assist the migratory passage of fish, and water pollution was made subject to heavy penalties. For the first time in a hundred years Atlantic salmon ran the Thames again in the early 1990s. The Thames was once so filthy that members of the Houses of Parliament had to hold scented handkerchiefs to their noses to avoid the stench. The Rhine is salmon-active in the 1990s too.

The Atlantic salmon had not been saved. Far from it. The 1996 catch figures were diabolical. But then the fish was under pressure from many more things than the Greenland and Faeroese interceptory fisheries. The salmon farming industry, employing large nets or 'cages' in estuaries and fjords on the west coasts of Scotland, Norway and Ireland, significantly contributed to the weakening of migratory fish runs (both of salmon and sea-trout), in many rivers, and to their virtual extirpation in some. The principal mechanism for this was a proliferation around salmon farms of parasitic sealice. These small brown bugs adhere to the salmon and suck nourishment from it. The cost of ridding salmon of sealice is one of salmon farmers' largest bills. They use very strong chemicals, different countries allowing the use of different treatments. The sealice, present in salmon farm estuaries in abnormal numbers, adhere to migratory smolts during the highly sensitive time when they emerge from freshwaters into saline estuaries. The smolts emerge tentatively, while osmo-regulatory changes adapt their metabolism to the salt. Sealice attach to them, eat off their protective mucous membrane and suck water from their bodies. In particularly bad situations sealice have been found almost completely covering the cadaver of the little smolt. The result is either that smolts die or they frantically return to freshwaters, missing their window of opportunity to get to sea and feed from the marine environment. In 1996 British government scientists found that sealice on salmon farms were often of different genetic strains; in other words, salmon marine aquaculture was breeding sealice. The last shreds of salmon farmers' protests that sealice were a natural hazard were stripped away. If smolts do not survive, clearly salmon stocks are destroyed at a primary stage in their development. Some sea-lochs in Scotland – Loch Fyne is an example – now have virtually no run of wild salmon into their lead-in rivers at all; the occasional salmon apprehended by the last persevering anglers is a farmed fish escaped from its cage in storms or when seals have rent holes in the net.

Irish and Norwegian scientists have proved the link between salmon-deficient rivers and salmon farms. In Britain, the world's second biggest farmed salmon producer after Norway, the government for a long time avoided involvement in any research which could harm the image of salmon farming, one of its flagship industries employing people on remote seaboards and an industry to which large and continuous streams of public supports had been made available. Through this time the environmental case against estuary and sea-loch salmon farming had been strengthening. Major calamities had occurred in Norway where whole stocks of farmed salmon had had to be destroyed after overstocking had led to viral infections and rampant disease. Sea-trout runs had vanished over wide areas, and the faeces from so many fish added to the detritus and waste food decaying underneath salmon cages which created a nasty chemicalised soup in what had once been one of western Europe's most unscathed environments.

The future of salmon farming – if feeding salmon in cages by pouring lower-grade fish into them proves sustainable and compatible with healthy whitefish-catching industries which also depend on this industrial foodfish – is in the open sea, or semi-protected minches with strong throughflows. Sealice do not adhere in open seas; the salmon farmers would enjoy much lower operating costs, and the fish would be fitter and less flabby from having to exert themselves against ocean currents. A few farms operate successfully at present in the open sea, seeing themselves as front runners in an inevitable exodus from the embattled theatre of the estuaries. However, sitting pretty in the estuaries is the comfier solution, especially in light of the high capital cost of making the move offshore, and the more expensive technology required to function in rougher conditions. The success of volume production has meant oversupply and the product price languishing in the doldrums for several years. A move offshore would increase the financial gearing of the salmon farm industry considerably at a time when expenditure is the last item on most aquaculturists' minds; on the other hand salmon growers would be released from the delimitations of estuaries already filled to overflowing.

The consequence of salmon farming has been the starving to death of many western Scottish, Irish and Norwegian rivers, failing to get back sufficient breeding salmon from the sea. Other troubles afflicting salmon include escaped farm fish occupying redds which wild fish would utilise, degradation of the upper catchment habitat where young fish commence life due to overgrazing and poorly designed afforestation, and, in some places, acidification of headwaters rendering small aquatic life insupportable. Orri Vigfússon's programme has to be measured against a host of counter-influences affecting salmon detrimentally.

Seals are part of the picture in a way already described. In addition, seals congregate around salmon farms and try to chase the fish into the corners of the nets where they can be bitten at through the mesh. As with sealice the presence of salmon farms concentrates a deadly predator just where wild salmon are most vulnerable. The NASF figure for the total population of North Atlantic seals, all species, is twelve million. That greatly exceeds, by about three times, the number of salmon. Although it is true that many of these seals are further north in the Atlantic than the Atlantic salmon usually reach, many are not; it is also true that seals kill many salmon in one day. When the first agreement with Greenland was struck in 1993, to suspend all except subsistence netting, it was one of the Greenlanders' fears that the surplus fish would merely be eaten by the hungry population of seals. There are a plenitude of reasons for the Atlantic salmon's low ebb.

ICES scientists say that ninety per cent of the salmon which leave their west Greenland wintering successfully re-enter rivers. Their safe sojourn off Greenland is essential. If the Greenland and Faeroese fisheries were still operating without restraint – Greenland turned down Orri Vigfússon's buy-out offer in 1996, fished, and ended up with profits considerably lower than the buy-out fee he had offered – the scenario would indubitably be in more distinct crisis now.

NASF has done more than just try and diminish the pressures on wild salmon. Its purpose is partly educational, and it has become an information centre for news on the North Atlantic. Fed by his extraordinary range of contacts, on fishing boats, in fish markets, in the Coastguards, in industry, amongst marine and environmental scientists, Mr Vigfússon can provide up-to-date information on a wide range of salmon and salmon habitat data. His office is awash with charts of sea temperature graphs, wind records, the movement of ocean currents, reports on catches of, and stock assessments on, capelin, sandeels, sprats, pout, and all the main commercial fish species. He charts American satellite information on ocean temperatures against the same readings as expressed by the Icelandic Marine Institute. Interestingly, he has an extremely close idea of any marine salmon poaching that takes place through his contacts in the markets. Serious-scale salmon poaching at sea is nearly nonexistent now, in his estimation. The reason he is provided with so much information, in some cases more than national governments would vouchsafe each other, is that he is personally a disinterested party, and he is manifestly trustworthy. The small, blue-eyed man with the square jaw has inspired many and received a heap of accolades, none of which mean much to him so long as his cherished goal – the restoration of the Atlantic salmon – is unachieved. His latest coup at the time of writing is to persuade the Royal Greenland Company, the biggest individual trader in North Atlantic seafoods, to commit itself to a certification programme on salmon, only buying fish from registered netsmen. Because of the dispersed and hard-to-monitor nature of Greenland's salmon network, this will mean more than meets the eye.

There are two sides to all stories, except fairy stories. Although Orri Vigfússon negotiated settlements with Greenland they were not to every fisherman's liking. The fishing community there was split. It was realised that previous agreements laboriously concluded with the Greenlanders had been violated in the remote, cut-off fishing communities themselves. Bruised by the collapse of the traditional harvest in seals, some of those with the longer view saw themselves getting gradually excluded from one of Greenland's few remaining resources. They saw that if they abstained from fishing for too long the outside world might argue their right of use had withered.

Einer Lencke, an attorney with the Greenland Home Rule Government, based in Copenhagen, and someone involved in the 1993 negotiations, said, 'The Greenland view is that no animal is sacred. Greenlanders think that if the resource can be harvested sustainably, they have a right to do it. There is a distinct non-utilisation risk with salmon. We have seen it happen with seals and whales.' Acknowledging that Greenland's salmon fishermen often make very little money from their salmon nets, he said, 'The importance is the lifestyle. And what of Orri's sack of money? Are they just going to drink it?' How frequently in peripheral communities, where native peoples are prohibited from continuing traditional lifestyles by high-minded urbanites basking in the luxury of unopposed moral declarations, is that cry heard!

Of course, the Orri Vigfússon solution expressly recognises the Greenlanders' underlying rights. In many regards the two perspectives are the same; he would like Greenlanders' lifestyles intact too. The drive to open new industries underpins this commitment. Until recently commercial salmon netting was a new lifestyle. The solution for Greenland's 50,000 inhabitants involves wider resource usage, not less. Salmon is irrelevant, ultimately. In 1992 salmon constituted only 0.2 per cent of exports. As a recent study, *Alternatives to the West Greenland Atlantic Salmon Fishery* from Dartmouth College in New Hampshire, shows, the paramount problem for Greenland is to get the US Marine Mammal Protection Act changed to allow Greenlanders to start exporting sealskin again, and hunting whales. To equate to the financial value of the 1992 salmon fishery, about 5,000 top-quality sealskins would have been needed and 29 minke whales. 1992 was a poor salmon fishing year, but the point is that neither minke whales nor seals are by any stretch of the imagination endangered. The Greenland-wintering salmon which home to Canada and

North America are the survivors of stocks and genetic strains many of which have already passed into extinction.

The Vigfússon case history is interesting because it shows the degree to which a great fish can inspire Herculean effort. Herculean is a word chosen advisedly: it is public knowledge that for some of his toughest sessions Orri Vigfússon was undergoing chemotherapy for cancer and only half-alive. In the wider frame the history of the NASF has more relevance. The Atlantic salmon may be a comparatively rare fish in the wild, but Orri Vigfússon's espousal of it has shown that one of his fundamental beliefs has great application: if you waited for governments and scientists it would be too late. He said in one of his speeches: 'We looked in the faces of scientists and politicians and decided it would be too costly and too late if we had to wait for them.' Many of those who have dedicated their lives to fisheries have felt exactly this despair about officialdom. Professor Warren Wooster, the distinguished founder of the Pacific equivalent of ICES, called the Pacific International Council for the Exploration of the Seas (PICES), expressed the same impatience about restoring Pacific salmon: 'I'm 75. I want to see something getting done,' he fumed, resignedly chuckling in wordly-wise conclusion.

Mr Vigfússon has a whole theory about this. In a speech made in Scotland entitled 'Farewell to Détente' he passionately lambasted the forces which prevailed against radical action. He made the point that scientists' requirement for unchallengeable truths meant, in effect, that they were blocking salmon rescue. He derided their reams of statistics when the essential facts were so clear. Civil servants, he said, performed more blocking. Politicians he berated for 'minimal action or none at all, since radical action tends to unbalance governments.' Urging anglers to participate in the debate and not surrender the resolution of their problems to biologists or other scientists, he called for an end to quotas if they were always inflated in political compromises, fewer meetings and entertainment evenings, proper compensation for retiring salmon netsmen, and for Scotland, one of the biggest users of the Atlantic salmon resource, to take up the cudgels and play a major rôle in its future management. The speech was by turns scalding criticism and focused hopefulness.

The point is: it could have been made about any fishery, not only Atlantic salmon. Mr Vigfússon's part in the 1990s in rescuing salmon is remarkable because it is private and because it has been effective in achieving some concrete aims it set out to achieve. Improved runs of fish to several parts of the western European coastline have been taken by the angling fraternity, after discovering that their timing coincides with the projected cumulative effects of the buy-out, as attributable to NASF action. Definitely in the case of salmon, governments would never have paid the Greenlanders and the Faeroese not to fish. One reason is that the Atlantic salmon, although mostly caught by commercial netsmen still (the proportion is fifty-five per cent; when NASF commenced it was eighty per cent), is perceived as a sport fish; those involved are seen as at play not at work; the salmon therefore matters less. What an absurdity! The truth is, it matters, for these precise reasons, more. West European governments have been remarkably slow to accept and institutionalise the vast economic significance of game sports tourism.

Orri Vigfússon's latest initiative, an effort to embrace in one movement the way of life of the North Atlantic as a whole, through an alliance of its various interests, has as one of its aims multispecies management in the sea. He sees clearly, as a few others are starting to, that this is the way forward for a completely rational, defensibly science-guided policy direction. Seabirds, marine mammals, man and fish must all be taken into the equations of resource allocation, and be called on to pay their dues in curtailment of predatory pressure. This theme is echoed by the European Bureau for Conservation and Development mentioned earlier. Maybe such broadbrush themes are too unwieldy for politicians; to be understood imagination is needed. Unilever's decision on sources of industrial fish, after all, may do more, as it is copied by others vying for the most environmentally-friendly image, than any CFP could achieve. And it is solidly based not only on market image and idealism but on the recognition that the safe future of other items in their products range, especially whitefish, hinges on a measure of ecological equilibrium. The 1997 formation of the Marine Stewardship Council between the World Wildlife Fund and Unilever, aimed at certifying that all fish products, not only industrial fish, come from sustainable sources, cements the process. Eco-labelling and general awareness about fisheries are coming fast. The independent panels set up by the Marine Stewardship Council have held, or are holding, workshops in six countries in 1997. Commerce and private individuals have a bigger part to play in saving the ecosystems in the sea than, up to now, perhaps either have realised. To an extent not seen before governments are either being sidelined or pushed from behind in the matter of ecology and environmental care. Orri Vigfússon was a front-runner in this process.

Chapter 11

The Future, By-catch and New Technology

'The merits of a fisherman can no longer be measured solely by how much he catches, but also on what he does not.' (Brent Paine, of United Catcher Boats)

'The public must become aware that catching fish for food in commercial quantities will require some by-catch.' (Steven K. Davis, LGL Alaska Research Associates, Anchorage)

The future of fisheries is going to be powerfully influenced by the question of by-catch and gear improvements and what can be done to match fishing's incidental fatalities to policies based on sound science. The 1992 high seas drift-net ban instituted by the UN is an example of what happens when powerful states are driven into a corner by pressure groups. To date the matter of discards and by-catch has, in general, been cursorily looked at in a simplistic fashion. The noise has been loud, illuminating and penetrating discussion slight. Because by-catch is susceptible to oversimplification it has often been forgotten that what matters in the end is the vigour of ecosystems. There will never be a way to eliminate by-catch totally and the public has to be prepared to understand that a level of by-catch is inevitable and, indeed, that in all harvesting operations, even say, of corn in the wheatfields or apples from the orchard, there is a degree of wastage. A fisherman at one fishing conference said the taxi on which he reached the meeting had a by-catch, in the form of all the insects on the radiator grille. The key point is not only how high is the degree of wastage, but how much does it matter in this place, at this time, in this stratum of this fishery.

Take the Bering Sea pollock fishery, the biggest fishery of a moderate sized finfish in the world. The by-catch figures look appalling. In 1992, studies determined that 300 million pollock were thrown back because they were undersized. Such a statement is typical of the type of statistical presentation made by conservationists fighting by-catch: it numbers individual fish. The full picture is in the wider-frame statistic: 300 million was only 1.6 per cent of the exploitable stock, even less of the stock including juveniles. The pollock stock is monumental. In most years despite the millions of fish discarded the by-catch rate is lower, around 0.5 per cent. For rock sole and flounders, trawling in the Bering Sea for pollock is more serious, although far fewer of them are being caught. It is important to note that fishery managers in the Bering Sea have recognised the high by-catch of young pollock and ruled that they are to be

220

deducted from the TAC. So fishermen pay for their mistakes, and pay hard. In the same 1992 season the value of all those young pollock discards in lost catch of adult pollock was $35 million. This is a big figure. The North Pacific has been an example of a very well-endowed fishery which has been managed, over the whole, with impressive seriousness of intent.

In some high-profile fisheries, where inadvertently-caught species are dear to the public heart, the level of by-catch has been more directly addressed with resounding success. One such is the tuna purse seine fishery in the eastern tropical Pacific; here a quarter of the world tuna catch is taken, in and around a stock of 4–8 million dolphins. Dolphins were the ideal marine mammals for which to launch a crusade: they jumped and pirouetted like ballet dancers, enjoyed interacting with people, even in the wild, and were warm-blooded.

The purse seine fishery for tuna started in the late 1950s. Concern about the entanglement of dolphins in the nets commenced in the early 1970s. Dolphin deaths were reckoned at over 300,000 a year, an ugly figure to live with. Facts gathered by the American National Marine Fisheries Service (NMFS) and the Inter-American Tropical Tuna Commission produced sums in 1972 which said 368,000 dolphins were brought aboard American vessels and 55,000 on non-American vessels. Although some would have been released alive, most would have drowned as the bottom of the purse closed and the net was pulled up. As in shark-finning a large animal was being routinely killed and utterly wasted.

Dolphins are medium-sized creatures and it is perhaps surprising that so many were caught when yellowfin tuna were the target. The explanation is that in the 1950s American fishermen twigged that dolphins and yellowfin tuna, for obscure reasons thought possibly to to do with dolphins' superior sonar-based prey-finding techniques which invited the tuna to tag along, lived in schools alongside each other. Fishermen paid out their one-and-a-half kilometre purse nets knowing that having seen the dolphins on the surface, there were likely to be tuna underneath. Dolphins betrayed the presence of the tuna. Tuna was and is a fish of great popularity in America; it awed and appalled tuna consumers to think of the ghost trail of dead dolphins that attended their mealtimes. The plight of the dolphins triggered a movement which resulted in the previously-mentioned Marine Mammal Protection Act of 1972.

Americans, once committed, know how to get on the warpath. Skippers and crews were trained in the safe handling of captured dolphins. Fishermen in tuna fleets were educated about dolphins and escape panels for dolphins were fitted into nets. Results were fast, amazingly fast. By 1977 American boats were down to a dolphin by-catch of 25,000; the figure was the same for the non-American boats. The American Tunaboat Association weighed in by making an annual award to the most successful dolphin releaser; in 1987 the winner encircled 137,000 dolphins and successfully released 99.9 per cent. By 1991 American boats had squeezed the dolphin figure down to a thousand; non-American boats were still catching 26,000. In 1993 American boats killed only 115 dolphins, foreign vessels 4,000.

The evolution behind the figures was as follows. To start with, American skippers had been allowed to catch dolphins inadvertently, without punishment. From 1976 there were quotas on the by-catch; these were progressively ratcheted down. In 1988 the Marine Mammal Protection Act was amended. Observers would have to attend tuna fishing operations, and fishing had to stop in the dark. In a rapid progression a 1992 special Act for dolphins took protection a step further. Any tuna put on the market had to carry certification; it had to be 'dolphin-safe', or caught in a manner which was consistent with dolphin escapement. The result over this time had been that most American boats had stopped fishing for tuna; this explains, partially, the freefall in the by-catch. It also explains the slower reduction in non-American dolphin by-catch. Other countries had taken up the slack in the eastern tropical Pacific. The main ones were Colombia, Ecuador, Vanuatu, Panama, Mexico and Venezuela.

The 'dolphin-safe' concept was America's reaction to being displaced in the fishery, as well as part of the ongoing dolphin protection crusade. It decreed no tuna could be sold in the States unless it came from a fishery with the approved rules. America being a large buyer of tuna, and close by to where many of them were being caught, this exerted muscular leverage over tuna-catching methodologies. Intermediaries in tuna-trading were targeted. In addition to the tuna-catching rules considerable funds were deployed on tuna research – $1 million in 1992 and 1993 – investigating how to fish without catching dolphins, finding out at what time the fish and the marine mammals separate, and researching their different diets. The use of sonar for separating tuna and dolphins was researched. Even using remote-control craft, banning purse seines, and using different net materials were tried. Observers were placed on vessels trying to qualify for the 'dolphin-safe' ticket. Foreign countries were forced, by steps, to reduce their dolphin by-catch to similar rates as the American boats. Finally America threatened additional sanctions, within fisheries, on nations which continued not to fish for tuna in the approved fashion. By 1990 all participants in the eastern tropical Pacific tuna fishery had adopted resolutions to safeguard dolphins. America had certainly exhibited a tough, co-ordinated,and comprehensive determination to secure the well-being of dolphins. They did, in fact, everything short of stop eating tuna. The success of the policy may have been even greater than the dramatic figures suggest. Early dolphin catches, before observers were counting them, may have been too low. Other tuna fisheries – off the Ivory Coast, Sri Lanka, the Philippines – recorded a heavier by-catch of dolphins per ton of tuna. It is perhaps only in America that this vigorous effort could have been attended by continuing pressure from conservationists to save the last dolphin. Lawsuits were pursued, attacking the NMFS for not being circumspect enough.

What has happened meanwhile to the subjects of the crusade? According to the Inter-American Tropical Tuna Commission, the international body established in 1949 and given a remit to consider dolphins in 1977, the recruitment rate for small cetaceans such as dolphins is 2 per cent. Marine mammals replenish their stocks much more slowly than fish. The highest take of any dolphin species by 1992 was safely below that, at 0.6 per cent. Dolphin numbers are slowly increasing. The Commission has examined alternative ways of catching tuna, but they all have a by-catch in other species such as sharks, billfish and turtles.

The fate of incidentally-caught dolphins had been at once the hottest subject in by-catch and the one which could boast, after thirty-five years of being addressed, that by-catch rates had dramatically tumbled. The problem had been solved but it left by-catch high on the agenda, not only in the public eye but in international fisheries conferences too. Populations of fish run down to nothing could not be photographed; they directly affect only those accustomed to eating them, selling them, or catching them. By-catch could be used as a stick with which to beat world fisheries. Photographs and film of fish being tipped back overboard, cascades of rejected silver protein, had the power to strike home on a wide front. The American by-catch expert, Lee Alverson, has referred to the 'press feeding frenzy' which developed over by-catch. By-catch has even become a word used to describe waste in fisheries in general. This has been a good thing in that it has focused on the need for proper harvesting in a world that can no longer be prodigal with its resources, and a bad thing in that matters have been oversimplified. Too often by-catch is talked of, as indeed I have done earlier in this book, solely in terms of rates of discard proportional to retained food-fish. These figures tell what part of the catch is surplus to requirements, not the meaning of the surplus in biological or ecosystem terms. Low by-catches of rare species – Atlantic salmon is an example – could have serious repercussions if that particular run of fish is a large part, or all, of the total run homing to a particular river, genetically unique in its adaptation to that natal habitat. The intemperate diatribes in newspapers and from conservation groups on 'the rape of the sea' refer to fish being wiped out: in fact, no marine fish has ever been wiped out by man.

By-catch has often been simplistically reported, statistically misreported, or reported without any consideration of its true environmental significance, and with even less consideration of the economics and politics which have contrived existing fishing systems in the first place. From a scientist's point of view in most fisheries the implications for fish's natural production of long-term climate change, and a changing environment, are of immensely greater significance. But long-term climate change is a turn-off, devoid of obvious perpetrators, and by-catch is a palpable subject on which folk can hone their indignation. Scientists are certainly perturbed by the waste entailed in by-catch, but they are perturbed more by the fact that fish thrown back are not tabled in the statistics. They are data gone to waste, not only fish fillets gone to waste.

Again, from a fishery scientist's point of view what matters is not what goes back into the water, but overall mortality. As mentioned in the discussion of by-catch and discards in the EU fishing area, the new evaluations of fish survival after escaping encounters with fishing gear are in some cases alarming. Enhancing the marketability of non-target species is obviously a good idea if by-catch cannot be avoided, but doing so can exert pressure on the by-catch stock by giving it a value, involve too-young specimens being caught, and interfere with other regulations on minimum size limits, etc. Specialist fishers for the by-catch fish are going to be thoroughly aggrieved if some outsiders have a reason to suddenly move the emphasis over to what were previously non-target fish. In one American halibut fishery where chinook are inadvertently caught as by-catch the fishermen keep the chinook to make money by selling it. Other systems which try and bring realism to bear provide incentives for retaining by-catch. If the fate of the by-catch is conversion to low-grade fish-meal some will argue it would be better returned into the food-chain to sustain other high-value stocks. The much-reviled shrimp fishery certainly has huge by-catches. However, in the southern Gulf States of the USA the by-catch is Atlantic croaker. They all die within a year anyway. Their loss is not commensurate with that of, say, a valuable flatfish. By-catch should be examined case by case, with pragmatism, occasionally, regretfully, involving the concession that some by-catch in a key fishery is unavoidable.

For perspective it is worth looking back over the history of fisheries. The early fisheries tended to target single species. In the early fishery for salmon in the North Pacific several species were not even fit for canning and were discarded. In the early cod fisheries everything was thrown overboard except the cod, including rejects that today have a high value. Present times have moved more in the direction of multiple use of fish and marine resources. Marketing has rendered our tastes more catholic. The French are pioneering new cuisines for deepwater species from the Atlantic; the New Zealanders have done the same for their deepwater fish. Velvet crabs and whelks are amongst a host of crustacea which are presently shipped from Britain to Spain, previously of no value, but which the Spanish are delighted to pay for. Elver fishers in the Severn Estuary in western England are selling live elvers to China for restocking Chinese rivers. In Japan virtually everything which hails from saltwater, even highly toxic puffer fish, is accommodated within a various and ingenious culinary preparation. Utilising these materials and putting a value on them will potentially bring them within the focus of management; their rôle in the ecosystem will be recognised complementarily in their rôle in human diet. These are all fair considerations attending the subject of by-catch.

The use of observers has proved effective in reducing by-catch. Treatment of the dolphin by-catch responded well to the presence onboard of outside monitors. Observer presence is easier, however, in a fishery where a few large vessels are docking in a limited number of ports. Where there are thousands of vessels docking in a multitude of small ports it is harder. Having observers counting fish on boats is costly. Automated technologies may answer this, and global satellite positioning, although not doing what observers do, could assist.

Observers have to be trained in the identification of fish. In the CFP area fish are routinely misdescribed to circuit over-quota situations. When observers were put on Korean and Taiwanese squid boats in the North Pacific they themselves were observed as not always being able to distinguish one by-catch fish from another. Unco-operative host fishing boats can find ways to distract the attention of observers; a young observer who had spent time on Taiwanese squid boats off the Falklands told me conditions were so awful no one could bear to go. In taut fishing situations like the CFP area in Europe, where the illegal proportion of catch is so big, one imagines the life insurance premiums for observers could be forbidding. Among the warring European fishing states there are presently no observers, nor talk of them. Norway on the other hand, the by-catch professionals, use all means to control their fishermen and gain knowledge of the fishery from the content of catches.

At present trap and pot fishers are able to argue their catch is in better condition than netted fish because it has not been crushed under the weight of its fellows; also that traps and pots can select the target age better. The potters and trap-men claim they are more ecologically-friendly, true in some cases, but as the Bering Sea crabbers show, not in others. Some associations of pot and trap-men are arguing they should have fishing-grounds reserved for them, for example, in western Scotland. Other countries – predictably Japan is a case – already have this arrangement, as of old. Indeed, in Japan some areas are reserved for hand collection of abalone by divers, a harvesting method of great antiquity which used to be performed by women. These female divers once became so highly-valued that the incidence of sex-selective infanticide fell.

Most attention, however, has focused on how to make trawls more selective. The House of Lords Select Committee Report advocated square mesh escape panels for round fish fisheries targeting fish such as cod, and diamond mesh panels for flatfish like turbot. It helps that much of the process of fishing can now be photographed. Sorting-grids have proved their mettle in this way. It is not only in the EU that gear scientists watch film of trawls approaching fish at different speeds and depths to devise escape hatches and other methods to scare off or release non-target fish. The day when the trawl would have an onboard camera in the cod end, enabling fishing to be halted if the wrong fish are headed into the net, is not inconceivably far off. Already acoustic alarms to deter porpoises, dolphins and whales away from fishing gear have proved successful on traps set for cod and on gill-nets set for groundfish. Different sounds, it transpires, are audible to fish and cetaceans. Whales, most obviously, can wreck large supplies of fishing gear; fishermen in New England where these experiments have been going for fifteen years had every reason to be fully co-operative with the scientists.

A limitless extravaganza of technical wizardry to curb the horrors of by-catch is not on the agenda. America, where more money has been spent on marine conservation than anywhere else, has deployed vast sums trying to restore Atlantic salmon to New England with negligible success. To achieve a twenty per cent reduction between 1991 and 1993 in the amount of halibut caught as a by-catch of groundfish fisheries the US Office of Marine Conservation spent $9 million. The cost of limiting by-catch cannot rise so high that fisheries become uneconomic. At the same time America has been responsive to public feeling. In the case of Atlantic salmon, dolphins, and sea turtles, the last of which America has been protecting by insisting on Turtle Excluder Devices (TEDs) since the late 1980s, protection and restoration programmes had strong public support.

The FAO stated in 1995 that a sixty per cent reduction in discards by the year 2000 was achievable. Given that the total for discards in all the world's fisheries is reckoned by Lee Alverson to be 27 million tons, representing a quarter of the global marine fishing catch, and that a major player, the EU, is in political deadlock on fisheries, it is a brave rather than realistic standpoint. Structural and organisational changes aside, the FAO reckoned the principal area to bring this turnaround about was more selective fishing gear. The FAO's targets were translated into solid lasting form later in 1995 in the Code of Conduct for Responsible Fisheries. This document, already part of the jargon in fishery management circles, calls for, *inter alia*, fewer discards, the development of technology to achieve this, a ban on fish dynamiting, poisoning and other neo-brutalist fishing practices, less lost fishing gear and 'ghost fishing', institutionalised reporting not only of target fish but of by-catch as well, protection and rehabilitation of special areas such as nursery and spawning areas, and the requirement of legislators to make regulations which take into account selective gears. This last in particular would introduce an excellent dimension into those fishery management regimes still conducted simply as a carving-up of the resource.

The UN's resolutions have not yet, as per the ban on high seas drift-nets, the full force of law. But documents like the Code put down markers; the best of them become reference points against which conscientious members can measure their performance, and they can become internationally-accepted standards of value to brandish under the noses of those who steer a course in isolation. There may be refinement of the Code; it is unlikely there will be a retreat from it. It was the fact that the UN took up the strong line on pelagic high seas drift-nets, admittedly shoved on by America, that enabled the USA soon after to justify the link-up between non-compliance and trade embargoes which proved so effective. Forget for the moment the justness of the drift-net ban and the whole question of certification, which can be an excuse for reviving old-fashioned trade protectionism: the role of the UN coupled with trade sanctions were completely effective, and quick.

Most nations cannot wave the big stick at other nations. Some, such as members of the EU, have surrendered the big stick powers to an overarching body anyway. The contempt with which its fishermen hold the CFP reduces the EU's standing in international negotiations; most fishing states at least have identifiable philosophies of fishing. There is one immediate solution for the by-catch problem which can be instituted anywhere without further ado, and it is frequently mentioned as a footnote to the technical intricacies of by-catch. It is effort control. This limits the time spent fishing. If fishing effort had been strictly controlled the scandal of by-catch would probably never have seen the light of day.

Where the worst by-catch disasters occur the skipper is having to hunt for his last kilogram of quota in a sea that has little left of what he is after. Consequently the net comes up with the wrong fish, which it may be illegal to land, and the whole lot goes overboard again, air bladders all expanded thus trapping the fish on the surface, a procedure which continues until he fills the last corner of the hold. One classic by-catch event which the British press latched onto as the heat was rising about the absurdities of the CFP involved a Scottish skipper in 1993 who photographed himself dumping an excellent catch of 2,000 boxes of large saithe, a haul which would have fetched £100,000 in the market if he had been allowed to land it. The points to be made were numerous: the target fish were scarce where the skipper was fishing; good quality saithe were plentiful, contradicting scientists' low stock projection and very restricted TAC; and why not have a quota system for numerous fish instead of one species, enabling this skipper to bring home and utilise the whole catch? The saithe were dead anyway. Where was the gain, under the CFP's conservation rules, in dumping them? The answer, often trotted out, is that the fishery administrators, dealing with a diverse fleet whose only unifying point is scorn for the regulatory system, fear their control over exploitation would be even less with multispecies quotas. It is not a strong defence of a lamentable waste. When the crunch comes in European fisheries, so will multispecies quotas.

The FAO cannot address the mess of Europe's CFP. Fortunately most fisheries are run with less strain and fewer contradictions. In addressing by-catch, however, one of the main contemporary issues about fisheries is being nailed. Admittedly by-catch remedies apply to some fisheries better than others. Admittedly fish escapement and fish survival are separate things. (Survival after being put back can vary hugely even for the same species. Halibut mortality from the long line fishery in the North Pacific can be as little as two per cent; for some pot and trawl fisheries it is one hundred per cent). By-catch remains a burning issue, if for no other reason than because it excites so much public distaste. The other issues are less technical and more abstract, more in the realm of politics and philosophy.

The world's capacity to catch fish is estimated to be around forty per cent greater than needed for the available fish. In some areas, for example where the fishing season has been cut to a few days, over-capacity in the fish-catching sector is clearly much greater. It is said we could catch all the fish available on a sustainable basis with the fishing fleets of the 1970s. Huge advances in fish-catching power have been made since then. Overcapacity matters to economists because capital is pointlessly deployed, and to fisheries because it becomes a political force pushing for reckless catches. Sticking to old-fashioned methods is possible only in unusual societies, either very affluent, or very disciplined and traditional. The overhanging question is: who is to drop out? This is the quintessential question for anyone who recognises what all, even fishermen, agree: that there is surplus fishing effort. Next, how is an exclusion policy to be justified? An exclusion policy is only justifiable, or only has a chance to be justifiable, within a coherent philosophy of fisheries management, which is applied with some rigour. Lastly, could the resource be increased? Obviously, if everyone stopped fishing tomorrow the stocks would bulk up. History shows that. But there are an estimated twelve and a half million fishermen, ten million of whom are 'artisanal', or low-tech and inshore, putting to sea in small boats. The myth that their efforts are ecologically benign has been proved as such: a myth. They provide livelihoods for 200 million people. Fish is the primary source of animal protein eaten by man. Stopping is not an option. Gaining more from the sea, however, is; I will come back to it.

This is the age of freeing-up of trade between nations. Increasingly tariffs and trade barriers are being assaulted and broken down in the stampede towards a global market. The overcapacity in fisheries, or the question of who gets pushed out, will, in the event, be decided on economic grounds. The countries like Japan which have strong economies and strong cultures and are able to lead the way fisheries are going, instead of reacting to events, are a small minority. The examples of New Zealand and Iceland, also Australia, and partially in the EU, in allocating fishery resources through ITQs, have spread and will continue to spread. What is dubbed 'the privatisation of the sea' will increase at the expense of open access. The coinage of Hugo Grotius's statement that the sea is a common resource is at last getting tarnished. As already intimated, the showdown in countries where small citizens have big rights will be mighty. The matter of what Serge Garcia and R.J.R. Grainger have called 'the gratuity of access to the resource', or the possibility of governments levying a user fee, will arise if quota holders start to look really rich in the context of strongly egalitarian societies, against a background of disgruntled, disenfranchised fishermen.

There are different sorts of fishermen today. Young men are leaving port in Europe in boats for which they have borrowed £10 million. These are high-risk ventures backed by harrowingly suppositional economics which would have made their fathers turn grey overnight. The bankers, on the other hand, sleep well. Big loans justify the arrangement costs, high-tech boats have a good re-sale value, and are versatile and capable of fishing in weather which keeps other boats in port. For the stakeholders in the right place in the game, profits, even in the embattled CFP area, can be big. That is partly why the CFP area is embattled. The Japanese market will pay figures in the region of £70,000 a ton for European lobsters, once a commonplace shellfish found widely round European coasts. American crab fishing boats, also capable of processing, fishing inside Russia's North Pacific zone at the peak of the fishery in 1993 hauled catches worth $20 million. Costs of the high-tech 170-foot crab catcher-processors could be paid off in eight months of frenzied operation, representing a phenomenal business opportunity. In the aggressive, everyone-for-himself sockeye salmon fishery in Bristol Bay, Alaska, where fast aluminium 1,000 horsepower boats are not shy of ramming each other and tearing through another fisherman's (or 'competitor's') net, the toughest young skippers can gross for their crew of four $300,000 in the four to six week season. A Bristol Bay sockeye old-timer said 'There isn't a polite fisherman left there anymore.' Half a kilo of live eels in China's depleted habitats (freshwater) is worth nearly $10,000 and fishermen have killed each other for them. A Japanese longliner fishing illegally in British Indian Ocean Territory in 1994 was punished in the courts by the loss of a catch of marlin, tuna and sword-fish, worth, along with the tackle needed to catch them, $1.5 million. This helps to explain why over-capitalisation seems a process with a momentum of its own. As the fish get scarcer bigger investments are made in cutting-edge technologies which will get this skipper to the resource in front of his fellows. It is called 'greed' by Greenpeace; it is staying in the game to the players.

The deficit in world fisheries is estimated by the FAO to be $22 billion yearly, a jaw-slackening figure. Subsidies in the form of vessel-scrapping incentives, investment grants, low interest rate concessions, fuel rebates and price supports are said to total $54 million across the world. At its most demanding, fisheries in the Faeroes islands took thirty per cent of the whole government budget for one year. The fisheries persist because, like the Atlantic salmon netsmen in Greenland or Britain say, 'It's a way of life', because governments see it as a duty to restrict imports, because votes are involved, and because from some perspectives it is better to have a man idling away his time on the sea than being paid unemployment benefit to do the same, or worse, on land. In the middle of this picture many fishermen go bust or contact their quota broker to sell out.

The temptation of tapping the natural riches of the world has a certain effect on men still. Empty nets are forgotten. Dreams power ahead of cold probabilities. It happened in gold rushes the world over. Where the resource is deep down and cannot be seen, the jackpot dreams perennially recur. To some extent the excitement of fisheries permeates the whole sector. In some indefinable form it is why I am writing this book. Until recently a woman aged a hundred was still working, intermittently, at a fish-processing firm in Buckie, a port in north-east Scotland traditionally linked to fishing boats and the sea. In what other sector are centenarians plugging on keenly? The change from the past is that today hard work, luck, and the bumper catch are not, in many places, what constitute success. There are fewer and fewer fishing-grounds to be pioneered. There are limits on catches, punishments for exceeding limits. There are more multinationals in fishing, particularly in the big fishing nations, while in the smaller states the skipper leaving port often leaves floating on a sea of debt. One might suppose the challenge, the gamble, had gone out of it. Perhaps in ways it has. The quota, a paper abstraction, is one of a fishing skipper's principal assets; it can be worth more than the boat. Within a system of transferable quotas the risk and gamble are still there, because you stake your money, and opt to buy more quota, on a future stock. Old fishermen say, truthfully, it is not what it used to be. However, fishing still has the power to magnetise young men and polarise national attitudes. In the overall scheme of things the economics of fisheries are a minor part of the great majority of countries' balance sheets. The future of fisheries is important because people feel in some way that if we cannot harvest a dividend from the world's last gigantic natural resource without harming it, we no longer deserve to be here.

229

Chapter 12

New Directions

'If you give a man a fish, you feed him for a day. If you teach him to fish, you feed him for life.' (Old saying)

'The only way to conserve fish is to kill fewer of them and to have a significant effect, you have to kill significantly fewer of them.' (Professor John Shepherd, Director of Southampton University's Oceanography Centre, providing evidence to the House of Lords Select Committee report 'Fish Stock Conservation and Management', 1996)

'Year after year all over the world, the long-term sustainability of fish stocks is being sacrificed in favour of the short-term protection of employment in the fishing industry. But the lesson to be learned from the collapse of the Grand Banks fishery is that ignoring the warning signs year after year can end in the loss of thousands rather than hundreds of jobs. In their heart of hearts, scientists, fishermen, managers and politicians must know that action must be taken now to prevent a repeat of the Grand Banks fiasco nearer to home. The question is, will they take it?' (Concluding sentence of the House of Lords Select Committee Report)

'Unless the issues of straddling and highly migratory stocks are resolved through the establishment of a robust international regime, there is a real risk of high seas chaos resulting from controversial unilateral action to extend coastal state jurisdiction beyond the 200-mile limit.' (Kevin Crean and David Symes, 'Sailing into Calmer Waters', *Fisheries Management in Crisis*, Fishing News Books 1996)

'. . . they [fisheries biologists] scurry back to their formulae regardless of new techniques; also they lose their objectivity and get wound up emotionally in the subject.' (A top fisheries biologist)

Ever since the early 1990s the FAO, in harmony with other overview international bodies, has reckoned the catch of fish from the wild has peaked. It used to be thought that world fish production would peak around 100 million tons. Then 1995 world catch figures of 109 million tons were announced, just below the record set in 1994. How then was the take bigger when everyone had known most fisheries were seriously over-exploited, fully exploited, under-exploited, or had collapsed, in the ominous designations of fishery scientists?

The reason is apparent in a closer reading of the figures. China is the world's biggest fish producer but half the fish comes from freshwater aquaculture systems, mostly of carp. Peru and Chile are the next largest, and although they have buoyant aquaculture sectors, their catch is largely of anchovies and horse mackerel, small fish used for reduction into meal. Of the world production total of seven million tons of fish-meal and one million tons of fish-oil, half the meal ends up feeding poultry, a quarter feeding pigs, fifteen per cent feeding fish; seventy per cent of fish-oil goes into margarine. The Peruvian anchovy fishery has already carved furrows in the brows of politicians and all in fisheries by its legendary stock nosedive: a reckless overkill of 12.4 million tons in 1970, accentuated in its effects by climatic disruptions from El Niño events, had translated by 1973 to 2 million tons, which had translated to a mere 100,000 tons by 1983. Industrial fisheries, the new growth sector, could collapse just like cod fisheries.

Stripping down the record fish production figure and focusing on fish caught in the wild for direct human consumption, a different picture emerges. Skipjack tuna, the first high-value food fish in the list of the top twenty catches by volume, is number ten. The catches of all tuna, by weight by far the most important family of food fish, have not increased from the figure of 4.5 million tons since 1990. The star tuna, bluefin, has been grossly overfished both on the high seas and in national EEZs. Atlantic bluefin numbers are said to have dropped from a population of 250,000 in 1975 to 20,000 in 1994. It has become a small stock, on the brink of calamity. Although new tuna stocks are still being found, many of the routes which used to be fished do not have them any more. Tuna are highly migratory and can be elusive, but there is every sign that the peak catch of wild tuna has been reached. Japan has proved it is always ahead of the game in fisheries and has reduced its distantwater fleet at great speed, the fleet that primarily targeted tuna. The doughty cod is the next big food fish in the top twenty and as we have seen, the world's cod grounds all have an established history of use, stocks being caught are dangerously young, scientists are sending out their SOSs, and the possibility of greatly increased harvest on a steady rising curve is remote. In 1973 cod was the number two fish in the world. Since then world fish catches have moved from a preponderance of high-value fish to one of low-value industrial fish like anchovies, pilchards and sandeels. This process can go no further unless we are to sieve the water with fine-mesh nets for abundant protein in the form of creatures like krill, a concept which has been considered but veered away from in the face of practical difficulties, and the potential international outcry about messing up Antarctica.

The wild catch plateau has been reached. The global fish-map is complete; no great undiscovered fishing-grounds await discovery and deepwater fisheries are coming up against the law of diminishing returns. World catches grew from the 1950s, when they were around 20 million tons, because more fish were caught in the usual places and because the intensive, factory-style fishing processes developed in the northern hemisphere swept down to the southern hemisphere. Fossil fuels, barring an occasional price upsurge, had remained cheap relative to other commodities. The last nooks and crannies of the oceans could be sought and found, and were. The overfishing of the North Pacific and North Atlantic spread out worldwide in half a century. The biggest fish catchers became southern hemisphere Chile and Peru; a total reversal had occurred.

Where good fishing zones had been found long ago and steadily revisited by artisanal fishermen over the centuries, they were now targeted by modern vessels which needed to cover the fuel costs of ranging so far by taking the maximum number of fish. Fishery management systems developed in the north, based on large stocks of small numbers of commercial species, were introduced into tropical waters where the marine ecosystem was characterised by fish of hundreds of species in smaller populations. There ensued a mismatch of systems to resource. Conflicts between distantwater fleets and local fishermen have a long history. The fishermen from afar can strip out a fishery and move on; it is the local fisher who waits for stocks, his outside larder, to reassemble. The conflicts became meaner as fishing fleets became more mobile and as distantwater fleets ran out of new water to fish. Today some warmwater coastal states build reefs to keep trawlers out. The 1982 Convention on the Law of the Sea formalised the EEZ declarations of the 1960s and 1970s. Current thinking recognises that preserving local fishermen's rights is in many cases a desirable aim. Mass migrations of peoples, powered onwards by the forces of conflict and starvation, are recognised by rich countries as a source of global instability.

In the most dramatic way Canada defended her local fishermen's rights when the Minister Brian Tobin fired on Spanish vessels in international waters, going on to develop the concept of the 'contiguous zone', an area where 'Canadian' cod were ranging outside 200 miles but still belonged tó Canada. Mr Tobin recommended a localised 24-mile extension of Canada's EEZ. This concept, not surprisingly, was taken up rapidly by a number of fishing states citing fishing hot-spots on their territorial boundaries in international waters, amongst them the 'loophole' in the Barents Sea, and the extension of coastal states' control on the pollack-rich 'Donut Hole' in the Bering Sea. Whose were these pollack anyway – Russia's, America's, or was the stock independent of both these states' coastal pollack, as claimed by the distantwater fleets? No one seemed to look at her example, such was her devastating and frightening fishing opportunism in far-flung corners of the world, but Japan had already considered and acted on these concepts of flexible zoning, although in inshore waters, many decades earlier.

Shadowing the spread of large-scale fisheries has been the rise in value of fish, and consequently the polarising of the market destinations. World market fish values have quadrupled since 1970. The developed world – Japan, Europe and America – has become a bigger and bigger importer of fish as poorer countries have become bigger exporters, and smaller consumers. The General Agreement on Tariffs and Trade (GATT), gradually being introduced worldwide during the 1990s, is likely to increase the volume of far-freighted fish as trade barriers are dismantled. More fish will flow into the markets of richer countries. This will increasingly hurt the world's coastal-dwelling poor for whom the fish cruising a skiff's journey away was a vital lifeline. As fish prices have risen poorer countries have sold this, often their biggest renewable resource, as a way of alleviating foreign debt payments. The scale of this transfer is in some cases considerable: in 1996 fifty-five per cent of all food exports from Morocco was from fisheries.

Development agencies often contribute to the resource transfers. Money is loaned or given for infrastructure developments – new wharfs, processing facilities, fishing boats, outboard motors – improving the supply of good quality fish. The international fish trading corporations step in, offer above domestic market prices for fish, local fish supplies dwindle, and the underlying resource is pushed down harder by hungry people. The fact that world fisheries are in chronic deficit with gross overcapacity is well-known, green organisations never tire of saying it. But they do not mention what is driving the overcapacity. The powerful groups in fisheries are the corporate buyers and big traders. An example is the role of Germany in the EU's CFP. Germany is highly influential in the development of policy in the CFP, a major fish processor, and yet a small catcher of fish. The producers, the fishermen, have become merely a digit in the equation. The same has occurred in modern farming, where production methods and practices are governed by the retailing multiples.

The price of fish is sky-high. Never before has wild fish nearly doubled that of red meat in the shops. It does in 1997 in Britain. BSE has affected the image of beef, salmonella did the same to chicken and eggs. Scare stories foisted on a credulous media which spoons them up with a dash of sauce to a gullible and ignorant public are actually often market-place mechanisms, deliberately instigated by traders, to bring toppy prices down, or get a new product leveraged into the market place as all else tumbles around it. Fish has a clean image. Stories about fish concentrating heavy metals never affected open sea fish breasting the life-giving brine.

The market for fish both in Europe and elsewhere faces one insuperable problem. The supermarkets which now dominate fish-selling, and fish-buying, want continuity of supply. In France one big supermarket chain has actually bought its own trawler fleet to secure it. However, fish are not available all the year round in the same condition. After spawning a North Sea haddock is half its normal weight. It follows that fewer fish would need to be caught if they were caught in correct condition, relaxing pressure on stocks. Food-retailing multiples want fresh haddock present on the chill-slab year-round. In Europe many fish are being caught, sold, and eaten in indifferent condition. Fish cognoscenti like the Japanese would not contemplate eating the stuff; to them freshness and condition is all. In Spain, too, the markets do not sell fillets but whole fish. You can see what sort of fish you are buying; possibilities for deception are edited out. The timing of TAC decisions in the Council of Fishery Ministers in the EU exacerbates the situation. They meet in December. After their announcements fishermen rush out to catch their allocation before someone else gets it, or before the fish get so scarce that steaming around finding them becomes expensive. From February to April demersal white fish in the North Sea are heavy with roe. By the time they are in prime condition, in mid-winter, fishermen have run out of quota. The quota year should start in May, to make the most of the resource.

One point of view among fish traders is that fish are unimaginatively presented. As one fish salesman put it: 'The traditional philosophy was, "Stack it high, sell it cheap. Hide it, fry it, don't let the child know it's fish." The future as I see it is in more discernment about the fish we catch. I can tell if a monk-fish has been at the bottom of a trawl, being towed along for four or five hours. Soon, if we make the most of our fisheries, everybody will be able to tell.' This particular fish merchant thought we should follow the Norwegian example, land everything, and develop new markets on to which to offload it.

The wife of an Irish fisherman remarked: 'Many people talk about fisheries, but few of them know what it is to fish.' Every fisherman would assent to that. The dangers, dirt, smell, exhaustion and isolation of being on a fishing boat used, at least, to be shared by the scientists. Now, mostly, they are not. Except in countries with national fishing policies, today's fishermen feel undervalued and misunderstood. Their pillorying by the media as ransackers of the environment is something they find hard to come to terms with. Privatising fishing opportunities has produced schisms in the fishing community; where quotas have been made transferable quota owners have benefited financially, not deck hands. There are fissures throughout the industry's structure. The EU's CFP is probably the apotheosis of decision-making in complete detachment from the fishermen. It is a lack of connection that has contributed to the destructive attitude of 'Devil take the hindmost', and cavalier attitudes to the legality or otherwise of catches.

As the politicking in the CFP shows, fisheries are a pawn on a bigger political chessboard. Fisheries are treated expediently, which has led, in a disturbing way, to them having become the bailiwick in the EU of the biggest participant, Spain. Officials working for the European Commission report that DG XIV appears to be staffed mostly by Spanish and Portuguese. My attempts to elicit the truth of this were thwarted; after prolonged rummaging around and being referred to a long cast of different personnel I was told that information on staffing in the Fisheries Department was not public information. So much for open government.

Spain has been another example of what seems a generalisation of fisheries: that if you are big at the beginning and fight your political corner, you can do well. Japan clung on tenaciously in the North Pacific, bringing fish back to a hungry home market at a profit while its negotiators clung to their seats at the negotiating table, manufacturing ingenious arguments to achieve delay. Like Japan, Spain has a huge home market for fish, and is a major importer. The Spaniards eat fish three times a week, other Europeans once. Japan and Spain demonstrate that he who pays for the product can govern the one who provides it.

Like Japan, Spain has reaped the sea's harvests for a very long time, although unlike Japan Spain has no generous continental shelf to succour a big inshore fishery, so had to scour the oceans. Like the Japanese the Spanish are the aristocrats of fishing: they treat fish carefully when caught; and have continued with the prosecution of a pole-fishing technique using small boats, in which a row of men hold fishing-poles over the side, a method which looks extraordinarily archaic and manpower-demanding until you notice how frequently, almost continuously when the going is good, the anglers – for that is what they are – reel in the tuna. Pole-fishing lands a beautiful unbruised tuna, and fish being caught individually means market gluts can be avoided.

Like the Japanese the Spanish have a big deepwater sector. Unlike the Japanese and everyone else, who are dismantling or re-deploying theirs, the Spanish, tapping CFP re-build subsidy schemes, are enlarging their deepwater fleet. As the Japanese did, the Spanish have been quick to occupy and take over boats which fly the flags of different countries. This is the 'quota-hopping' scandal of the mid-1990s, when large numbers of British boats were bought by Spain, and some other European states, for their quotas. Nearly half of some sectors, such as beam trawling, became foreign-owned while still flying the British flag. It made an obvious nonsense of national quotas, the cornerstone of the CFP's conservation policy.

Spanish administrators and politicians have been caught out in public using the terms 'fishery development' and 'opportunities to open new fisheries' synonymously. Spanish public officials are on record advocating fishery policies based on fishing a stock down so low it becomes uneconomic, then zeroing in on another while it recovers. The Spanish have an appalling record for over-fishing and a worse one for breaking the regulations. Like the Japanese they regard paying fines when they are caught infringing conservation rules as part of the routine cost of fishing. The Spanish predilection for eating young fish wreaks havoc on stocks.

This relish for the babies has come into particularly high profile in 1997 in the Malaga region of sourthern Spain. The fry of anchovies are being sieved from the sea in small-mesh nets to satisfy the Spaniards' ingrained tradition of serving the young anchovies as 'chanquetes' at social occasions. No amount of public pressure against this fishery violation, including public campaigns using slogans such as 'Don't eat the future', and the confiscation of offending vessels, has succeeded in stopping it. For reasons like these the fishing catches in Spain's Mediterranean waters are in freefall in every species; which explains Spain's relentless lobbying to ensure that after the CFP review of 2002 Spanish access to the remaining waters from which it is presently excluded is secured, and that the attempts by third world countries to regain control of their fisheries is thwarted by EU strong-arm diplomacy.

The Spanish are suspected of being major offenders in the capture of charismatic billfish as a by-catch in high seas longlining. Up to 90 per cent of the great billfish are estimated to meet a tragic end, untargeted, on long lines. Spain is known to under-report this by-catch, or not report it at all. Spanish violations are indeed far-flung. Spanish-owned boats, often re-flagged to other countries, have been principal offenders in the notorious free-for-all, probably peaking in 1997, for the large and slow-growing toothfish in the Atlantic high seas off New Zealand. This opportunistic onslaught in a place hard to police is being conducted in defiance of the internationally-constituted Commission for the Conservation of Antarctic Marine Living Resources.

Like the Japanese too the Spanish care for fishing as an art. Older fishermen still dexterously tie some of the longline fish lures out of ears of real corn. Fishing is also part of culture. The labyrinth of nets, configured over many years, to trap migrating tuna in the Straits of Gibraltar, and the way the divers set the maze of netting in February, check it regularly, and then dive again around April, expectantly awaiting the arrival of the tuna shoals, shows a fishing culture at its most sedulous. The tuna, nosing their way towards the anchovies, sardines and mackerel burgeoning in the eastern Mediterranean, are cornered in the trap at the end of the maze. As the net is lifted and a huge splashing announces the presence of hundreds of tuna, including fish of nearly a thousand pounds, young men abseil off the boat decks and stand waist-deep in the thrashing fish, striking their gaffs into the enormous, fat, silvery bodies, then handing up the gaffs for others to heave aboard, and taking another. In this scene, for all its gore and death-drenched drama, is visible a people at work for whom fishing is more than just picking vegetables. The Spanish, whatever else may be said, fish from the heart.

To the criticisms inside Europe that they wreck other people's stocks the Spaniards have a retort which stings. They point to Spain's inexhaustible fishing energy, with fleets positioned in the Indian Ocean, the Pacific, South Atlantic, and round South America, far further spread than other European states. Their crews are willing to stay half a year away from home, exhibiting a commitment which marks them off.

Zeal, however, in the modern world is not enough. Indeed this old-fashioned rapacity, and swashbuckling sectoral self-interest, is out of kilter with contemporary thinking on shared resources. Spain's example is interesting because it shows a developed country adhering to archaic and historical habits, running right up against the buffers of sustainable ideologies. The world is moving from harvesting ideologies to husbandry-type ideologies. Of course, the unreconstructed aboriginals still exist (Sid Cook of Argus Mariner saw a man in the market-place in Kunak, Borneo, rushing about brandishing a stick of dynamite, explaining a hundred species of fish could be killed at once this way, and shouting 'This is the future') but fishery administrators worldwide have put on their thinking-caps for fear that some day, maybe not far off, they could be hanging up those hats permanently. There will be more major stock collapses, more eviscerated fishing communities and boats pulled up to rot, before sense is seen and politicians stop playing games. However, times are changing quickly, and everyone knows it. There are many examples of the husbandry approach. One which is being considered in many fishing states with, or even without, a good continental shelf, is the construction of artificial reefs.

In the case of artificial reefs, or the generality of 'fish-aggregating devices' (FADs) to which reefs belong, Japan's inevitable lead is a modest 200 years or so. They were constructing reefs in the 1700s. The Americans were the next to cotton onto the fact that you could enormously boost fish stocks in inshore fisheries by creating the right habitat; in their case it took an aircraft crash off Alabama and some perceptive local fishermen and charter captains to note and act on the fact that around the wrecked plane fish numbers multiplied. Now, excepting in the EU where pro-active stratagems are impossible, most fishing states in both temperate and tropical waters know about reefs and have built them, or intend to.

The theory is simple. It is evident to any wreck diver. A collapsed structure on the seabed soon starts a chain reaction. Weed and small organisms colonise the multiple surfaces. The outgrowths attract phytoplankton and zooplankton. The cavities, tunnels, caves, and arches of seabed structures provide that essential ingredient for marine life – shelter. Remember: the most dangerous event for a fish is the appearance of the wrong sort of other fish. Holes in the reef become fish houses. Congers lurk in the tunnels, parrotfish stare from the caves, young lobsters sit securely in the crevices as food filters down to them. Constructed reefs are merely man's effort to imitate coral reefs, nature's great FAD. Coral reefs are living organisms capable of concentrating fish populations in a remarkable way known to anyone who has dived near them even with simply a face-mask. The whole ecosystem becomes concentrated including, as those same weekend divers will know as they peer over their shoulders or paddle nervously round the spires of coral-heads, big sharks. Reefs have been blown up, picked at, and hewn away by the magpie human race precisely because they are the richest, densest aggregations of ocean life. Now, for reasons unknown, reefs of coral all round the world seem to be packing up and dying.

The Japanese have researched reef building with their customary professionalism. The government's reef structure and design guide has 68 prototypes. A whole section is devoted to tides, waves, currents, and sediment dynamics. The Japanese know from the shape and size of the reef which fish will be attracted. The positioning of a reef is critical; badly-placed ones fail to attract fish, or get broken up and washed away. Trials done in 1957 showed that one reef had increased fish numbers by 4 per cent after a year, while another of identical design in a different place had clocked up a 2,000 per cent gain. Fish were counted by divers. Present at 36 pounds per acre of seabed before the successful reef was installed, after 200 days fish were present at densities of 1,500 pounds per acre! It is hard to think of a system in nature which builds so fast. It makes aquaculture look positively plodding.

Reef design and positioning theories differ, but many think reefs perform best when at right angles to currents. This brings regular doses of organic material to the reef to enrich it on the wave side, and provides shelter on the other. Reefs are best placed where other natural structures are, or once were, or where young fish congregate. Young fish shoals give away the invisible nutritional benefits of favoured spots. Local fishermen can often tell best where reefs will prosper.

The structure of the reef is more important than what it is made of. Using submarines, underwater video, and acoustic recording through tagged fish, the Japanese have researched the eyesight of fish in relation to underwater fixed structures, and the ranges different species will move between structures. Fish mostly have poor visual resolution although their field of vision is wide.

Therefore the horizontal dimension is the most important, and reefs should be densely assembled. Although the Japanese have both high and low reefs for different sites, the higher-profile reefs being for deeper water and bigger fish, the low reef no higher than five metres from the seabed is the standard. The wake patterns behind reefs being hit by tides is a determinant in how far apart reefs should be. Of course, different species will travel different distances between reefs, demersals like soles, flounders, and dabs, apparently being content regularly to move 600 metres between structures. Maximum abundance of rock-dwelling fish was ascertained to be at distances between reefs of 400 metres.

In America artificial reefs came into being less as a result of enlightened policy but because her flexible system and adaptability rose to the occasion of happy findings. The Alabama fishermen who benefited from that unfortunate air crash began to their amazement to land red snapper, a novel species. State and sport fishermen's funds financed laying and securing 1,500 derelict automobile bodies on the seabed. They then sank old wartime troop transporter craft. Fishing boomed. Reef builders became so enthusiastic, dumping stuff everywhere, that the Coastguard stepped in and halted it. By now sport anglers thought of reefs as 'gardens that grow fish'. The American Corps of Engineers, which has jurisdiction over inshore seabed structures, joined in the act and gave permission for an area of nearly 1,000 square miles to be developed.

There arose the phenomenon of the grocery-cart reef (grocery-carts seem to have a bedded role in fisheries: remember the grocery-cart salmon fishing net as used by Indians on the salmon-spawning beds of British Columbian rivers). The aluminium cages were cabled together. As crustacea and other marine organisms colonised them, the shopping trolleys exfoliated. They were to carry more food and faster than they ever did in the supermarket. It was deduced over time that the more complex, heavier and bigger the structure, the more fish would grow on it.

Eventually regulations were devised, detailing what materials could and could not to be used – cars had to have working parts removed and the bodies steam-cleaned etc. – and commercial businesses began to manufacture fish-friendly structures which they called modules, of ecologically acceptable materials, which often resembled giant sponges. Reefs had to be marked with buoys. Reef experts also exfoliated. They came up with the delightful figure that an ex-Army tank could produce 1,600 pounds of fish a year against the performance of troop carriers which could avail sport anglers and bait fishers of an impressive 5,000 pounds of reef fish every year. They even learned that uncrushed car bodies hosted a different species assortment to crushed car bodies. For vermilion snapper it is a good flattened car, but for red snapper the car-frame as it came off the assembly-line is the ticket.

Artificial reefs had eloquent self-justifications. In 1995 despite its relatively short coastline the state of Alabama had a third of the recreational catch of red snapper, a fish once barely recognisable among all the southern USA Gulf states. A local tourist chief said the artificial reefs generated over $50 million a year for the local offshore economy in just two small cities, Orange Beach and Gulf Shores.

Artificial reefs seem to be a way of restoring, replacing and creating habitat to which there are few objections. The Third World has many reef projects up and running and stimulating offshore economies. Local materials can be used, or junk and rubble materials which have served their time. Materials of opportunity will often do. In the Gulf of Mexico redundant oil rigs have been felled and modified to become reefs. Fishermen know the same could be done to effect in the North Sea, if everyone could agree, since they noticed that the legs of oil rigs became fish sanctuaries, fishing boats not being allowed close, and feeding grounds. Attempts to put this idea on the agenda have grounded in the quagmire of sectoral self-interest; trawling grounds would be obstructed, for example. In the tropics the stumps of coconut trees have served as a good base material for reefs. Air bottles are tied to ropes which then point upwards and act as the scaffolding on which organic life settles and grows. Reef building is not beyond the resources of anybody. In most tropical countries the reefs are run on a community basis to provide local food.

One point, common to all reef fisheries, has to be reaffirmed. There must be a social discipline in the way reefs are used. They cannot be harvested prematurely. Baby lobsters, for example, can be planted out into the crevices of reefs. However, they take many years to grow to marketable size; they must be left until they yield their best. This involves restraint and community support. In America the reefs are common property. The charter captains instituted their own disciplines to ensure the fish gardens stayed in flower. Charter skippers did not sit on the reef all day picking off its citizens; they moved on when the sport angler had had a go. Artificial reefs test the concept of the fishing community.

Some modern technology takes the construction of an artificial reef as a secondary aim. Technology developed in America has looked at creative ways of addressing coastal erosion. This involves fixing in erosion-prone places materials composed of multiple fronds. The densely-packed fronds, of various compounds, simulate sea grasses and as they are hit by wave-action, accrue algae and build up an accumulation of micro-organisms, at the same time dissipating the force of the wave. Studies showed that in one area where the materials were deployed, after a hundred days, an original fish density of ten fish to every hundred square metres had turned into a thousand-fish density. At another site, though in freshwater, the materials had filled with crayfish. If beach erosion problems could be solved in tandem with fisheries enhancement, public funds could be saved and resources created for private use. The frond system is not restricted to erosion situations and other joint-project experiments are underway in Scotland and Spain for the increase of algae production in order to attract sea urchins, abalones and periwinkles, which then form the base of a chain attracting finfish.

Implicit in the concept of sea-husbandry is discipline over harvest. As touched on, there are divergencies about stock assessment in the scientific community, and divergences thereafter about appropriate rates of harvest, rates of replenishment, and so on. One of the many problems for scientists in producing 'clean' figures is that catching efficiency in fishing boats is increasing continuously. Logically all fisheries controlled by limiting the efforts of fishermen should measure the increase in catching efficiency and reduce fishing effort accordingly.

The core problem for most fisheries is obtaining reliable data from fishing boats. It is seldom that catches and the rate of catches are not factored into stock assessments; usually they are the scaffolding statistics. Alternative methods of assessing fish numbers, such as measuring the quantity of eggs found in the plankton, a method used for mackerel in the CFP, are expensive and not suited to all stocks. In the EU's CFP area, where the black catch is so high, discards are measured by some countries and not by others, and fish are deliberately mis-identified to get round quotas, the scientists have only approximate ideas of catches. Their indices of stock abundance remain inferior to, and less real, than actual catches. Because they miss information on catch numbers they also miss out on vital data about the age composition of the stock. In their mathematical models they may include guesstimates for these factors; these remain guesstimates. Without knowing what is being caught in the fishery, stock assessments and the consequential TACs are inevitably haphazard. This is often behind the repeat scenario of fishermen saying scientists' calculations are up the spout and the proof is in the net; they are finding far more fish than scientists predicted (they are slower to draw attention to fewer fish being found). Of course, an underestimate of true catches will lead to an underestimate of stocks, low TACs, and frustration for fishermen. This opacity in true catch figures has caused such distortions that the Mallaig and North-West Fishermen's Association in Scotland has hired its own scientist, a former fisherman, to whom the members will give true figures. One wonders what the fishery police, and fishery scientists, would do to get that scientist's notebooks, and what the fishermen would do to prevent it! Things have come to a pretty pass when official and unofficial scientists are gathering data alongside each other.

In addition to the absence of a solid base on which to make assessments, data gathering by the two sources ICES and the EU's own scientists, in the view of Mike Holden in *The Common Fisheries Policy*, is neither transparent, accessible to outside inspection, nor reliable. He says '. . . there is severe pressure within the scientific community to conform.' He goes on, 'The consequences are potentially disastrous.' This comes from someone once at the top of CFP administration. Scientists admit their figures are often far adrift. They concede that for most species stock assessment is an imprecise art.

More alarming is that the House of Lords Select Committee report found that fishery science was actually deteriorating. Too much effort was taken up with routine sampling, too little with developing an overview of what is happening. Because of failure to co-operate by some nation's statistical offices ICES has abandoned some of its fishing effort data. For the science to be in retreat as the need for it increases seems folly, presumably born of a despair at the future of the CFP and the diminution in importance of an industry in shrinkage.

Then there is the matter of inclusivity. About a dozen major species in the CFP are under quota. However, the number of species for which catches are recorded is 128. Over a hundred species have no TAC set for them, from which quotas would be divided between different fishing states. Dogfish, lemon sole, brill, John Dory, turbot, skate: all are without quota. Any state can fish for them. Most shellfish have no TAC, although there may or may not be technical measures such as size limits, seasons, licences, and so on, to control catches.

TACs are recognised to have merits in pelagic fisheries where surface-swimming fish, like herring and mackerel, congregate in huge shoals. In mixed demersal fisheries, when the net is pulled up full of different sorts of fish, and the fishermen may only have quota for one of them, the TAC and quota system, in the rudimentary form in operation today, is open to many criticisms. In a paper by fisheries analyst Brian O'Riordan, of Intermediate Technology, an international non-governmental organisation based in Rugby, England, the critical assault is severe: 'The system of TACs and quotas has played little or no role in the conservation of fish stocks, and has proved ineffective as a management tool.' Harsh words, perhaps, and fishery scientists might argue that European fisheries, in ramshackle state though they are, do still persist. However, the questions of inclusivity, of whether more fish should be subject to stock assessment and TACs, of when direct effort control is going to be adopted in the armoury of fishery managers trying to keep catches under control, and doubts about the adequacy of base data used by fishery managers, means the criticisms will not go away. European fish stocks hang on a wing and a prayer.

The multispecies management concept is edging onto the agenda. The concept of multispecies scientific assessment was first mentioned in the 1950s, but multispecies effects on the life history of a particular fish were measured as a constant. That was simplistic. A cod which had grown well could eat bigger fish than another of the same age which had grown slowly; and itself escape predation more successfully. More food was available to fish in some years than others, affecting growth. More critically, recruitment fluctuated wildly, just as it did for the single species under scrutiny.

During the 1980s scientists at the Lowestoft Laboratory tried to develop models which would include sizes of stocks and the condition of the fish in them. It was found that in longer-term scenarios multispecies models sometimes predicted the opposite of what single species models predicted, given an assortment of fisheries regulations such as bigger meshes in nets, lower catches, and so on. Scientists at Lowestoft have spoken of the 'formidable mathematical difficulties' of multispecies assessments; even single species are far from straightforward. Lower catches of important predator types like cod, whiting and saithe could increase mortality in the multispecies model. High catches of those species means the surviving individuals grow faster, availing themselves of more food, and seems, in a way not understood, to lead to a biologically-led attempt to re-balance the age classes. If cod, the most versatile predator, had not been a major commercial species there would have been an argument for fishing it down to nothing! The permutations facing fishery scientists are painfully complex. Dr John Pope of the Lowestoft Laboratory has written, 'Within North Sea modelling studies there is a continuous tension between the need to increase realism by adding details and the need to increase comprehensibility by using simplifying assumptions.'

The nub of the matter of the CFP is its complexity in every aspect: the fishery science is complex; the policy itself is complex; above all, the politics, which opportunistically intrudes, is complex. The view of virtually everybody informed about European fisheries is that the CFP and a common fishing area was a bridge too far. In the European context fisheries were an unsuitable vehicle for federal treatment. Their ad hoc inclusion on the agenda right at the beginning was an over-hasty error. Because of the EU's decision-making process, and the right of one nation to veto major treaty alterations, it seems to be a communal hammer and chain from which the frustrated participants cannot escape. It is the least stable fishery policy in any major fishing ground. Instability means imminent change, and between the time of writing and this text appearing in print much may have changed.

The CFP is governed, ultimately, as most fishery policies are, by wider politics. On top of that, also like most fishery policies, it is in thrall to a prevailing urban socio-cultural atmosphere opposed to a scientific and rationalist approach. Most people would agree that inclusivity in stock assessment is a desirable thing. Most would also urge that the industrial fishery, which accounts in the North Sea for a larger catch than the food fish catch, should be included in the science and in the management. Indeed before long it is likely, if politics permit, that the catches of pout, sandeels and so on will be under science-based quotas. There should be a management plan, and limited access to the fishery. Manifestly, it is the correct approach. However, inclusivity and the term multispecies also means consideration of the role of sea mammals. This will send shivers up the spines of knee-jerk conservationists.

It should deter nobody. The great taboo on sea mammals is logically indefensible. Greenpeace have moved to the absurdist position of saying the industrial catch should be controlled to provide more food for sea mammals. Just because we identify with seals, and Californians identify with sealions, does not mean that their rôle in the fishery should be out of bounds, nor that management of them should be out of bounds.

There are four ways fish stocks are reduced. The effect of fish on each other is by far the largest determinant. Man has his take, as we have seen, increasingly of smaller, industrial fish. Seabirds have their take, particularly of small fish in larval form, and when fish eggs float amongst the surface plankton. Seabird numbers in many places are increasing, almost certainly a trend connected to the volume of discards which provides a ready-caught supply of fresh fish floating for the taking. Lastly, there are marine mammals, the totem of environmentalists, but nonetheless huge participants in the fishery.

At the turn of the third millennium man has mostly reached a point of ceasing to harvest sea mammals. One of the four participants in the fishery has been pushed off into a separate compartment, beyond consideration, and the key turned in the lock. A logical and rational, properly science-guided fishing policy would include a harvest of sea mammals. Animal rights people are so worried about the possibility of scientific data being accumulated about the ecosystem role of sea mammals that some of them even object to the hunting of whales for further research by traditional whaling countries like Norway, on the grounds that the whale hunts are disguised commercial ventures. Those who cry out loudly about ecosystems and ecological approaches should realise what they are saying. An ecosystems approach involves first recognising how the components interact – and marine mammals are top predators – and then managing them. Abandoning this logic is an admission that resource allocation is now to be governed by socio-cultural considerations, and that science, fact, and the realities of natural systems demand courses of action we have become too squeamish, too detached from the truths of nature, to countenance.

Already, by and large, seabirds have been abandoned as a source of food. This is not for reasons of flavour: puffin is a dark meat which tastes agreeably like pheasant leg. It is that we regard eating seabirds as somehow archaic. Animal rights activists protested vociferously in 1996 when a boatload of Scottish islanders set off for their traditional harvest of gannets. There is no conservation issue. Gannets are numerous. Logically there is no reason not to eat gannet any more than there is not to eat wild duck. The issue, in truth, as fishermen are beginning to realise, is quite different. The issue for extreme greens is that we should be starting to question why we eat anything from the wild at all. Marine mammals and seabirds are off menu; now they want fish off as well. The fact that this would mean starvation for millions of poor coastal dwellers matters not a jot; they live in the purities of their own abstractions. The same, or similar, people argue that fish should not be used for reduction to meal for farm animal, bird and fish feeds. Aquaculture of marine fish, their pre-ferred option for the fish element in human diet, needs fish-derived feeds. Seafish will not thrive eating vegetables.

Whales are protected nowadays, so analysis of their diet from stomach con-tents is difficult. Undeterred, a British whale scientist, Christina Lockyer, tried to work out how much great whales ate from old southern hemisphere whaling statistics, and also from measurements of smaller specimens at a whaling station in Iceland. Handling animals of this size, in addition to the fact that, like ruminants, they have multiple stomachs with three chambers, presented her with a daunting task. It was known that blue and fin whales spend around 120 days a year on their feeding grounds in the Antarctic during which time their weight gain is astonishing. A blue whale, as the old whalers learnt, could increase its weight by half over the brief season. Christina Lockyer's tentative data suggested a blue whale ate about 2,200 pounds of krill four times a day. As the whales grew, presumably they ate more. Many whales eat larger fish directly. Nor are all whales, as is the general perception, rare. For example, the population of minke whales, the smallest of the rorquals, has been censused at around a million, with big areas of the range so far unstudied.

The rôle of whales in fisheries has been computed here and there by brave scientists. Some perspective needs to be put on their gargantuan consumption. Assuming sperm whales eat five per cent of their body weight daily it has been calculated, using International Whaling Commission figures, that in the North Pacific this cetacean alone consumes more fish and squids than the total catch by man. This is in the most productive fishery in the world.

The present public disquiet about whaling is the legacy of the uncontrolled slaughter of the past. In sanctifying whales we are atoning for these excesses. Whales have been shown too to have appealing characteristics. They are often family or 'pod' based, they suckle their young, they are capable of tremendous communications over a vast range (blue whales communicate on extremely low frequencies across ocean basins for distances of 5,000 miles), and they have an intelligence and sensitivity which is now part of popular culture. The immensely popular film *Free Willy* humanised a killer whale, capitalised on killer whales' proven gentleness in captivity and made the species irreconcilable with their equally well-attested ferocity as a marine pack-hunter. The more we lose touch with our own status as another biological race, with the same blend of sociability and single-minded self-replication, the more perfectly normal characteristics of animals appear to the fractured perceptions of modern man as subjects to be viewed with wonderment and awe. In all probability other life-forms, cobras for example, have touching and impressive features too, but we have not sought them out. Is it really enough to say that whales are valuable sentient beings which we must protect indefinitely from the forces of logic, from true ecosystem management? As said, a harvest of twenty-nine minke whales would, if the products had been marketable, have given the Greenlanders enough money to equal the value of the whole 1992 Atlantic salmon drift-net catch. As it happens, courtesy of Orri Vigfússon and his sponsors, the salmon catch is being bought off, or there is a good chance it will be bought off. But would it not have been better to have taken the twenty-nine minke whales?

The same arguments apply to seals, which in the opinion of some have substantially replaced whales in marine ecosystems since the over-harvest of whales in the past. Seals are more destructive eaters than whales, taking bites from fish and killing much that remains uneaten. Whales – the killers' habit of ripping the oil-rich tongues out of bigger whales is an exception – swallow fish whole. Numbers of seals have rocketed while many whale stocks will never re-attain antediluvian proportions. The Cape fur seal doubled its numbers from one million to two million between 1970 and 1990, and the figure is expected to double again by the year 2000. In several countries (South Africa, Norway, Scotland) it is estimated that seals are eating more fish than the nation's fishermen are landing. Where is the logic or justice in excluding them from management? The alternative is to admit that fisheries are to be managed only partially, because we still feel the need to atone for the gruesome excesses of the past, because we are too sentimental to concede that inclusive quotas, embracing all interdependent species, would be a progression.

The sea covers two thirds of the world's surface and contains by far its largest natural food resource. Already we have substantially restricted what we may take from it. Meanwhile, human numbers and needs are rising. If consistent policies about extractions from the sea are adopted the pressure to produce food from the land will trigger land-use changes which could be even more controversial, less sustainable, and put cherished land patterns at risk. Determining that sea mammals are to be excluded from the rules of exchange that apply to other natural forms will exact a penalty, and the price will be paid in another natural system.

Entire communities in the far north of the northern hemisphere were devastated when trade in seal products was prohibited. Those that have been in the far north and seen the sad results of this prohibition, and who understand the traditional rôle of sea mammals and fishing in these hard-pressed societies, are persuaded of where the injustices lie in double-quick time. Affluent societies find it hard to envisage, as products from all over the world are wheeled into supermarkets, what it means for a society to subsist entirely on one or two entirely natural products, then to have the traded value of them entirely eradicated. Anyone in doubt should ask an Inuit if the provision of canned and frozen basics in the store has been a fair exchange. The fact that a restoration of seal harvests and trade in seal products would rationalise the management of world fisheries makes the argument forceful. The emergence of bodies like the European Bureau for Conservation and Development, espousing the principle of multi-species management, demonstrates that the logic is getting more widespread recognition. An Irish EU fisheries official told me that proposals for fishery projects were increasingly required to have a multi-disciplinary approach, adding 'There must be a sea change in perspectives on sea mammals; in fact, it is happening already. The eco-sentimentalism of the 1980s is discredited now.'

This was confirmed at the 10th Meeting of the Conference to the Convention on International Trade in Endangered Species of Wild Fauna and Flora (more commonly called the CITES convention) held in summer 1997 in Zimbabwe. Over 1,100 delegates from 128 countries attended, and to the surprise even of many there, when Norway and Japan proposed downlisting of three species of whales – grey, minke and Bryde's – off the highly endangered list, votes were evenly balanced. The movement of a species from one listing to another needs a two thirds majority, so the status quo for these whales was retained; nonetheless delegates had effectively voted for the re-opening of commercial whaling. Commentators on the conference noted a new swing in opinion away from extreme protection in favour of sustainable use.

For the peoples of the far north there is still a tenacious hope that the trade ban will be lifted and that the world will again look more kindly at controlled marine mammal harvests. This is mirrored in a personal experience. In my jacket resides a slim leather-bound address book. Purchased over twenty years ago I discovered on trying to get the same leather in another item that it was sealskin. The leather is as pliant and durable and elegant as when it was new; on the evidence of the address book sealskin is one of the best leathers that exist. It is hard to accept that on the premises we have rehearsed it will never enter circulation again.

At the International Whaling Commission (IWC) in Cambridge, England, the Secretary, Ray Gambell, thinks that with regard to whales the far north people have a right to be optimistic. He thinks there has been a shift from the hardline attitudes of the 1980s. Accepting the arguments of those former whaling nations like Japan, Norway, and Iceland that scientific stock assessments of some whales have now been refined to within five per cent accuracy, and assuming extremely conservative harvest levels that there is no reason to abstain from whaling any more, he puts the evolution of feeling about whaling into the context of the 1982 UN Law of the Sea Convention, the UN Conference on the Environment and Development in Rio de Janeiro in 1992 (since known as the Rio Summit), and the crisis in commercial fish stocks. In his words several governments are increasingly taking the view 'We can't make whales so special they stand out from the general run of things.' He talks of the developing feeling of self-confidence in asserting aboriginal hunting rights amongst the native communities of Scandinavia, Greenland and Canada. Viewed through a wider cultural prism: 'The habit of imposing one culture on another is becoming less acceptable.' The efforts to ban all whaling forwarded by some of the 39-nation IWC membership, usually, like Switzerland, from countries with no history of whaling or of use of whale products, is thus against the trend towards management of the natural world on an ecosystem or 'wise use' basis. It is also against the interests of commercial fishing and jobs in fishing and processing, and heavily punitive on just the type of aboriginal or native peoples for whom public sympathy is rising. What is being proposed is not a reversion to the opportunist whaling of the past, which neglected to utilise whole sections of whale, but rather a science-led cull only of whales, like the minke, which are numerous and have been censused. This would be carried out only in areas where they have been censused, by local communities in what is called 'small-type coastal whaling' operations, in which the animal is put to optimum use.

To keep whale populations in trim with fish stocks other nations, like Norway, Iceland, and Japan would harvest whales too, above board, without recrimination, without threats of trade retaliation. The number killed would be irrelevant to the health of stocks. Those opting to buy sea mammal products would look on the choice as a way of apologising to the people of the far north for a historic wrong. Revived uses could be found, for example for the leathers, which would reduce the need for farmed skins which so distresses welfarists. Sea mammals would return to the fold of natural and renewable resources.

It is vital to see that behind the determined attitudes of Inuits and peoples of the far north today, pleading for the right to harvest and market traditional prey, lies a painful history of bitter ironies. In the late nineteenth century when the whalemen who had depleted western arctic whales turned to killing walruses for their oil, they commenced to deplete the Eskimos' last staple. For the Eskimo hunts for seals, whales and walruses were interdependent. Whaling harpoons were attached to 35 fathoms of line made from walrus-hide; attached to the line were inflated seal-skins called 'pokes' which exercised drag on the line exhausting the whale. The whale trailed as many as twenty floats, enabling the hunters to close for the kill. The jointed-together, therefore flexibly-framed umiaks, or rowing-boats, from which they pursued whales were icefloe-proofed with the tough hide of the bearded seal.

The aftermath of the huge walrus hunts was a pitiable human tragedy: returning whalers found arctic winters had decimated Eskimo communities, sometimes wiping out the last living soul. Without walruses, and with whales gone, they had starved (the walruses had provided more than their own meat; from their stomachs Eskimos often ate freshly-swallowed clams from the seabed). The Eskimos' emaciated corpses, sometimes stacked like cordwood outside the rudimentary settlements, made a telling impression on many whalers, some of whom refused any more to hunt walruses. Today, over a century later, those slowly reassembled communities, surrounded by a population of walruses which many consider over-burdensome on the present ecosystem, are deprived of a market for their sea mammal products. Armchair sea mammal preservationists should consider this, and the fates of the often-shattered subsidy-dependent far north peoples. It is perhaps then more understandable that the Inuits of Greenland play fast and loose with the carefully-stitched agreements negotiated by Orri Vigfússon with their far-off government. The Europeans and their descendants have delivered them some terrible testimonials over sea mammals.

The Canadian scientist Dr Chesley Sanger in his post-doctoral thesis on the origins of Scottish whaling developed a model of fisheries development which he demonstrated applied both to the traditional old whaling, for easy whales with simple gear, and to the modern-era whaling, which he defines as starting with the improved technology of the 1860s, ultimately embracing every whale species. The model shows seven stages of progression. The first is initial discovery of the stock, followed by exploitation. Then comes success, and profit, fuelling the inevitable fifth phase, over-exploitation. The two closing phases are the decline of the resource and then collapse. Dr Sanger shows both phases of whaling unfolding in identical stages, except that with modern whaling the sequences sped along more quickly. Underlying the model is the fatal fallacy of the common property resource. Scoping more widely he extends his model to all fisheries; those profiled in this book could serve as a sample. The turning-point in fisheries is occurring now because, in the main, there are no initial discoveries left to be made. The finitude of the possibilities is apparent to all.

There is no option in fisheries to return to a primal Arcadia where fish swim again in the numbers and compositions they did before the ancestors of man crawled from the swamps. The sea of hundreds of years ago is gone. Climate and fishing operations have changed it forever. Some animals will never recover from the human assaults on stocks. The blue whale is the best example, thought to have had a population of 250,000 before its fishery started, but now down to 460 in the southern hemisphere and uncensused, though probably rising, in the northern hemisphere. There are plentiful examples of fishery effects on the composition of fish populations. Before both foreign and American fishing boats started the enormous harvesting pressure wrought on the fertile Georges Bank off north-eastern USA, gadoids like haddock and silver hake composed more than half of the fish population. Flounders and skates were present too, a few other species, and a tiny percentage was dogfish. By 1986 when the fishery had been severely depleted by over-fishing, dogfish had taken over as the kingpin predator, with the gadoids in low numbers. Nature, it is correctly said, abhors a vacuum. This explains another sort of effect, when a key predator is removed. The decimation of sea otters resulted in an explosion of sea urchins, a favourite sea otter food. In the next stage of the chain reaction heavy kelp and inshore seaweeds were eaten away away by the sea urchins; these frond-like plants had harboured many plant-forms and protected shorelines from erosion. When sea otter numbers were restored, so was the kelp. An ecosystem basis of management attempts to hold in equilibrium the whole range of living material. Huge climate changes like El Niño, the prolonged cyclical warming of the eastern Pacific associated with meteorological disturbances and thought to have been occurring since about 3000 BC, change fish stocks and ecosystems faster than imbalances created by humans, for they affect life at the bottom of the pyramid, at the microfauna level. Mankind, in effect, clings to the wheel of these phases, trying to understand them, maybe hoping to predict them, unlikely ever to change them.

Human management of fisheries is, rightly, spreading outwards. In 1995, in addition to the UN's Code of Conduct for Responsible Fisheries, there was an agreement on straddling stocks and highly migratory fish stocks. This emphasised governments' duty of care to leave fisheries for future generations in as good a state as the present, and the need to control numbers of resource users, or fishermen. Along with various other predictable bromides, such as the need for precautionary approaches to TACs, the user-pays principle was mentioned. Free fishing of fish in the far-off ocean is heading for the exit. In time, users will pay for all their fishing, the 'sports' as well. To managers, and theoreticians, this looks nice and tidy. What they may not have calculated is that once fishermen have to pay a user fee they acquire user rights. They will have the right to demand that fisheries are managed in a defensible way, in proper equilibrium, properly researched, properly policed and properly distributed.

Conclusion

'I never eat farmed salmon. I rod-fish wild salmon. It's quite another quality.' (Axel Eikimo, Director of Control, Norwegian Fisheries Ministry)

'Nothing will happen until the stocks totally collapse.' (The most commonly-heard statement on European fisheries)

'Fishing: it's a way of living, not a life.' (Govey Cargill, retired fisherman from Gourdon, Scotland)

'After the war the fish were so big the sizes would have scared you. Skate were 12 feet across; there were 16-stone halibut. We cleaned it up in three years. Just with lines.' (Govey Cargill)

'We will solve the fisheries problem – by fishing them out.' (Iain Sutherland)

'We do not subscribe to the apparently rapidly developing mythology that modern twentieth-century capitalist economics are incapable of managing wild fisheries – and we certainly do not subscribe to the late twentieth-century version of the noble savage myth that fisheries are only able to be managed by artisinal fishers.' (Alastair MacFarlane, Deputy Chief Executive of the Fishing Industry Board of New Zealand)

The subject of this book has been the capture of wild fish. The question is always asked: to what extent could aquaculture expansion take the pressure off the harvest from the wild? Personally, I do not think the world population, and therefore food needs, will rise as quickly as projected, nor that aquaculture can be a substitute for the wild catch at a greatly significant level. Presently by far the world's biggest fish farmer is China, but most of the fish are vegetation-eating carp in freshwater ponds. Marine aquaculture has attracted a wide range of extremely capable people, and grown roughly in parallel with wild capture fisheries, since the 1950s. Now it is continuing to expand after wild fish catches have peaked. However, its record is patchy. The slash-and-burn marine equivalent to uncontrolled peasant agriculture has become notorious in coastal shrimp farms and the stories of over-stocking, rampant disease, local pollution and financial collapse are legendary. Mangrove swamps, which are superior habitat for fish and all marine life, have often been cleared to make way for shrimp farms which then have exhibited the economic trajectories of a firework. Several countries including Thailand and India are furiously back-pedalling on their policies on shrimp farming, trying to discourage more aquaculture, and to halt further environmental losses.

The question is not how many species can successfully be reared in cages; if it was, the answer is many (presently some 300 finfish), and more to come. The

249

rates at which fish convert feed into growth exceeds that of domesticated stock like pigs and sheep. Small crustacea called daphnia, in an enriched environment, can grow about 20 tons of flesh per hectare in under five weeks. Aquaculture has great possibilities. The question is, can it be developed without harming the surrounding environment, without dangerously depleting stocks of other valuable parts of the food-chain for fish, at a profit, and producing fish fit to eat? Salmon farming is the nearest thing to successful marine aquaculture and, as has been described, the environment and wild fish stocks have paid a heavy price. While some countries (like America) are encouraging more and more of it, others with longer experience (Ireland, Norway) are placing tight restrictions on locations of farms and husbandry practices, and forcing some operations to close. If marine aquaculture could develop in open sea sites, if it could compete on price at lower stocking densities, its overall contribution would be in less doubt. The technology for open sea farming, on a larger scale, already exists. The other route out of the embattled estuaries is in the opposite direction, onto land, where the development of recirculation systems promises the possibility of fish farming using sterilised, recycled seawater with minimal discharges and minimal parasitism. These are the areas of potential technical advancement. What seems harder to resolve is the fundamental, underlying difficulty of finding a feed for farmed fish which does not punish wild fisheries at a lower level.

Many types of marine fish farming, having accelerated the processes of nature by converting low-value fish into high-value ones, actually result in a net loss of fish by weight. An extreme example is the quantity of pilchards fed to bluefin tuna in the southern Australian tuna farming industry. In this unusual system, a wild capture/aquaculture hybrid, young tuna are netted in the wild, and then caged and fattened for the Japanese market. The problem is that 17 tons of pilchards are needed to convert into one ton of tuna. This procedure is only operational because of the high value of bluefins. It illustrates how the equations of marine fish farming may be inherently unstable. The ecological consequences of such a conversion rate have not been looked at and thought through; hitherto such operations have been conducted at the primitive level of simply being successful, economically, in the short term. Many questions arise at this point. For the time being, however, Nature's potential to replenish the oceans on a sustainable basis overshadows the puny efforts of man to concentrate and intensify the process and shorten the growth cycle.

Fish farmers have benefited from one of fisheries' less obvious characteristics, its internationalism. From the beginning, nearly five centuries ago, the need to move fresh fish to the consumer, or cure them, stimulated tremendous exertions in trade. It does so even more today. The movement of fresh fish, overflying continents, is a remarkable transfer of goods across the globe. A trader told me: 'It's an international trade which nobody knows about who is not in it.' Because of the demand for freshness no other commodity requires, as a precondition, such speed of delivery. The future is in freshness specification. Already, long lines using single hooks in fisheries vying to supply the valuable

Japanese market have mini clocks on them; when the fish strikes the little clock stops. The buyer knows to the minute how fresh the fish is. The new development of freighting fish live has accelerated. Fish-catching novelties include using remote-controlled fish to lure shoals into waiting nets, the act of fishing moving from a highly active one to a passive one, the hunter mounting an ambuscade.

The green movement, which has done so much to alert the world to the waste of by-catch, has got one thing badly wrong. It is fisheries that are exterminated, not fish. Long before the last pair of breeding fish are caught, the fisherman has run out of the means to go on seeking them out. The Newfoundland Grand Banks were not annihilated utterly. The sea is too big. Furthermore, fish, at the last extremity, seem to adapt their behaviour better to survive, a subject whose science is at an early stage. Now in 1997 the glimmerings of revival in the vital Grand Banks' northern cod are visible. The fishery is re-opening.

The reason fisheries point the finger at us all when they fail is that it represents a failure of human institutions. If we get fisheries wrong what chance is there that we will be able, for example, to adequately protect Antarctica; or to explore space further, without destroying ourselves? Today more and more people are piling into the tumult over sustainable fisheries and those with experience and vested interests, interests in the future not only the present, may get drowned out. The adjudicators, the ringmasters, in a scenario of this sort should be the politicians. It is their look-out to ensure fishery policy is not driven by socio-cultural forces, but by reasoned debate involving those with something tangible to contribute. If good science is pushed aside in favour of anthropomorphic prejudices the future will be unpredictable. The precautionary principle must not become a doctrinal tenet because the impacts of fishing on a fish stock are impossible to predict prior to actual development. Conversely, rapid-action ransacking of freshly-discovered stocks, as has happened with some deepwater fisheries, must be avoided, and free-for-alls taken in hand from the early stages. Fish stocks should be treated to management recipes which reflect their individual differences and their place in the wider ecosystem of the sea. Put another way, fisheries should be, and deserve to be, managed with commitment. Ultimately there is no excuse not to aspire to manage the sea with as much care as the land.

Scottish fishermen used to quote the Bible, saying if herring deserted the locality it was because of 'the wickedness of the people'. In other words the presence of fish off the coast equalled health in the community. Many would hold to that today, particularly those living in the empty fishing-ports, opposite seas which have been mismanaged. Healthy fisheries have the potential to revive these places.

Sitting on the mantelpiece of my study is an egg. It is smoothed and polished and made from a substrate of the seabed which was once solid plankton. It also has a smell. The smell is of something very old, without adornment, very thick in the fragrance of dense living matter. It is the smell of the biggest natural resource on the planet.

Bibliography (Books)

Surveys of Fish Resources, by Donald R. Gunderson, John Wiley and Son (1993).

The Living Ocean, by Boyce Thorne-Miller and John Catena, Island Press. (1991)

Sea Change: A Message of the Oceans, by Sylvia A. Earle, Constable (1996)

Nature's Keepers: A New Science of Nature Management, by Stephen Budiansky, Weidenfeld and Nicolson (1995)

Controlling Common Property: Regulating Canada's East Coast Fishery, by David Ralph Matthews, University of Toronto Press (1993)

Fish Ecology, R.J. Wootton, Blackie (1992)

1973 Seminar in Maritime and Regional Studies, ed. Clark G Reynolds and William J. McAndrew, University of Maine Press (1974)

Global Fisheries: Perspectives for the 1980s, ed. Brian J. Rothschild, Springer-Verlag (1983)

The Icelandic Fisheries: Evolution and Management of a Fishing Industry, by Ragnar Arnason, Fishing News Books (1995)

The Provident Sea, by David Cushing, Cambridge University Press (1988)

The St Lawrence Basin, by Samuel Edward Dawson, Alston Rivers (1905)

The Seas, by Sir Frederick Russell and Sir Maurice Younge, Frederick Warne (1975)

The Textual Life of Savants, by Gisli Pálsson, Harwood Academic Publishers (1995)

The Forest and the Sea, by Marston Bates, Museum Press (1961)

Fisheries Management in Crisis, ed. Kevin Crean and David Symes, Fishing News Books (1996)

Capitalism from Within: Economy, Society and the State in a Japanese Fishery, by David L. Howell, University of California Press (1995)

Ocean Forum: an Interpretive History of the International North Pacific Fisheries Commission, by Roy I. Jackson and William F. Royce, Fishing News Books (1986)

The Atlantic Salmon, by W.M. Shearer, Fishing News Books (1992)

British Trawlers in Icelandic Waters, by Jon T.H. Thor, Fjolvi Publishers (1992)

The Mackerel, by Stephen J. Lockwood, Fishing News Books (1988)

The Crystal Desert, by David G. Campbell, Minerva (1992)

The Common Fisheries Policy, by Mike Holden, Fishing News Books (1994)

The Dorling Kindersley Encyclopedia of Fishing (1994)

The Icelandic Capelin Stock, Marine Research Institute (1994)

Scottish Fishing. Photographs from the George Washington Wilson Collection, Aberdeen University Library (1982)

From Herring to Seine Net Fishing on the East Coast of Scotland, by Iain Sutherland, Camps Bookshop, Wick

Sunset Playgrounds, by F.G. Aflalo, Witherby and Co. (1909)

Antarctic Isle: Wild Life in South Georgia, by Niall Rankin, Collins (1951)

Fishermen: A Community Living from the Sea, by Sally Festing, David and Charles (1977)

The Open Sea: Its Natural History, Fish and Fisheries, by Sir Alister Hardy, Collins (1959)

The North Sea, by George Morey, Frederick Muller (1968)

Great Waters, Sir Alister Hardy, Collins (1967)

Wick of the North, by Frank Foden, North of Scotland Newspapers (1996)

The Inexhaustible Sea, by Hawthorne Daniel and Francis Minot, Scientific Book Club (1954)

Bibliography

Fishing Villages in Tokugawa Japan, by Arne Kalland, Curzon Press (1995)
The Romance of Fish Life, by W.A. Hunter, A. and C. Black (1931)
Sea Fishing as a Sport, by Lambton J. Young, Groombridge (1872)
Dead Reckoning: Confronting the Crisis in Pacific Fisheries, by Terry Glavin, Greystone Books (1996)
Sea Fishing with the Experts, by Jack Thorndike, George Allen and Unwin (1956)
A Sea of Small Boats, ed. Jack Cordell, Cultural Survival Inc. (1989)
Beyond the Tides, by Philip Street, University of London Press (1955)
Field Investigations of the Early Stages of Marine Fish, by M.R. Heath, Academic Press (1992)
Rational Fishing of the Cod of the North Sea, by Michael Graham, Edward Arnold (1948)
The Silver Darlings, by Neil Gunn, Faber and Faber (1941)
Fisheries Ecology, by Tony J. Pitcher and Paul J.B. Hart, Chapman and Hall (1982)
The Old Man and the Sea, by Ernest Hemingway, Cape (1952)
Fishermen and Fishing Ways, by Peter F. Anson, Harrap and Co. (1932)
The Fishermen: The Sociology of an Extreme Occupation, by Jeremy Tunstall, MacGibbon and Kee (1962)
Seal Cull, by John Lister-Kaye, Penguin (1979)
Creatures of the Deep Sea, by Klaus Gunther and Kurt Deckert, George Allen and Unwin (1956)
The Fish Gate, by Michael Graham, Faber and Faber (1953)
The Economics of Fisheries Management, by Lee G. Anderson, John Hopkins University Press (1977)
The Cruise of the Cachalot, by Frank T. Bullen, Smith Elder and Co. (1898)
Administration and Conflict Management in Japanese Coastal Fisheries, by Kenneth Ruddle, FAO (1987)
Crisis in the World's Fisheries, by James R. McGoodwin, Stanford University Press (1990)
Whales of the World, by Nigel Bonner, Blandford (1989)
Population Production and Regulation in the Sea, by David Cushing, Cambridge University Press (1995)
The British Fisheries Society 1786–1893, by Jean Dunlop, John Donald (1978)
World Sea Fisheries, by Robert Morgan, Methuen (1956)
Fishery Development and Experiences, by W.H.L. Allsopp, Fishing News Books (1985)
The Politics of Fishing in Britain and France, by Michael Shackleton, Gower (1986)
Lovely She Goes, by William Mitford, Michael Joseph (1969)
Environmental and Natural Resource Economics, by Tom Tietenberg, Harper Collins (1996)
Captains Courageous, by Rudyard Kipling, MacMillan (1897)
British Trawlers and Iceland, by Jon T.H. Thor, University of Göteborg (1995)
The Herring and Its Fishery, by W.C. Hodgson, Routledge Kegan Paul (1957)
The Herring, by Arthur Michael Samuel, John Murray (1918)
Of Whales and Men, by R.B. Robertson, Richard Clay (1956)
Beautiful Swimmers, by William W. Warner, Little, Brown and Co. (1976)
Salmon: The World's Most Harassed Fish, by F. Anthony Netboy, André Deutsch (1986)
The Castle of Lies, by Christopher Booker and Richard North, Duckworth (1997)
Distant Water: The Fate of the North Atlantic Fisherman, by William W. Warner, Little, Brown and Co. (1977)
Fish and Fisheries in the Baltic. Essays, Olsen and Olsen (1994)
A Sea-Faring Saga, by James Slater, Commercial Fishing Enterprises (1979)
The Economics of White Fish Distribution in Britain, by R.A. Taylor, Duckworth (1960)

The Ocean World, by Louis Figuier, Chapman and Hall (1869)
Sharks and Other Predatory Fish of Australia, by Peter Goadby, Jacaranda Press (1959)
Fisheries Mismanagement: The Case of the North Atlantic Cod, by Rögnvaldur Hannesson, Fishing News Books (1997)
A River Never Sleeps, by Roderick Haig-Brown, Collins (1948)
The Diversity of Life, by Edward O. Wilson, Allen Lane (1992)
The Sea Around Us, by Rachel L. Carson, Staples Press (1951)
Moby Dick, by Herman Melville
A Century of Fisheries in Northern America, ed. Norman Benson, American Fisheries Society, Washington DC (1970)
Fishing: The Complete Book, by Tre Tryckare and E. Cagner, David and Charles (1976)
Fishing from the Earliest Times, by William Radcliffe, John Murray (1921)
The Cod Fisheries: The History of an International Economy, by Harold H. Innis, University of Toronto (1940)
North Cape, by F.D. Ommanney, Longmans, Green and Co (1939)
Whales, Ice and Men, by John R. Bockstoce, University of Washington Press (1986)
The Great North, by Felice Bellotti, Andre Deutsch (1957)
Fish or Cut Bait, ed. Laura Jones and Michael Walker, The Fraser Institute, Vancouver, (1997).
The Living Ice, by Pol Chantraine, McClelland and Stewart (1980).

Other Publications

The Code of Conduct for Responsible Fisheries, FAO (1995)
Tools For Sustainable Fisheries Management, by Brian O'Riordan, Intermediate Technology (1996)
The Pelagic Fishery in the UK, by David Cleghorn, Sea Fish Industry Authority (1996)
The Common Fisheries Policy: End or Mend, by Austin Mitchell, Campaign for an Independent Britain (1979)
Monitoring the Common Fisheries Policy, European Commission Report (1996)
Fisherfolk Safeguarding Aquatic Diversity through their Fishing Techniques, by Brian O'Riordan, Intermediate Technology (1996)
Catching for the Market, conference organised by the Sea Fish Industry Authority (1994)
Alternatives to the West Greenland Atlantic Salmon Fishery, by Nicholas E. Flanders, Flemming Enequist and Oran R. Young, Dickey Centre Institute of Arctic Studies (1995)
The United Nations Resolutions on Driftnet Fishing: An Unsustainable Precedent for High Seas and Coastal Fisheries, by William T. Burke, Mark Freeberg and Edward L. Miles, University of Washington (1993)
Study of Deep-Water Fish Stocks to the West of Scotland, by J.D.M. Gordon and J.E. Hunter, Scottish Association for Marine Science (1994)
Overfishing: Causes and Consequences, The Ecologist (1995)
A Global Assessment of Fisheries Bycatch and Discards, FAO Fisheries Technical Paper 339 (1994)
Multispecies Models Relevant to Management of Living Resources (Symposium held in October 1989), International Council for the Exploration of the Sea (ICES) (1991)

Bibliography

The Biological Collapse of Atlantic Cod Off Newfoundland and Labrador, Jeffrey A. Hutchings and Ransom A. Myers

The North Atlantic Fisheries, ed. Ragnar Arnason and Lawrence Felt, Institute of Island Studies (1995)

What Can be Learned from the Collapse of the Renewable Resource? Atlantic Cod, Gadus Morhua, of Newfoundland and Labrador, by Jeffrey A. Hutchings and Ransom A. Myers, National Research Council of Canada (Volume 51 No 9) (1994)

The Influence of Gulf Stream Warm Core Rings on Recruitment of Fish in the Northwest Atlantic, by Ransom A. Myers and Ken Drinkwater, Journal of Marine Research (No 47) (1989)

Why Do Fish Stocks Collapse?: The Example of Cod in Atlantic Canada, by Ransom A. Myers, Jeffrey A. Hutchings and Nicholas J. Barrowman

Was an Increase in Mortality Responsible for the Collapse of Northern Cod?, by Ransom A. Myers and Noel G. Cadigan, National Research Council of Canada (Volume 52 No 6 (1995)

The Collapse of Cod in Eastern Canada: The Evidence from Tagging Data, by R.A. Myers, N.J. Barrowman, J.M. Hoenig and Z. Qu, ICES (1995)

Fishery Statistics. Falklands Islands Government Fisheries Department. Volume I (1989–1996), Falkland Islands Government (1997)

Fisheries Management and Sustainability: A New Perspective of an Old Problem, by S.M. Garcia and R.J.R. Grainger (Paper prepared for the 2nd World Fisheries Congress in Brisbane) (1996)

Global Impacts of Fisheries on Seabirds, Birdlife International (1995)

It Can't Go On Forever, Greenpeace International (1993)

SAMUDRA magazine (Issues 10, 11 and 12) (1994/95)

Stability and the Objectives of Fisheries Management: The Scientific Background, by J.G. Shepherd, British Ministry of Agriculture and Fisheries, MAFF (Leaflet No 64) (1990)

Aide Memoire on Scientific Advice on Fisheries Management, by J.G. Shepherd, MAFF (Leaflet No 70) (1992)

Why Fisheries Need to be Managed and Why Technical Conservation Measures on Their Own Are Not Enough, by J.G. Shepherd, MAFF (Leaflet No 71) (1993)

No Place to Hide: Highly Migratory Fish in the Atlantic Ocean, by Ellen M. Peel, Center for Marine Conservation, Washington DC (1995)

Industrial Hoover Fishing: A Policy Vacuum, Greenpeace Report (1997)

Fish Stock Conservation and Management, House of Lords Report by the Select Committee on Science and Technology, Her Majesty's Stationery Office (1996)

A Guide to the Deep-Water Fish of the North-eastern Atlantic, by John D.M. Gordon, Ewen M. Harrison and Sarah C. Swan, Scottish Association for Marine Science

RSPB's Vision for Sustainable Fisheries, by Euan Dunn and Nancy Harrison, Royal Society for the Protection of Birds, RSPB (1995)

Interactions between Fisheries and Marine Birds, by Euan Dunn, RSPB (1994)

Fishermen and Fisheries Management, by David B. Thomson, FAO

Report from the Commission to the Council and the European Parliament on the Establishment of a Satellite-Based Vessel Monitoring System for Community Fishing Vessels, Commission of the European Communities paper (1996)

An Investigation of Seal Predation of Creels in Orkney, by John Crossley, Scottish Natural Heritage report (1994)

Ecological Effects of the North Sea Industrial Fishing Industry on the Availability of Human Consumption Species, by Jack Robertson, Jacqueline McGlade and Ian Leaver, Robert Gordon University (1996)

255

Brief Description of the Fishery Industry in China, by Globefish, FAO, (Volume 41) (1995)

Fisheries Development, Fisheries Management, and Externalities, by Richard S. Johnston, World Bank Discussion (Paper No 165 (1992)

North Sea Quality Status Report 1993, ICES, Oslen and Olsen (1993)

Conservation: An Alternative Approach, The UK's National Federation of Fishermen's Organisations (1992)

North Sea Fish Crisis, Greenpeace (1996)

The Economic and Social Effects of the Fishing Industry – A Comparative Study, by H. Josupeit, FAO Fisheries Circular (1981)

The Salmon Net (The Magazine of the Salmon Net Fishing Association of Scotland) Nos XXVI (1995) and XXVII (1996)

Memorandum from the RSPB to Select Committee on Science and Technology Enquiry: *Fish Stock Conservation and Management*, RSPB (1995)

FAO and Fisheries Development, by Patrick McCully, The Ecologist (Volume 21 No 2) (1991)

Modelling the Predation, Growth and Population Dynamics of Fish within a Spatially-resolved Shelf-Sea Ecosystem Model, by A.D. Bryant, M.R. Heath, N. Broekhuizen, J.G. Ollason, W.S.C. Gurney and S.P.R. Greenstreet, Netherlands Journal of Sea Research (No 33) (1995)

Transfer of Erosion Control Technology to Aquaculture and Fisheries Enhancement, Enviro-Marine, Scotland (1996)

The Kewalo Research Facility, 1958–1992: Over 30 Years of Progress, by Richard W. Brill, National Marine Fisheries Service, Honolulu (1992)

Report on Sea Fisheries: Commission of Enquiry into the Resources and Industries of Ireland (1921)

Our Living Oceans: Report on the Status of US Living Marine Resources 1995, USA Department of Commerce (1996)

The ICES Multispecies assessment Working Group: evolution, insights, and future problems, by J.G. Pope, ICES (1991)

Chaos and Its Implications for Marine Resources Science, by J.M. McGlade, Marine Environmental Management, Review of Events in 1993 and Future Trends (Volume 1 Paper No 15) (1994)

Integrating Social and Economic Factors into Fisheries Management in the European Union – An Issue of Governance, by Jacqueline McGlade, Environmental Management, Review of 1994 and Future Trends (Volume 2 Paper No 15) (1995)

Solving Bycatch: Considerations for Today and Tomorrow, University of Alaska (1995)

Seasonal Fishermen's Unemployment Insurance Benefits in Canada: Continuous Debate: 1935–1995, by William Schrank, Memorial University of Newfoundland (1995)

Canadian Government Financial Intervention in a Marine Fishery: The Case of Newfoundland, 1972/73–1980/81, by William Schrank, Blanca Skoda, Noel Roy and Eugene Tsoa, Ocean Development and International Law (Volume 18 No 5) (1987)

The Cost to Government of Maintaining a Commercially Unviable Fishery: The Case of Newfoundland 1981/82–1990/91, Ocean Development and International Law (Volume 26) (1995)

An Inshore Fishery: A Commercially Viable Industry or an Employer of Last Resort, by William Schrank, Noel Roy, Rosemary Ommer and Blanca Skoda, Ocean Development and International Law (Volume 23) (1992)

Employment Prospects in a Commercially Viable Newfoundland Fishery: An Application of 'An Econometric Model of the Newfoundland Groundfishery', by William Schrank, Noel Roy and Eugene Tsoa, Marine Resources Economics (Volume 3 No 3) (1986)

Extended Fisheries Jurisdiction: origins of the current crisis in Atlantic Canada's fisheries, by William E. Schrank, Marine Policy (Volume 19 No 4) (1995)

Bibliography

Case Studies and Working Papers Presented at the Expert Consultation On Strategies for Fisheries Development, FAO Fisheries Report (No 295)

Fish for Food and Development, FAO, Rome (1991)

Falkland Islands Economic Study 1982, report chaired by The Rt Hon. Lord Shackleton, HMSO (1982)

Recruitment Variability and the Dynamics of Exploited Marine Populations, by Michael J. Fogarty, Michael P. Sissenwine and Edward B. Cohen, Trends in Ecology and Evolution (1991)

Resource Productivity and Fisheries Management of the Northeast Shelf Ecosystem, by M.P. Sissenwine and E.B. Cohen, chapter five from *Food Chains, Yields, Models, and Management of Large Marine Ecosystems*, ed. Kenneth Sherman, Lewis M. Alexander and Barry D. Gold (US Government publication)

Marine Fisheries at a Critical Juncture, by Michael P. Sissenwine and Andrew A. Rosenberg, Fisheries (Volume 18 No 10)

Achieving Sustainable Use of Renewable Resources, by A.A. Rosenberg, M.J. Fogarty, M.P. Sissenwine, J.R. Beddington and J.G. Shepherd, Science (Volume 262) (1993)

Achieving Long Term Potential from US Fisheries, by Steven L. Swartz and Michael P. Sissenwine, Sea Technology (1993)

US Living Marine Resources: Current Status and Long-Term Potential, by Michael P. Sissenwine and Andrew A. Rosenberg, Conserving America's Fisheries, ed. Richard H. Stroud, National Coalition for Marine Conservation Inc. (1994)

Sources of Uncertainty: A Case study of Georges Bank Haddock, by Michael J. Fogarty, Andrew A. Rosenberg and Michael P. Sissenwine, Environmental Scientific Technology (Volume 26 No 3) (1992)

Climate Change 1994. The Intergovernmental Panel on Climate Change, Cambridge University Press (1995)

Fast Fish and Sober Captains, by Orri Vigfússon (speech from a Fraser Institute Conference) (1996)

Decadal-scale regime shifts in the large marine ecosystems of the North-east Pacific: a case for historical science, by Robert C. Francis and Steven R. Hare, Fisheries Oceanography (1994)

Pacific Salmon Production Trends in Relation to Climate, by Richard J. Beamish and Daniel R. Bouillon, Canadian Department of Fisheries and Oceans, Nanaimo, British Columbia (1992)

Evolving Principles of International Fisheries Law and the North Pacific Anadromous Fish Commission, by Yvonne L deReynier (unpublished)

Chondros, the newsletter 'Dedicated to Rational Use and Conservation of Sharks, Skates, Rays, Sawfishes and Chimaeras', published at 1003 Hermitage Drive, Owensboro, KY, USA 42301 (various issues)

Various issues of *Shark News*, the Newsletter of the IUCN Specialist Group

Selected Advantages Conferred by the High Performance Physiology of Tunas, Billfishes, and Dolphin Fish, by Richard W. Brill, University of Hawaii (1996)

The Impact of Grey Seals on Fisheries: The Need for their Management, by Basil Parrish, The Scottish Fishermen's Federation (1993)

ITQs In New Zealand: Bureaucratic Management Versus Private Property. The Score after Ten Years, by Tom McClurg (conference paper from 'Managing A Wasting Resource: Would Quotas Solve the Problems Facing the West Coast Salmon Fishery?' Conference held in Vancouver in 1996)

Sustainable Management of the Chatham Rise Orange Roughy Fishery, the Parliamentary Commissioner for the Environment, New Zealand (1992)

ITQs In New Zealand: The Era of Fixed Quota in Perpetuity, by Michael P. Sissenwine, Fishery Bulletin, US (1992)

The New Zealand Sea Food Exporter 1993/94, The New Zealand Fishing Industry Board
Trends in Finfish Landings of Sport-boat Anglers in Texas Marine Waters, May 1974–May 1992, by Thomas A. Warren, Lee M. Green and Kyle W. Spiller, Texas Parks and Wildlife Department (1994)
Saltwater Finfish Research and Management in Texas, Texas Parks and Wildlife Department (1995)
Management of the Red Drum Resource in Texas, by T.L. Heffernan and R.J. Kemp, Texas Parks and Wildlife Department (1978)
Red Drum in Texas: A Success Story in Partnership and Commitment, by Lawrence W. McEachron and Kevin Daniels, Fisheries (Volume 20 No 3)
History and Management of the Red Drum Fishery, by Gary C. Matlock, Texas Parks and Wildlife Department (1978)
Eutrophication, Fisheries, and Productivity of the North Sea Continental Zone, by R. Boddeke and P. Hagel, from *Condition of the World's Aquatic Habitats*, proceedings of the World Fisheries Congress, Theme I, ed. Neil B. Armantrout
Thorsteinn Pálsson's address at the conference *Iceland's Fisheries and the European Union* (1996)
Fisheries Management in Iceland, by Kristjan Skarphéðinsson, Ministry of Fisheries, Iceland (1991)
Close to the Sea, Ministry Of Fisheries, Iceland
Report on Legal Remedies in Relation to Sea Trout Protection and Salmon Farms in the West Coast of Scotland, by Walter G. Semple, Bird Semple, Glasgow (1996)
Fisheries Research and Management of the North Sea; the Next Hundred Years, by John Pope (from eight lectures for the Centennial of the Danish Institute for Fisheries and Marine Research) (1989)
Chronicles of Marine Fishery Landings (1950–94), by R.J.R. Grainger and S.M. Garcia, FAO (1996)
Canadian Royal Commission Report on Seals and Sealing in Canada (1996)
Is deepwater a dead-end?, by Phil Aikman, consultant to Greenpeace
The Physical Interactions Between Grey Seals and Fishing Gear, EC Report: PEM/93/06

Index